D1214002

RADIANT SCIENCE, DARK POLITICS

A Memoir of the Nuclear Age

Martin D. Kamen

Foreword by

Edwin M. McMillan

UNIVERSITY OF CALIFORNIA PRESS

Berkeley Los Angeles London

University of California Press
Berkeley and Los Angeles, California

University of California Press, Ltd.
London, England

Library of Congress Cataloging in Publication Data

Kamen, Martin David, 1913–
Radiant science, dark politics.

Includes bibliographical references and index.
1. Kamen, Martin David, 1913– . 2. Biochemists—United States—Biography. I. Title.
QH31.K277A37 1985 574.19'2'0924 [B] 83-13510
ISBN 0-520-04929-2

Printed in the United States of America

1 2 3 4 5 6 7 8 9

To Virginia Swanson Kamen, whose steadfast encouragement made this book possible

Contents

Foreword
by
Edwin M. McMillan

I first met Martin Kamen in 1937, when he came to Ernest Lawrence's Radiation Laboratory at the University of California in Berkeley as a newly minted Ph.D. in Chemistry from the University of Chicago on a research fellowship. To set the stage for that event I should say something about what the laboratory was like then. It bore no resemblance to any modern research establishment. It had one building, an old wooden structure that had been scheduled for demolition until Lawrence prevailed on Robert G. Sproul, then president of U.C., to save it for him. In this building worked a small group of dedicated people, some of them faculty, some graduate students, some visitors and fellows, under Lawrence's direction. There was no formal organization and no set program of research, and a newcomer like Martin pretty much had to find his own way. The fact that he was a chemist in a group dominated by physicists was no problem; there was plenty of chemistry to be done in the identification and purification of radioactive isotopes, and anyhow the kind of chemistry that he had learned under William D. Harkins at Chicago had a large component of nuclear physics mixed with it.

Then, too, there was Martin's friendly and cheerful personality, with its many facets and some eccentricities, such as his refusal to learn to drive a car. His stories of life in the seamier parts of Chicago added a touch of exotic interest, as did his tales of life as a graduate student with some dictatorial professors of the old school. Life at the Radiation Laboratory had its relaxed moments, such as the summer vacations at Don Cooksey's place in the Trinity Mountains and the parties at Di Biasi's restaurant, and at these Martin was a star performer with a line of stories that became famous.

I have not mentioned his musical interests, which were so important in his life, nor will I go on about his later career and achievements, but I do wish to emphasize what I consider his most important single contribution to science, the discovery of carbon 14. This took place before I left the Radiation Laboratory in November 1940 to do radar research, and I was in Berkeley and interested in the work while it was going on. The use of radioactive tracers to follow the course of chemical reactions was a subject of great interest and high promise at that time, and new radioactive isotopes were being discovered almost daily. The most desired of all was a long-lived isotope of carbon, the key element in organic chemistry.

Already known was carbon 11, with a half-life of twenty minutes, and Martin with his colleagues Sam Ruben and Zev Hassid were already trying to study the process of photosynthesis using this as a tracer, a frustrating procedure at best, since the tracer isotope was decaying away rapidly while they were carrying out the required chemical manipulations. The obvious cure was to find a long-lived carbon isotope. With the strong encouragement of Professor Lawrence, Kamen and Ruben set out on a concerted search for this bonanza, and they found it. The consequences of their discovery are without limit, and others have received high awards for researches made using carbon 14, while its discoverers have not been rewarded on a commensurate scale.

The reason for this I believe lies in the sad circumstance that one of the two co-workers suffered a fatal accident while working with phosgene in the laboratory, so that what might have been a joint award was no longer possible. I also believe that an award to Kamen alone would not be inappropriate. I knew both men well, and in fact have published joint papers with both of them. It was Kamen who had the spark of originality and the initiative to propose lines of attack, and Kamen with whom one had the stimulating discussions. In my mind, at least, I think of Martin Kamen as the discoverer of carbon 14, as well as being a man of wide talents in many fields, and I am honored to have him as a friend.

Author's Preface

The English clergyman, scientist, and political thinker Joseph Priestley, whose experiments inaugurated the modern era of research into photosynthesis, began his memoirs, published in 1806, with the following modest declaration:

> Having thought it right to leave behind me some account of my friends and benefactors, it is in a manner necessary that I also give an account of myself, and as the like has been done by many persons and for reasons which posterity has approved I make no apology for following their example.

Let that also serve as preamble to this account of my own first four decades. As one who was present at the launching of the Nuclear Age and the advent of Big Science, which it ushered in, I was closely associated with many of the seminal figures of the time, and knew many others—among them E. O. Lawrence and J. Robert Oppenheimer, Otto Warburg and James Franck, Arthur H. Compton, Max Delbrück, Harold Urey, Irène Joliot-Curie, Georg von Hevesy, Edwin McMillan, Alfred Hershey, Glenn Seaborg, Linus Pauling, Arthur Kornberg, Melvin Calvin, Robert Mulliken, Emilio Segrè, Luis Alvarez, and Niels Bohr. With the notable exception of Oppenheimer, each of these won a Nobel Prize. They are clearly worth writing about.

Moreover, my own experiences in the early application of isotopic tracer methodology to the study of photosynthesis, as well the hitherto unwritten complete story of the discovery of carbon 14, should be of interest to those historians of science who shoulder the burden of providing definitive studies on the impact of the constructive and revolutionary developments in chemistry and biology that came in the wake of the atomic bomb blasts at Hiroshima and Nagasaki. I owe this to the record.

The story of my life as a musician is the second thread that runs through the pages that follow. Curiously, I became a scientist almost by accident. Indeed, I had been expected to become a musician, and music has always been a happy counterpoint to my scientific work. Among the friends and benefactors of whom I hope to leave some account are many musicians—prominent among them Isaac Stern and Henri Temiauka.

The third, sinister strand to my story is the struggle in which I found myself caught up, quite against my will, to defend myself against the assaults of the House Un-American Activities Committee and others who sought to prove on the basis of false evidence, and sometimes of no evidence at all, that I was a spy and a traitor. An account of how this fight was fought, over many painful years, and finally won, seems almost a civic responsibility. Certainly I must record the devotion to the cause of justice and efforts on my behalf of some of my friends and benefactors, which were crucial in preventing my career as a scientist from being brutally terminated almost at its outset.

The source material had to be sorted, organized, and digested from a sprawling mass of documentation—letters, memorabilia, articles, and the like—accumulated over nearly half a century. The task of selection was onerous, and much that digressed too far from the main story had to be omitted. Even so, the text in its original form stood in need of tightening up, and I am indebted in this respect to Peter Dreyer, whose editing helped produce a coherent final version, while effecting necessary cuts and emendations. Careful reading of the manuscript by Edwin McMillan and others, including Emilio Segrè and Glenn Seaborg, has eliminated a number of errors of fact. Suggestions on matters of content and style by James H. Clark, director of the University of California Press, proved most helpful. And the efforts of Peggy Earhart and Laurie Sprecher produced a readable typed copy from the original handwritten manuscript. I also thank Maurice Lecovre for generously providing the photograph of the Viking Mission ^{14}C Life Detector element. Above all, this task would not have been attempted or brought to completion without the steady encouragement and support of my wife, Virginia Swanson Kamen, to whom this book is dedicated, and who devoted many long hours to reading and correcting page proofs.

1

Beginnings
(1913–1929)

"MOST CHILDREN have the research spanked out of them by age four," Julius Nieuwland, the famed chemist-priest, is said to have remarked. I experienced no such conditioning. Russian-Jewish refugees from the repressions and cramped prospects of the Old World, my parents encouraged anything that could be imagined to foretell a successful professional career for their firstborn and only son. The immigrant wave of the early 1900s was to produce a heavy crop of scientists, artists, and writers.[1]

But refugees seldom have time to spend assembling family records while fleeing for their lives. Nor do periods of turbulence that disperse families promote safe storage of memorabilia. The story of my origins must rely on the recollections (judiciously weighed) of parents, relatives, family friends, and even casual acquaintances, supplemented by a few albums, letters, and official documents.

My father was born Aaron Kamenetsky on May 5, 1884, in the town of Slonim in White Russia. He was the second of three brothers. As the family owned one of the largest bakeries in this fairly large town, the boys enjoyed a solid upper-middle-class education. Aaron became a part-time teacher of Russian and also secretary to the local police chief, who for some time remained unaware that he was also a member of the local SR's, or Socialist Revolutionaries, whose chief was Aleksandr Kerensky, eventual premier of the provisional government that preceded the Bolshevik takeover. In due course, Aaron's connections became suspect, and at the age of eighteen or nineteen, he found it expedient to flee, using a forged passport obtained by a friend.

As the first step on the way to the New World, he arrived in Finland just as that country was beginning to enjoy de facto independence and was welcomed there as a refugee from hated Imperial Russia. He even participated in an independence celebration, riding on a horse supplied for the occasion. Soon thereafter he left for England, arriving in London without the slightest notion as to what he might do for a living. He became attached as a helper to an itinerant peddler of crockery and other household items, acquiring in the process a working knowledge of English. On one occasion, he came to Stratford-on-Avon, which he remembered excitedly as William Shakespeare's birthplace. Still, he had his heart set on going to the New World, and at the first opportunity, using the meager savings he had accumulated, he shipped out in the steerage of the S. S. *Kensington* on a summer's day in June 1906. He arrived in Quebec at the age of twenty-one, describing himself some years later in his U.S. naturalization papers as "5½ feet in height, with fair complexion, brown hair, accompanied by no one and destined to no one." He was a slight man of only about 135 pounds, a weight he maintained the rest of his life.

Aaron met his future wife, *née* Goldie Achber, in Toronto shortly thereafter. She had come there to join some of her immigrant family after a short term as a textile worker in New Hampshire, where an older sister was just established with her already large family, consisting of a husband (a Mr. O. Bean) and several children. Goldie had been born in 1889 in a village somewhere in Lithuania or Latvia, where her family were probably small-town farmers and merchants. Eventually, they had dispersed in various directions—the oldest boy migrating to South Africa, where a prosperous branch of the Achbers was to become entrenched, while the rest (six sisters and two brothers with their parents) took up residence in Toronto and New Hampshire.

Aaron drifted into this family circle soon after his arrival in Toronto, where he had taken up photography as a possible livelihood, being attracted to this profession by his natural artistic bent. He and Goldie became engaged in 1909, and moved to Chicago after their marriage in 1911. There they entered into partnership with a Mr. Lasswell, who owned the Hyde Park Studio on the near South Side of Chicago in the ward that also housed the University of Chicago and was the locale for an impressive collection of artists, journalists, and intellectuals. This studio already enjoyed a leading place among the photographic establishments on the South Side and was to retain this position for half a century, until my father's retirement in the 1960s. It no longer exists, having been torn down to make room for urban redevelopment projects of the late sixties.

Many famous sitters had their portraits taken by my father. These included Eduard Beneš, the successor to Tomáš Masaryk as president of Czechoslovakia; Arno Luckhardt, the famed physiologist; Fanny Bloomfield-Zeisler, the outstanding woman pianist of her time; and a renowned retired contralto, Mme. Rosa Olitzka-Sinai. Perhaps his most glamorous client was a young girl named Mary Philbin, who entered a Hollywood contest and won her place in movie history on the basis of the portrait made by my father. I believe her most famous role was in *The Phantom of the Opera,* which starred Lon Chaney.

Goldie learned to take and process photos so that my father could range far afield in drumming up more business than could be developed in the Hyde Park area alone. Now known as "Harry Kamen," thanks to a clerk in the naturalization bureau downtown in the Cook County courthouse, he often left home as an itinerant photographer, working with associates in various small towns in northern Illinois and Iowa.

Goldie was still in the learning stages, however, and just five months pregnant early in 1913 when Harry suddenly came down with appendicitis, a dreaded affliction in those days. A young doctor, Roger T. Farley, who was to be our beloved family physician all through my childhood, was called, and declared an emergency operation essential. Dr. Farley had a staff position at the Illinois Central Hospital, a mile from the studio, and there he and Harry repaired on the trolley, leaving Goldie distraught and faced with running the studio. The operation was successful; but for the next two months Harry was immobilized while convalescing and Goldie bravely bluffed her way, meeting customers, taking their pictures, and filling orders. Sometimes the photos came out minus heads or feet, but somehow she managed to arrange new sittings until she had acceptable results. She lost no customers. By early summer, Harry was back on his feet and Goldie left in the ninth month of her pregnancy to have her baby in Toronto. Her sisters there had insisted she come back so that she could be tended by family, rather than by strangers. My birth in Toronto on August 27, 1913, was accompanied by confusion as to what might be the English equivalent of the Hebrew name I was given—Menachem David. Quite irrelevantly, it came out "Martin David," although I subsequently discovered that the name on my birth certificate was "Nathan"! Goldie came back to Chicago in December, when I was three months old, but again confusion reigned—this time at the Canadian-American border, where the authorities duly noted her presence but not mine.

The family album was a special long-term project, following naturally from the professional activity of my parents. Harry had become a skilled portrait photographer and enjoyed demonstrating his skills on his son. Almost daily sittings were tolerated well at first, but as I approached my teens, I became more resistant, and finally large enough to terminate the series. The attenuation of the record was also hastened by the arrival of my sister Lillian when I was six years old. She took on the chore of model until she, too, rebelled a decade later. The experience my father acquired in dealing with our recalcitrance he put to good use in dealing with other children, acquiring a well-deserved reputation as a portrait photographer of babies and the younger set.

The studio, which was the center of family life throughout my childhood, was reached from street level by a flight of stairs lined with portraits and advertisements. The family, including my Aunt Ethel, lived in a cramped space connected to the studio reception room on the second floor by a dark, dingy corridor crammed with all manner of discarded furniture, records, and photographic equipment. The living space consisted of one large room partitioned into a kitchen, bathroom, and two bedrooms. The tiny quarters frequently brought on confrontations between my parents, and especially between Harry and Ethel, who in Harry's opinion did not contribute her share to the upkeep of the family. Goldie found herself squarely in the middle of these arguments and needed all her considerable charm and good sense to keep the peace. The stress of such conflicts undoubtedly had an unsettling influence on me, because I still recall becoming frightened and crying during some of the arguments that were a frequent feature of our family life.

A traumatic and lasting bit of character building occurred when I was about five years old. I had accumulated a small hoard of about a hundred pennies, by doing small tasks around the house and being a "good boy" in general. My mother gave me a small cotton purse with a drawstring in which to keep my collection, at the same time sternly warning me not to show it off to anyone, particularly to the neighborhood kids. Nevertheless, I went out with my bag of riches and was immediately surrounded by a dozen or so ragamuffins, some of whom towered over me. The crafty leader of the gang distracted my attention with a shiny dime that seemed ever so much more desirable than my drab pennies. A fast shuffle took place, leaving me alone holding the thin dime and a sadly shrunken bag with a few forlorn pennies. Attempting to feign nonchalance, I slunk back into the house. My mother was not fooled and quickly extracted a confession. Angrily, she called my father and explained to him the need for stern measures, even a spanking. My father made a try. Pulling me face down in his lap, he raised his hand to administer punishment to his terrified and repentant son. Almost immediately, he fell into tears, as did my mother and I.

The effect on me was lasting. For years afterward I was a miniature miser, so much so that I was relied upon, much to my father's chagrin, to keep him from wasting money on foolish purchases whenever he went to fairs and other public events unaccompanied by my mother.

Our neighbors in the tenement complex were a polyglot mixture of first-generation settlers, including some of Irish and Swedish origin. We were the only Jews, but no racial enmity was evident. The hard struggle for existence seemed to knit us together into a community, socializing among its members being facilitated by a porch in back overlooking a rear alley. The only objects of prejudice were the blacks a mile away across Cottage Grove Avenue to the west, who were perceived as a threat to economic wellbeing and hopes of future prosperity by the whites in the eastern portion of Hyde Park.

As family fortunes improved, we were able to move to a more spacious apartment across the porch with an address on the adjoining street, Blackstone Avenue. This was fortunate, because with the arrival of Lillian more space was imperative. We could still reach the studio by walking to our old quarters across the porch. Next, we acquired a new and marvelous status symbol—a four-door Buick sedan. My father was, and remained, the only driver. This is worth emphasizing, because I never did learn to drive. I don't recall ever wanting to. In later years, as a result of much observation, I reached the conclusion that this skill is not essential in a civilization where there is always someone available to provide a ride. To put it more colorfully, "There is always a sucker who owns a car!"

Shotwell Hall across the street was a center for recreational gatherings, sports events, and all manner of neighborhood fiestas, providing fascinating glimpses of adult activity from my bedroom window. I learned about the marriage rituals of Central Europeans, whose uproarious nuptial celebrations often lasted far into the night. I still recall the vigorous jumping and dancing of inebriated celebrants stripped down to their red underwear. Riots calling for police action were frequent.

I do not remember how I learned to read, but I was reading at age five before I went to school. One of my earliest recollections is trundling downtown with my father to haunt the bookstalls and return with armloads of books on all subjects, but mostly on history. I had a strong visual memory and could commit whole pages of whatever I read for instant recall, even when—as in the case of Tolstoy's *War and Peace,* read at age eight—I had little understanding of what I was reading. This learning process was reinforced, starting when I was eight, by the presence in the studio of an old Scottish gentleman named Mr. MacClear (my father's spelling), whom my father had hired on a part-time basis to help with the studio accounts. Mostly, Mr. MacClear would sit in the big overstuffed chair in the reception room and hold forth to me, an entranced audience, on a wide variety of topics, including discussions of whatever books I might be reading at the time. Whenever he mentioned a word or a thing I did not know, I would run to the dictionary to find its meaning without waiting for him to explain. MacClear was an indigent intellectual with practically no income except that which my father provided. I have no knowledge of how he had acquired his learning, and neither did my father, who continued to support him to the day of his death, despite my mother's objections that we could not afford such altruism. My father contended that it was good for me to be in the company of such learning, and I, for my part, loved the old man. I recall taking him his pitifully small salary when he was no longer able to come to the studio, seeing him with swollen limbs and in pain in his terminal illness. Even then, he would inquire what I had read lately and seemed to forget his condition in the excitement of making sure that I had not missed some important point.

When I reached six, I began my academic career at the Ray School, a public grammar school that kept the students off the streets during school hours up through the eighth grade. The principal, an imposing figure of irreproachable character, had the misfortune to be named Arthur O. Rape. He was really the epitome of respectability, as was Hiram B. Loomis, the principal of the next higher institution of learning I attended—the Hyde Park High School. Incongruously, both ruled over institutions regarded citywide as notorious for juvenile crime.

Coming from a sheltered family existence, I was terrified by my first exposure to the hard facts of life in the school yard. I compensated for my small size and lack of aggressiveness by superior scholastic achievement, however, and happily for me, the school bullies decided that I needed their protection. I used to walk home with Wally Herman, a muscular and athletic youth of formidable prowess as a schoolyard brawler, and I often joined him in chasing and harassing terrified girls and other natural prey. Once, I had my comeuppance when Wally suddenly departed out of sheer boredom, leaving me to take over as official tormentor. The girls, three of them, and all bigger than me, deposited me unceremoniously on my back and left giggling as I strove to recover some dignity.

Pressure to stay at the head of my class came from two sources. First was my father, who subjected me to an inquisition almost every night about what had happened in school during the day. He expected report cards to show nothing but "superior" or "excellent" (S or E). A grade of merely "good" (G) was unacceptable and a "poor" (P) unthinkable. Second was the expectation of the boys in the class—including the school bullies—that I would uphold their superiority over the girls in scholastic contests such as arithmetic competitions and spelling bees.

Nor was this all. I was also required to attend Hebrew school after normal school hours and perform in a superior fashion there as well. The estimable Rabbi Benjamin Daskal had visions of my becoming a famous theologian and was supplying me with advanced commentaries by ancient Hebrew sages.

Finally, since it had been discovered that I possessed considerable musical talent, I was given violin lessons, first by a Miss Bailey at a local music school, the Wilson Conservatory of Music, and thereafter by a rough, cigar-smoking type named Edward Roeder, a Belgian violinist of some ability, whose ideas on discipline were extreme. Seeing possible fame and fortune for himself in my future as an infant prodigy, he drove me relentlessly to achieve precocious virtuosity. Having a remarkable ability to sight-read almost any score and a facility to carry a performance through in a reasonably acceptable manner, I was soon giving full-fledged concerts in and around the South Side. Another boy, named Meyer Kaplan, was in a similar fix. When I was nine and he was ten, we played a concert that was duly reported as follows in the July 27, 1923, issue of *Music News*:

> Two little boys who played at this time were certainly convincing evidence of the fact Mr. Roeder has a new, unusual and extremely rapid

> method of teaching beginners. These two are Martin Kamen and Meyer Kaplan, aged 9 and 10 years, respectively.
>
> Martin has studied just one year and Meyer only two years and yet they both display remarkable advancement, playing with absolutely sure intonation and good musicianship, as well as with technical excellence, to say nothing of the fact that both very evidently, on Friday evening, understood amazingly well the content of the difficult numbers they played.

The notice went on to express amazement that I had memorized in the short space of three weeks the J. B. Accolay Concerto. (Written by a long-forgotten nineteenth-century musician, this was a popular piece for all aspiring violinists.) This feat of memory had been facilitated by my strange ability to visualize a page and recall its contents as though it were actually in front of me.

In sum, I was under extraordinary pressure as a child because of my father's resolve that I realize my full potential and amount to something. My sister Lillian, being a mere girl, received little notice. Although there is no doubt that he loved us both dearly, this did not come through to me because of the daily nagging I received, or to her because of his assumption that her welfare was entirely my mother's concern.

The pressure of these expectations was mitigated by a rich family life, organized by and centered on my mother, who provided all the love and sympathy I was not getting elsewhere. At least once a month, sometimes more often, the extended family—mostly aunts, uncles, and cousins on my mother's side—would congregate for an incredible feast wherever our home happened to be. My mother cooked enormous quantities of all the traditional dishes, sometimes forgetting just how much she had prepared. I remember the end of one such feast, when after several dozen members of the family as well as a few friends and retainers had collapsed in the living room, she suddenly cried out, "My God—the roast! I forgot the roast!"

Some of these reunions were accomplished even in the era before the Buick, when we had no transport other than the streetcar line to carry us many miles to the relatives on the far North Side. After we "got wheels," it was still an all-day affair to travel to the North Side and back, leaving only a few hours for the family reunion. For me, these affairs were something of a trial because I would almost invariably be put on exhibition as the coming young Heifetz, making me highly unpopular with the numerous cousins present—all of whom were reminded by their luckless parents that they lacked my grace and talent.

I had hardly reached my early teens before I began to sense that authority is not always infallible and enlightenment not invariably associated with goodness. The earliest lessons came from an unlikely source—the Boy Scouts. Not surprisingly, I had become keenly anxious to be a Boy Scout and had joined as a Cub just before entering high school. I could hardly wait to reach twelve and become a full-fledged Scout. After a year of painless activity, during which I passed a few merit tests, I prevailed on my parents to let me attend a summer camp. A uniform was purchased and a photograph duly taken. I obtained a Scout knife, axe, and water jug and prepared for the summer, which began with my joining a large crowd of eager Scouts who had converged downtown on the municipal pier. I think the year was 1926. We were then taken from the charge of our families and shipped to South Haven aboard a first-class steamer. The trip across Lake Michigan was pleasant, as the lake was in one of its occasional calm periods, but we were rudely awakened from our dreams of a carefree summer when we landed and started for the camp, which was a few hours' drive east of town. The trip was made in overcrowded trucks, in which we stood all the way over a jolting, rough road. On arrival, we found that the camp was a complex of ramshackle canvas tents, grouped in sections, each named after a Scout commandment. I found myself in "Camp Courtesy." The connection between the name of the camp and the character of its inhabitants was nonexistent, as I soon discovered, noting that "Camp Reverence" housed the most unregenerate gang of young thugs in attendance.

Our camp day began about thirty minutes before dawn with reveille. To be sure that we awoke, the trumpet was consigned to a boy whose experience with it was minimal. He produced an incredible cacophony, which startled everyone awake. Stumbling out of our bunks in the damp cold, we groped our way barefoot in the dark to the parade ground for roll call. This being completed, we were driven by exhortations, coupled with force where necessary, into the lake for a morning dip. The lake bottom was rocky and full of mud holes and ill-tempered snapping turtles, and the water was always near freezing.

The morning was spent in busy-work, some of it instructive, such as knot-tieing, cookery over open fires, and other activities applicable to advancement in Scouting. Occasionally, special projects were assigned in the afternoons. The climax of our stay was an overnight fourteen-mile hike, advertised as a joyous experience, including sleeping under the stars. The reality was considerably less exhilarating, largely because rain dampened the fires needed for cooking. Moreover, we had to sleep to windward of a silver fox farm.

An indication of the quality of the food provided can be obtained from the fact that a few days after our arrival a truck loaded with rotten bananas, a rejected shipment acquired by the Scouts in town, penetrated only a few yards into camp before it was engulfed by a horde of famished boys wolfing down the moldy, evil-smelling fruit, wholly disregarding the many spiders swarming over it. Lodging was also inadequate. The tents were open to rain and there was no proper drainage. The result was that we lived in a state of constant dampness.

It can be appreciated that we hailed the day of departure as a deliverance from hell. The relief at the prospect of returning home quite cancelled the pain of the jolting, crowded truck ride and even of the miserable voyage back across a rough Lake Michigan in a boat famed as the "Sea Sick Carolina"—officially the S. S. *Carolina.* This unhappy craft lacked stabilizers and wallowed in the swells of the stormy lake in a manner that guaranteed nausea for all. When we debarked at the pier in Chicago, hollow-eyed and indescribably filthy, we must have presented an unsettling sight to our anxious families. It took several baths to remove the accumulated grime, and several weeks of home cooking to restore the weight I had lost at camp. I had, in addition, managed to inflict a nasty cut on the forefinger of my left hand, which threatened my career as a violinist. I also had received an examination at camp that resulted in an off-hand diagnosis of a hernia. A visit to Dr. Farley was required to disprove this and allay my parents' concern.

This sad tale should not be taken to mean the Scout movement was all that bad. For one thing, the whole camp program was provided at a ridiculously low cost (something like thirteen dollars, all expenses including transportation being covered). So one might say we got our money's worth! For me, it was a disaster, and the first step in my evolution as an anti-group type. Experience through the years of adolescence and later did nothing to strengthen my respect for policemen, ministers, and other figures of authority.

Meanwhile, I dutifully attended school, got top grades, played concerts, and performed as expected. But I was chafing at my isolation from other children my age, of whom only normal performance was required. Nor was the turbulence of life in a big city in the twenties, with the sensationalism attendant on race riots, gang wars, and the Loeb-Leopold case, conducive to cultivation of an inner sense of security.

One Christmas Eve, when I was about ten years old, it had snowed and I was out coasting on a sled down Blackstone Avenue in front of our apartment. At the corner there was a drug store. When I reached it, I turned and flopped back on the sled to coast back to our apartment. Then I heard several loud reports. Looking across the street, crowded with Christmas shoppers, I saw a man who had been running clap his hand to his head and fall to the sidewalk. A policeman with gun in hand arrived on the scene. Glancing up, I saw a bullet hole in the showcase of the drug store just where I would have been standing if I had not at that precise moment fallen on my sled.

Perhaps I should not leave the reader with an exaggerated impression that life was a constant gamble. As was remarked by a famed pundit on an occasion I shall refer to later, "Many were killed, but not all!" In fact, I only saw one other incident in which there was a crime committed in all the time I lived in Chicago, and that did not involve a shooting.

One should not underestimate the cultural richness of the Chicago scene as a counterfoil to the urban squalor. Memorable concerts I attended with my father were frequent. One of these, in 1928, the last Pablo Casals was to give in the United States for nearly a half century, was especially notable. I believe that out of the five or six hundred auditors there to worship at the feet of this master, my father and I were possibly the only non-cellists. I had not reached a stage of sophistication beyond the impression that the greatest music ever written was Tchaikovsky's "March Slav." I even resented the imposition of the Chicago Civic Opera on me in substituting Rossini's *Barber of Seville* for Gounod's *Faust* at the last minute (the fact that the cast included Tito Schipa, Feodor Chaliapin, and Virgilio Lazzari I only appreciated subsequently with the wisdom of hindsight). At the Casals concert, he received the "cooperation" at the piano (so the program stated) of a tall, thin, bushy-haired, and very competent musician, one Nicolai Mednikoff. They made an odd pair, as Casals was short, fat, and bald. The impact of Casals's playing was inexpressible, not only on me but on the whole audience. I remember that the most impressive offering, even to me at the age of fifteen, was Casals's opening with the Bach Unaccompanied Partita in D, which was received by the audience in stunned silence.

I was smaller and less athletic than my schoolmates and so had suffered from a sense of isolation for which I compensated by constructing a fantasy life in which I was a star athlete, even imagining whole leagues of baseball teams and lineups for each team. (Later, in high school, I was manager of the school soccer team, for which I earned my one and only athletic letter.) Like any normal South Sider, I was a rabid supporter of the White Sox, and had only contempt for the other major league team, animals known as the Chicago Cubs, who represented the enemy—that is, the North Side. To be a White Sox fan required a degree of faith wholly inhuman. As the "Black" Sox, they won one pennant in 1919, when I was six. I had to wait forty years before they won again. The revelation of the chicanery of my heroes in selling out the World Series in 1919 was a giant step forward in the promotion of a basic cynicism I was to develop in years to come.

A consequence of all this interest in baseball, coupled with my tendency to memorize everything in sight, was a knowledge of the personnel of every team in any of the major and most of the minor leagues in the country. (My head is still full of the lineups and statistics for both Chicago teams during the first half of this century.)

When I reached thirteen, preparations for a Bar-Mitzvah were in order. Thanks to our growing prosperity, no expense was spared. Nor was I. The beadle of the synagogue, who was rarely sober and whose appetite for garlic and onions was always noticeable, informed me that I had had the good luck to draw the longest Bible reading of the year. It was the famous portion of Isaiah beginning "Arise, shine—for thy light has come." I learned it by rote in incomprehensible Hebrew. It is still a mystery to me why, in seven years of intense study, I was not taught Hebrew as a spoken language.

In addition to the session at the synagogue, I also was scheduled to give a speech of thanks to my parents at a big reception held in the evening, besides playing some selections on the violin. The speech was a kind of takeoff on the Gettysburg Address. The affair was held at a popular dining hall—Sam Gold's Florentine Room on the near South Side—and was attended by a paralyzingly large mass of family, relatives, and retainers. A major feature—from my standpoint—were the baksheesh they brought as presents. After the dust had settled, I found myself the owner of innumerable pen-and-pencil sets, with a straggling of religious artifacts and books.

The significant results were twofold. First, I came out in the open—as a non-believer, that is. Second, one of those who attended the affair and heard me play was the cantor of the synagogue, a Mr. Kittay, who informed my father that I needed competent instruction. He had penetrated through the facile and superficial nature of my performance and seen that my talent had been grossly exploited. Indeed, because I had the disability of being so good a sight reader and could get along with minimal amounts of practice, I had not acquired a soundly based technique. Kittay recommended that I seek instruction from Ramon B. Girvin, one of the two leading violin pedagogues in Chicago.

A few weeks later, through Kittay's good offices, I had an audition with Mr. Girvin. A number of his students (who were to become lifelong friends later) tell me Girvin was astounded by my virtuosity and that I was regarded with awe by the students, who had received advance notice of the phenomenon about to descend on them. But Girvin was not so dazzled as to fail to notice that I was really faking most of what I played. It seemed to him that I had enormous unguided talent, and that with application and proper direction, it would flower. He maintained this optimism against all odds through the next six years, helped by a fervent belief in Christian Science and in the essential goodness of humanity. Regrettably, I had been irreversibly turned off by my experiences as a child prodigy and had no desire to stay in the limelight or put in the long hours of systematic practice required. I retained my habit of getting through lessons with as little effort as possible.

However, having finally been exposed to the richness of chamber music as revealed at the Girvin Conservatory, I became interested in playing the viola, an instrument unlikely to keep me in the limelight. At thirteen, I entered Hyde Park High School, where I became concertmaster of the orchestra. Spurred on by my growing desire for anonymity, I volunteered to learn the viola when Miss Finley, the music teacher, spoke about the need for a viola player. She was dismayed at this development, for she needed a concertmaster more than a violist, but I was adamant, though I softened the blow somewhat by remaining first violin in the high school quartet.

That summer, I took the school viola home and learned to read the viola clef and play the instrument after a fashion. From then on, I lost interest in the violin, and Girvin's expectations were doomed. He persisted in violin instruction for me, refusing to take my interest in the viola seriously, so that I had little formal training on the viola then, and none after leaving his school. Borrowing Girvin's viola, however, I busied myself learning the chamber music literature—an effort that was to prove crucial in establishing one of two bases for much happiness later in life. The other—science—was yet to come.

I was well embarked on my self-education in the chamber music literature and playing in numerous ensembles at the Conservatory and elsewhere. It was clear that I was on the way to a professional career as a violist. A tryout soon for the Chicago Civic Orchestra, the feeder organization of the Chicago Symphony, was in prospect.

I should mention what little more I know about my origins, even though this knowledge was acquired years later. All through my childhood, my father kept impressing on me that learned scholars and famous rabbinical commentators abounded in his family. I tended to put all this down to an overactive imagination and a desire to magnify unremarkable individuals who may have had only pretensions to high scholarly status. One day he pointed to a photograph on the wall showing a distinguished-looking gentleman with a beard. Chin in hand, sunk in thought, he appeared to be staring off into space reflecting on some knotty philosophical problem. This, it seemed, was a grand uncle who had taught in Vilna (now Vilnius, in Lithuania), and had known twenty-six languages. If I didn't waste time with "those bums on the street," he declared, I might turn out to be as learned. Some fifteen years later, while at the University of California in Berkeley, I had occasion to go to the main library to track down some publications of the physicist James Franck. In so doing, I chanced upon a reference to one Israel Frank-Kamenetzky (born 1880–died?) who had written a doctoral thesis at the University of Königsberg in 1911 on the works of an ancient Arabian sage who predated Mohammed by some hundreds of years. A copy had come to Berkeley because the two universities had an exchange program in comparative religion. I went into the stacks and retrieved it (the first person to do so since it had been received). It turned out to be a 58-page document, almost entirely in Arabic with learned footnotes in Hebrew, Greek, and German. A curriculum vitae, in German, was decipherable and from it I learned that the author had taught in a gymnasium (high school) in Vilna and had had a career that seemed to accord completely with the data my father had provided to his incredulous son that day back in the twenties. What was more, there was a thesis in the same area of theology by another Frank-Kamenetzky, obtained at the University of Berne some years previous to that of Israel Frank-Kamenetzky.

It is possible, therefore, that I am related to the Frank-Kamenetzkys who have been prominent scientists (physical chemists) in the Soviet Union. I have never had an opportunity to check on this. Israel Frank-Kamenetzky died in a typhus epidemic in World War I, and most of my relatives in Eastern Europe probably did not survive World War II. My father kept a lock of brown hair inside a pink silk handkerchief with a notation in his handwriting, "Souvenir from my aunt Olga, sent to me from Russia, who had been killed by the Germans during the war with Russia. October, 1940."

In the early fifties, Isaac Stern, whose friendship was one of the many wonderful results of my involvement with chamber music, was entertained at a cocktail party at the home of a prominent doctor after a concert in Buenos Aires. His host asked if he knew a cousin of his in the United States who played the violin. To Isaac's amazement, the cousin's name was Martin Kamen! On returning Isaac informed me, to *my* amazement, that I had a cousin in Buenos Aires. I later received a letter from this cousin, a Dr. Novik, stating that he had arranged with the Argentine Atomic Energy Commission for me to give some lectures on nuclear chemistry there. I was aware there were relatives in the Argentine, but had no notion who they might be. One of my father's sisters emigrated there and married a José Novik. They had prospered greatly and owned a ranch near Rosario.

Later, in 1968, I received a letter that had taken some six months to reach me in Paris, where I had gone to start a new program on the biochemistry of photosynthesis at the French government's scientific establishment at Gif-sur-Yvette. This letter had been written by one Julio Kaplan, informing me that he was a distant cousin and was hoping he could see me when he came to Chicago to study business administration at what I took to be a small college in the downtown area of the city. It had been addressed to me in Chicago (which I had left some thirty years previously) and had made its way through all the places I had been in the intervening years before finally turning up at Gif-sur-Yvette. I answered cordially, inviting him to visit me when I returned to California, hardly expecting to hear more. The next winter I received an honorary doctorate at the 1969 fall quarter convocation of the University of Chicago. As I idly turned the pages of the program, I was flabbergasted to read that this same Julio Kaplan, who had earned an engineering degree in Santa Fe, Argentina, in 1962, was to receive a master's degree in business administration from the University of Chicago at the convocation. When he was called by name to receive his diploma, he strode up and past me. Although I was readily identifiable as his cousin, being the only occupant of the stage receiving an honorary doctorate, he made no sign of recognition. Moreover, he seemed to bear no resemblance to me or any of my family. However, after the ceremony, he came running to intercept my wife and me as we were leaving the chapel and said he was indeed my cousin.

Another wildly improbable event, which occurred in New York in the forties, indicated that I might have cousins from a branch of my father's family that had fled to Italy. At the time I was dating Beka Doherty, a research reporter at *Time,* who was eventually to be my wife. She had met a wonderful Italian correspondent sent to New York by the Milan paper *Corriere della Sera*. This journalist, introduced to her as "Mike Kamenetzky," was famous for having worked in the Italian underground during the German occupation under the alias Ugo Stille. When we were introduced, there was no question about a resemblance, not only in physiognomy but even in our rapid-fire manner of speech.

The improvement in our financial status in the late twenties supported a move across the tracks to a prestigious new address on Cornell Avenue in an area of upper-middle-class families. We now lived in a relatively luxurious eight-room apartment in a building we "owned"—that is, on which there was a mortgage being retired. My father had speculated in real estate through the twenties, and his operations had continued to pay off in spectacular fashion, though my mother still clung to the belief that there was something wrong with money acquired in this easy fashion. In fact, we were rich—like so many others—only on paper. But the conviction that prosperity was permanent gripped everyone as the fever of speculation mounted. Hyde Park was expected to have a future unrivaled in the city, and even the world, for "residential and business advantages." No one was aware a crash was in the offing.

My knowledge of the outside world did not extend beyond the confines of the square-mile area in which I had been reared. The notion of going to faraway places, coupled with an interest in oriental studies (surviving from my early days in Hebrew school) prompted a vague suggestion on my part that I might seek a career in archaeology, a subject appropriate to study at the University of Chicago, with its Oriental Museum and distinguished Department of Near Eastern Studies under the aegis of J. H. Breasted. The fact that this was hardly a career for making money was of no moment, as it was clear that there would be no great need to worry on this score, with the future of the family fortunes ever bright. Optimism persisted in the face of unmistakable signs that the boom times were ending.

At my graduation late in 1929, as a feature of the program, I played the first violin part in a duet with one George Wilson, a fellow graduate. (This turned out to be my last public appearance as a violinist.) There was still a conviction that I would have a career of some sort in music, although probably not as a soloist. My mother was particularly unenthusiastic about my continuing in this vein, because she knew the traumas I had endured as a result. On one occasion, while performing Sarasate's well-known "Zigeunerweisen" at the Eighth Street Theater in downtown Chicago, before an audience of over a thousand people, I had blanked out halfway through the piece, and only became conscious of what I was doing during the last few bars. Apparently, I had carried through solely on reflexes. I was told I played brilliantly, although I remembered nothing at all about the performance, except its beginning and end.

At Hyde Park High School I had chiefly taken courses in the humanities and classics, exhibiting no interest in, or aptitude for, things mechanical or scientific. However, some glimmer of a talent for logical thought appeared to exist, as I had taken all the mathematics offered. Hyde Park High was known to have a highly competent mathematics faculty, which had built an impressive record producing students who won a major share of the math scholarships offered in citywide competitions by the university.

Because of my grade skipping, I finished at Hyde Park High at the end of the fall and winter semester early in 1930 and could not enter the university before the spring quarter, which began a few months later. There was thus some time to discuss what my program in college should be, but no definite conclusions were reached as I prepared for university life.

2

College Years
(1930–1933)

WRY REFLECTIONS on the vagaries of vocational guidance were very much in order in the spring of 1930, when I found that the university administration had assigned an organic chemist, J. W. E. Glattfeld, to be my program advisor. I was sure I would not be majoring in any aspect of science, whatever my academic career might be. How wrong I was eventually became evident, particularly to Dr. Glattfeld, who found me returning to see him only eight years later for advice in his field of expertise—the chemistry of C_4-saccharinic acids. But back in the spring of 1930 our concern was the expeditious disposal of a general science requirement. Dr. Glattfeld proposed an apparently painless possibility—the introductory course in geology. As a tentative probe of mathematical leanings, based on the fact that I had taken all the relevant high school courses available and had done well in them, he also recommended the spring quarter offering in college algebra. As the beginning of my prospective major in the humanities or classics, I enrolled in the beginning course for English majors—rhetoric and composition, and looking toward the fall quarter, I planned a year of German.

Meanwhile, my involvement as a chamber musician and violist was deepening. Downtown at the Girvin Conservatory I became a member of several chamber groups. I earned teaching certificates in theory and harmony under the tutelage of a noted German authority, Dr. Albert Noelte.

More significantly, my social horizons were broadening as I gained entry into intellectual and cultural circles because of the demand for violists in amateur chamber groups. I found myself traveling all over Chicago and its suburbs, penetrating even the socially elite northern areas such as Wilmette, Winnetka, and Evanston. On one of these occasions, at the home of an eminent brain surgeon and accomplished cellist, Dr. Adrian Verbrugghen (the son of Henri Verbrugghen, then conductor of the Minneapolis Symphony), I had my first encounter with hard liquor—a highball made with Canadian Club rye whiskey. I found it repellent and could not imagine why Prohibition was having such tough sledding.

Another devoted amateur cellist, Dr. Otto Saphir, head of a pathology division at Michael Reese Hospital on the near South Side, recruited me into his group. Sessions at the Saphirs were always occasions for much good music and camaraderie, not to mention a most elegant buffet. Occasionally, Otto would disappear, called to the hospital for an emergency consultation or autopsy, but these mischances were not allowed to interfere with the party in progress.

The devotion to music and the arts of the Saphirs and the Verbrugghens was exhibited on a somewhat less elevated level at an "engagement-breaking party" organized by a Chicago Symphony violinist who had been exploiting the hospitality of a prosperous Gold Coast Jewish family that considered him a fine prospect as future husband of their nubile daughter. As pressures mounted, he saw the need to be extricated and suggested a party at which the family could meet his friends. Delighted, they arranged a gigantic buffet for the great occasion, whereupon a mob of some thirty Bohemians, including a gaggle of woodwinds and strings, ranging from violins to basses, not to mention a child prodigy cellist, complete with mother, descended on them. The group was rounded out by an Ethereal Type of indeterminate sex, who recited Sanskrit poetry, accompanied insubstantially by a partner at the piano. Everyone had a voracious appetite. As the night wore on, massive quantities of delicatessen, ordinarily enough for any normal group, kept disappearing, creating emergencies that the harassed family had to meet by endless calls to the caterers. I came away in the early dawn in the throes of an incipient gastritis, and facing the grim prospect of the jolting, hour-long ride back to the South Side on the No. 1 streetcar. The object of the whole exercise—to break off the engagement—was achieved, but at no small cost! Another objective—breaking apartment leases—was achieved by the same kind of exercise on a few other occasions.

The Gary Philharmonic Orchestra was organized and directed by a young University of Chicago student and gifted oboeist, Robert Buchsbaum. He prevailed on a few of us professional outsiders from Chicago to come to his hometown to help his struggling orchestra present an occasional concert. In return he offered transportation and meals, the latter supplied by his devoted mother and family. The ride to Gary at a dizzying speed on the long since defunct South Shore Railway was an exciting experience, especially as we had read of unfortunates whose cars had been hit while stalled on the tracks and hurled hundreds of feet off the right-of-way.

Usually, we returned early the next morning, but once we decided to stay over and do all eighty-two Haydn quartets at a single sitting. This turned into a marathon effort over two days and a night of continuous effort, with several shifts in personnel and an intermission for a dip in the lake. Toward the end, the haggard group resorted to playing only fast movements.

Forays into the hinterlands near enough to make a round trip from Chicago and back in a day were occasionally provided for the Girvin Conservatory orchestra. On one memorable trip, we went to the little town of Three Oaks, Michigan, where we presented a program of light classics to a farm audience. We were scheduled to leave during the intermission to get back to town before midnight but barely made it because we heard the beginning of the program after the intermission. The performer was a rugged, seamy-visaged old gentleman with a shock of straight white hair that fell over one eye. Sitting in a chair strumming a guitar, he held his audience transfixed with a steady stream of folksy witticisms interspersed with recitations of poetry and renditions of farm ballads. I remember speaking with him just before he went on and could not leave until I had heard some of his program. The other members of the orchestra had the same feeling. So fascinated were we that we almost missed our bus. I was thrilled to discover later that the balladeer was none other than Carl Sandburg.

Particularly important then, and through the years that followed, were my friendships with fellow students at the Conservatory. One of them was Milton Preves, who became first viola in the Chicago Symphony in the late thirties and has remained in that prestigious post for decades. Others included Leonard Sorkin and George Sopkin, who later were to organize the famous Fine Arts Quartet, Jimmy Hansen, Oscar Chausow, and Eddie Gradman, to name a few of those who came to fill important positions in symphony orchestras. Clearly, Dr. Glattfeld could not be faulted for supposing my interest in any academic program he might devise to be peripheral.

Back at the university I discovered to my great delight that the instructor in geology—one King Hubbert, later to achieve prominence as one of the era's leading geologists—was a devotee of classical music. We spent much time listening to and discussing music. I spent much less time on his course, for which I received a C. Possibly the essentially descriptive character of introductory geology, lacking the rigor I admired in mathematics, was a factor in my mediocre performance. Happily, I did better in my other subjects, and salvaged a creditable B average despite the geology fiasco. (Neither of us would have imagined that years later I would join Hubbert as a member of the National Academy of Sciences.)

The experiences of the spring quarter seemed to underscore conclusively that my future did not lie in science, but events took a new turn. The family fortune had deteriorated alarmingly with the Depression in full swing. My father's chronic optimism was shaken, and our hopes of financial security were dimmed. We had lost all of our real estate holdings; and prospects in the photography business, never very good, were even less promising as the inroads of the Kodak camera were becoming apparent. My father kept looking for get-rich-quick schemes in magazines such as *Popular Mechanics* and *Popular Science.* He was so engaged one evening in the summer of 1930 when a discussion took place concerning my future studies. My parents wanted to know what I expected from a career in English, the major they understood I had chosen. I could not make a good argument for a future livelihood as an English teacher, or whatever, nor did a career in music look reasonable, especially as we knew of many established musicians who had been reduced to beggary. My father happened to see an advertisement in the magazine he was reading in which some correspondence school promised that one could "be a chemist and make millions." He asked me if I could see anything wrong with trying chemistry. I agreed it might be worth considering. So, in the fall, I requested a new program, based on a major in chemistry.

I now came under the aegis of Dr. Adeline de Sale Link, who had the responsibility for constructing such programs. The first sequence was a year-long introductory chemistry course, in which she collaborated with the eminent inorganic chemist H. I. Schlessinger. The latter was famous for his dramatic lectures, which on a clear day could be heard a good distance away from the auditorium in the old Kent Laboratory in which the class of some hundreds of students met.

All that was left of my previous program was a year of German—now, ironically, taken to get rid expeditiously of the humanities requirement for graduation. Here again I found an appealing and friendly instructor, Dr. William Kurath, with whom I struck up a warm interaction, such that he even entertained hopes that I could be rescued and brought back to a career in the humanities, presumably in German scholarship. However, I began to find chemistry all-engrossing.

Hitherto I had read omnivorously in philosophy, history, and literature and had acquired a habit of rapid but general, rather than slow but specific, absorption of the printed message. Now I found it necessary to pay attention to details and facts, rather than skimming and occasionally daydreaming as my attention wavered. I realized that I had an enormous gap to bridge and much time to make up, so I used every moment possible to delve into texts and source books, not only those recommended or required in courses, but others that I dug out of the departmental library and book stores. As my mathematical sophistication increased, the level of difficulty I could overcome rose, so that by the end of my undergraduate studies, I had read and mastered adequately, if not completely, many basic works of the great scientists. In addition, I began to appreciate the humanistic values in the early writings of such seminal figures as Boyle, Priestley, and Franklin. I formed the habit of copying text material, so that I was forced to take time over each sentence. I thus accumulated a large store of useful knowledge, which became part of me rather than remaining on the page.

Now that I seemed to have a definite orientation, it was the family decision that I proceed with all dispatch to a quick implementation of the program by doing what I had done in high school—that is, take extra courses in the summers as well as through the regular year of study. I therefore undertook an accelerated schedule, attending the university all four quarters of each year whenever appropriate courses were offered, while at the same time keeping up my activities as an amateur, or sometimes professional, musician.

In addition to the excitement of Schlessinger's course, there were also the equally well organized and effective beginning college physics courses of Harvey B. Lemon, who was aided by a corps of hardworking young assistants. These took three quarters, covering all aspects of classical physics from mechanics through electricity, magnetism, sound, and light. At unannounced times we were treated to talks by a famed, crusty old professor of physiology, Anton J. Carlson, as well as the up-and-coming bright star of physiological chemistry, A. Baird Hastings, who was soon to leave and take a prestigious appointment as professor of biochemistry at Harvard. Unusual efforts were made to keep the lecture and lab material exciting. Gimmicks abounded, based mainly on surprise. Thus, one day, during an unavoidably dull passage on some aspect of mechanics, the classroom was suddenly shaken by the thud of a large object falling to the floor. The professor beamed at the obvious discomfiture of the class. He had pushed a big steel ball off the table. The object of this exercise was to fix in the consciousness of the class Newton's second law, according to which the steel ball should have stayed put.

The Physics Department was accounted one of the great strengths of the University of Chicago. Founded by the legendary Nobel laureate Albert A. Michelson, its activity centered on his special area of precision interferometry. Its base was broadened by the acquisition of Arthur H. Compton, a young Nobel laureate who was creating an actively productive group in cosmic-ray studies. Many other stars also crowded the faculty, among them Carl Eckart, who was beginning his illustrious career as a theoretician with fundamental contributions to the bases of quantum theory.

There were also areas of distinction in chemistry, but with some indications of a need for new blood. The head of the department was Julius Stieglitz, the brother of the photographer Alfred Stieglitz. A product of the late nineteenth century, his active career in organic chemistry had been one of distinction, but his learning essentially looked backward. Organic chemistry had been strengthened by the presence of Morris S. Kharasch and my first advisor, J. W. E. Glattfeld, as well as by some promising younger investigators. There was considerable activity of significance in inorganic chemistry, with Anton Burg, then a young instructor working as Schlessinger's assistant, beginning his notable contributions in boron hydride chemistry. A strong group in ammonia system chemistry gathered around Warren C. Johnson, a product of the great school of inorganic chemistry founded by Charles A. Kraus at Brown University. In physical chemistry, the ranking professor was William D. Harkins, whose contributions to surface chemistry, as well as to nuclear systematics, maintained a tradition going back to that established by the fabled Herbert N. McCoy's researches in radiochemistry. Two younger associate professors, Thorfin R. Hogness and Thomas F. Young, had been recruited from the nation's great center of physical chemistry at Berkeley, established by Gilbert N. Lewis, and were considered highly promising, as was a young instructor and assistant to Harkins, David M. Gans.

My preparation for a career in science (culminating in a doctorate some six years after entry as a freshman in 1930) was to be entirely confined within the mile-square area where I had received all my previous education. This relatively short transition from fledgling humanist to scientist did not go smoothly. In my sophomore year, I got two C's in the first two quarters of intermediate physics largely because of a lack of rapport with the instructor. These were in contrast to an A I earned in the last quarter of the same sequence with another instructor. Although I did extremely well in all my other courses, these C's ruined my chances of a scholarship, then more urgently needed than ever, as the family fortunes continued their decline.

In my first encounters with the quantitative aspects of chemistry, a succession of courses crucial to continued success as a chemistry major, I had the good fortune to experience the firm, but kindly, guidance of Edward Haenisch who presided over the "quant lab" under the aegis of the gentlemanly Willis C. Pierce, who later played an important role in the establishment of a strong chemical curriculum at the Riverside campus of the University of California. The two A's I earned in quantitative inorganic analysis were a major factor in convincing me that I might someday become a chemist.

On the other hand, there was the advanced qualitative inorganic analysis course presided over by Mary M. Rising, apparently devised to discourage all but the most determined from becoming chemists. It consisted of ten "unknowns"—devilish mixtures concocted so as to present every known complication likely to arise in the "wet" analysis system taught by the department, based on an ancient text by Stieglitz. The gossip mill held that Dr. Rising was a particular favorite of the chairman's, which was confirmed years later when the widowed Stieglitz married her. The obvious affection between her and the chairman was noted cynically by the faculty, an attitude transmitted to everyone. As a consequence, she was defensive and hostile much of the time in her class and faculty contacts.

The assistant caught in the middle between instructor and students was a lanky, earnest graduate student, Calvin Yoran. An adversary relationship was in any case guaranteed by the nature of the course. The laboratory was a decrepit, miasmic enclosure on an upper floor of the old Kent Chemical Laboratory, which in itself would have delighted Dr. Caligari. When the exercises were at their height, the atmosphere in the laboratory was a murky, bluish haze compounded of the reactions between hydrogen chloride, hydrogen sulfide, hydrogen cyanide, and ammonia fumes, through which one might descry the blurred figures of the young alchemists going about their work. On one occasion when the chairman thought to favor Dr. Rising's young charges with a visit, he retreated abruptly with an anguished expression of wonder that anything human could exist in such an atmosphere.

The young chemists-to-be numbered several who were preternaturally bright and accomplished at solving the composition of unknowns by clever questioning of the assistant. One day Dr. Rising made a personal visit to see how the class was doing. Yoran watched in horror while she took the notebook with the answers for the unknowns into the lab to receive questions from the students. In the wake of her turn around the laboratory, which lasted a mere thirty minutes, Yoran received several dozen perfect analyses. He was later able to enlighten Dr. Rising, whose suspicions about what the students might be saying were thereby aroused, generating an adversary relationship that resulted in undeserved penalties to students genuinely interested in learning.

On one occasion, a remarkably precocious young chemist, David Ritter, working as an assistant in Anton Burg's research group and already an authority on silica chemistry despite his undergraduate status, was called upon to recite the reactions involved in the qualitative test for the presence of silica, based on the use of hydrofluoric acid to volatilize silicon as a fluoride. It was Dr. Rising's bad luck that she had picked the one student in the class who was completely up to date on knowledge concerning the basic chemistry of silicon. To her amazement, Ritter began a highly technical discourse, which dismayingly bore little relation to the obsolete explanation given in the text. Ritter explained that the silica "skeleton" seen in a bead of water held over the container in which the silica was present was hydrofluosilicic acid (H_2SiF_6), whereas the Stieglitz text described it as silicon hydroxide ($Si(OH)_4$). Dr. Rising interrupted Ritter in full flight to inform him that he was in error, a statement to which Ritter took vigorous exception (a bad mistake on his part, as it turned out), expostulating that analysis showed two atoms of hydrogen, one of silicon, and six of fluorine. Dr. Rising was quite unprepared for this revelation and hastily took refuge in a remark that what happened in a beaker did not necessarily happen in a drop of water, thereby knocking the props out of Ritter's argument—and all of wet qualitative analysis as well! As Ritter kept protesting, his grade dropped. In later years, he went on to forge a career of distinction in the chemistry of paper.

As my academic career unfolded, my musical activities continued, fortunately as it turned out, because I was able to finance my studies by occasional engagements as a jazz violinist playing in the numerous havens of alcoholic renewal up and down 55th Street, while Prohibition continued its ineffectual sway. I joined a small "combo" consisting of drums, clarinet, piano, and violin. I fail to recall who was in it, but remember we had a limited repertoire of four pieces, including "Tiger Rag" and "The St. Louis Blues." Our clientele was usually too far gone to realize how dismally limited our program was. We often finished in the early morning hours with as much as four dollars each, the result of the besotted generosity of our auditors as we passed the hat. (This was a considerable sum for a few hours' work in those Depression days.)

During my sophomore year, a development of major significance for me occurred—the creation of a Department of Music, rumored as owing to the influence of the celebrated soprano Claire Dux, wife of a trustee, Harold Swift. An energetic young musician named Carl Bricken, supposed to be a protégé of Mrs. Swift's, was installed as acting chairman.

Bricken was most ambitious for his new department, but aware of the opposition to its founding. He set about selling critics at the university on its new venture by inaugurating a vigorous plan of development. A symphony orchestra was organized, liberally salted with professionals from the downtown music schools. I was recruited as first viola. As soon as Bricken could raise funds, I received full tuition at the university. This was a godsend because the family finances were by now in complete disarray. Departmental activity was augmented by creation of a university quartet made up of first chair musicians from the orchestra.

Also of significance to the university was the adoption of the New Plan, a venture in undergraduate education begun during the brief tenure as president in the late twenties of Max Mason, a physical scientist. "I believe that opportunity and not compulsion should be the spirit of the undergraduate college. The university has the opportunity of abandoning the childish game of marks and grades and of emphasizing the fact that education is fundamentally self-education," Mason observed. A new plan was created intended to air out the musty recesses of the old curriculum. A vigorous implementer was needed, and he appeared in the form of the thirty-year-old Robert Maynard Hutchins, hailed as the Boy Wonder of American Education.[1] A glamorous figure as the youngest dean of the Yale Law School (he had been appointed when he was only twenty-seven) and married to the beautiful and statuesque Maude Phoebe Hutchins, an accomplished artist and writer, whose poems were deemed good enough to appear years later in the *New Yorker*, Hutchins hit the campus like a whirlwind. Indeed, it soon appeared that the trustees and faculty had gotten more than they had bargained for, in that the new president's entourage included his friend, Dr. Mortimer J. Adler from Columbia University. Adler had acquired considerable notoriety as a critic of modern educational philosophies and an exponent of ideas later to be developed in the "Great Books" program at the university.[2]

I received honorable mention for excellence in junior college work in the middle of my junior year (at the end of 1932), by which time I had negotiated the two-year sequences in chemistry and physics, as well as mathematics through college calculus. As for the humanities, I had abandoned them completely once my year of German was over. The German sequence was, however, the only one in which I managed to achieve straight A's. In the sciences and mathematics, an occasional B intruded.

Beginning in the summer of 1932, I began in earnest on my eventual major in physical chemistry. This involved general physical chemistry, atomic and molecular structure, "contemporary" theoretical chemistry, and advanced experimental physical chemistry, to which I added advanced physics courses in electricity, magnetism, light, sound, line spectra, and atomic structure.

Meanwhile, the Depression was deepening and bringing to the campus an ever growing spiritual malaise. The university, regarded as a hotbed of radicalism by the downtown Establishment, as represented by the Hearst press and, in particular, by Colonel Robert McCormick and the *Chicago Tribune* (by its own claim "The World's Greatest Newspaper"), was increasingly becoming the target of hostile publicity. Notables among the faculty such as Robert Morse Lovett (English) and Maynard Krueger (Political Science) were especially active in leading socialist activities and were convenient foci for press attack. Many students, reacting to these provocations, and naturally rebellious against the System, as it appeared to falter and come apart under the economic pressures following the collapse of the stock market, actively took up political protest, which found an outlet in agitation against the rise of fascism in Europe in the early thirties, with a corollary sympathy for communist Russia. I was beginning to be drawn into these activities, occasionally attending meetings of groups affiliated with, or belonging to, organizations such as the Young Communist League, the Socialist Club, and others that found their way onto a list of organizations and groups proscribed by the attorney general in later years. A natural disinclination to run with the mob kept my participation at the innocuous level of merely verbal protest and the signing of petitions, which circulated abundantly around the campus. While I cannot recall ever actually joining anything, peer pressures were strong for some participation. Hardly anyone I knew was not passionately involved in protest against what was perceived as the steady downward slide of the nation toward chaos and catastrophe.

There was unrest in academia as well. The transition from the old order of Newtonian mechanics to the still undefined new system of quantum mechanics confirmed even for the supposedly stable natural sciences that the values of the past needed drastic reevaluation, if not abandonment. Awareness of this was largely confined to those of us in chemistry and physics, who were hearing disquieting attacks on simple causality as the basis for understanding natural phenomena for the first time.

Our heroes in the arenas of conceptual science were J. Willard Gibbs, Ludwig Boltzmann, and Ernst Mach, prominent among whose contemporary disciples were the physicist Percy Bridgman, at Harvard, and a remarkable mathematician at the University of Chicago, Arthur C. Lunn. I had begun avidly reading the works of Mach, prominent exponent of the hard line of "operationalism," a system that abandoned pictorial representations and emphasized that physical observations could do no more than provide correlations between observations and models that, it was hoped, might have predictive value. In its extreme form, operationalism denied the existence of atoms. (I admit I found it difficult to maintain this view later when I was working for my doctoral thesis and saw the actual tracks of atoms and electrons visible as vapor condensation trails.)

By early 1933, when I was in my senior year, I had joined a group (we liked to consider ourselves an elite few) who had discovered Lunn and thought of him as a kind of Moses who might lead us out of the wilderness. There was a strong drive to identify with him as another victim of the Establishment. Early in the twenties, Lunn had, quite independently, and considerably before the groundbreaking articles of Louis de Broglie and Erwin Schrödinger, proposed an elegantly precise formulation of wave mechanics. He had treated particles as expressions of "beat" phenomena—that is, discrete solutions of wave equations—in analogy to the Rayleigh theory of sound. His ideas were obviously far ahead of their time and appeared wholly unphysical to the then editor of the *Physical Review*, a Professor Fulcher, to whom Lunn had sent his manuscript. The paper was rejected as of no interest to the readers of the *Physical Review*. Enraged, Lunn reacted by withdrawing into a shell from which he never emerged as far as further publications were concerned. Thus, from being an active contributor to mathematical physics, beginning with his critically important work on the Stokes's Law corrections needed to validate Millikan's classic determinations of the electrostatic charge on the electron, Lunn turned into a recluse, working at home and in his office on his ideas about the basic nature of physical theory and, in particular, on applications of group theory and symmetry considerations.

Lunn came to be regarded as "too mathematical" for the physicists and "too physical" for the mathematicians. Notwithstanding, his powerful gifts as a deductive theoretician and master of classical physical theory earned him promotion to a full professorship in the Department of Mathematics, then and thereafter one of the strongest departments in the university.

When our group was first drawn into his orbit, he had ceased to teach elementary subjects and was assigned largely to handling topics of his own choosing at the graduate level. These included the classical theories of thermodynamics, sound, electromagnetism, and vector and tensor analysis, as well as a course on relativity. Lunn's remarkable ability to see all of these areas in an interconnected logical framework provided a basis for understanding the nature of physical science that we were unable to find elsewhere in the university, although we took many courses in physics and chemistry that were supposed to cover the same ground. For example, there were courses in thermodynamics in both the Chemistry and Physics Departments. The former, taught by T. F. Young, was a narrowly conceived exercise in handling various cycles in heat engines, which continued at the graduate level in a soul-deadening effort to work out precise values for activity coefficients of sodium chloride in dilute solutions. In physics, Arthur Dempster taught a conventional and uninspired version of nineteenth-century thermodynamics, replete with its ancient symbolism of lowercase Greek letters. To him, as well as to most of his colleagues in the Physics Department, the system developed by Gilbert N. Lewis and other physical chemists was something alien, or even obscene, and the Third Law an unmentionable abomination. Thus, the beauty of thermodynamics as a logical system, complete in itself, was imparted to us only in the presentations given by Lunn.

Lunn's method of lecturing was refreshingly different from any other we encountered, as might be expected from such a maverick. He never organized his talks around the theme his course was advertised to involve. Instead he would shamble into class, typically attired casually in a worn old suit, settle into a chair, nibble at a piece of chalk, and begin a discourse on whatever he had in mind that moment. He needed no notes, such was his mastery of the material. One could count on basic penetration of the subject matter whatever it might be. Thus, over a sequence of half a dozen courses, his students learned the important features of the subjects covered even if there was no immediate correlation between the talks and the course content advertised. He never made assignments and gave no tests. As far as he was concerned, his responsibility ended with the presentation and it was the student's task to listen and learn.

Lunn had not, however, been so casual in his approach to pedagogy early in his career. One day, so the story goes, he was flattered to receive a deputation made up of a trustee and the department head, who proceeded to compliment him on his excellent teaching. That such august persons were making this extraordinary effort to extol the work of a lowly assistant professor was heady, but the effect was dissipated when the trustee remarked that the football team captain was not achieving a passing grade in Lunn's class. The chairman hoped Dr. Lunn could see the light. Lunn solved the problem, thereafter, by giving only A's.

Lunn was also an excellent pianist and organist—in fact, a musician of professional calibre. His father had been a piano tuner and instrument repair specialist, and the young Lunn had accompanied him when he went on his rounds tuning pianos and organs. Discoursing on the theory of sound, Lunn recalled that on one such occasion he was idly munching a ginger snap while his father was sounding a particular organ stop in a local church. Thirty years later, while walking in the same area, he heard the organ stop sound from the church nearby, and immediately felt the sensation of tasting the ginger snap again! (One might speculate on how long it will be before an explanation in terms of molecular or developmental biology is offered for such a phenomenon.)

Having discovered that we had similar interests musically, Lunn and I began regular weekly sessions at his house, going through the violin and piano sonata literature. We developed a close relationship in this way, which was an enriching experience for me not only in terms of friendship with a great individual but also in expanding my own knowledge of this aspect of chamber music. Moreover, in a wholly privileged way, I learned more about science on a personal basis than would have been possible as an auditor in class. From my senior year through graduate work until I left the university with a doctorate in late 1936, this friendship continued as a beacon in the general darkness of the troubled thirties.

The anti-science faction, led from the president's office, continued to burgeon. Pro- and anti-Hutchins sides formed, culminating in a celebrated debate between the leaders of the two camps. The sciences were represented by a doughty old warrior, physiologist Anton J. Carlson, and the opposition by the smooth-talking dialectician Adler. On the afternoon of the confrontation, Mandel Hall was stormed by hundreds of impassioned student partisans clamoring for admission to an auditorium that could hold only about 800. We chemists managed to get seats in one of the front rows flanking Arthur Lunn, whom we had talked into attendance.

Adler opened with argumentation likely to ensnare Carlson in semantic quagmires, but the latter refused to notice him. Instead, when his turn came, Carlson advanced to the center of the stage, fixing a beady-eyed stare on Hutchins sitting in a balcony box, and growled in his heavy Swedish accent, "Ay didn't come here to shadow box!" He then read a list of ten pertinent questions for the president to answer, and sat down. Adler arose and held forth at length, but with no answers. The debate had become a pair of monologues. As Adler rambled on, a thin smile was seen to flit across Lunn's features. On paper supplied him by his ardent disciples, he scribbled: "The index of progress in science is the ratio of ideas to words!"

However, there was no doubt that the anti-Establishment pronouncements of Adler and Hutchins appealed strongly to many of the brightest young undergraduates, strong on brains and weak on judgment. They flocked to courses by Hutchins and Adler on the "History of Western Thought," which, we science students were given to understand, did not progress beyond the thirteenth century and the commentaries of St. Thomas Aquinas.

As times grew worse, my mother's indomitable cheerfulness and devotion kept the family intact. Although our only means of livelihood was the meager income from the studio, she managed to make ends meet. My father still struggled to find some way to fortune, and hit upon the idea of merchandising rings and bracelets with photographs implanted in them, providing both sentimental gifts and means of identification. He found a partner, a Mr. Scharfman, who had concocted a process for pasting small photos inside plastic and shaping them to fit in a ring. The two of them established a factory downtown, where my father installed Scharfman and some apparatus to copy pictures sent in by customers for processing into photo rings and bracelets. This business gave promise of prospering, particularly as my father could leave the studio entirely in my mother's hands while he promoted the new venture.

Pilgrimages to my mother's family went on as usual, as did routine trips to the North Side to visit various relatives. I had begun to absent myself more and more from these gatherings owing to the demands on my time at the university, as well as to a growing disinclination to be involved in family affairs. Much of this feeling arose from my perception that the family seemed to exist only to make demands on my mother, who always responded uncomplainingly and generously.

Around this time, catastrophe befell my mother's sister Rachel, who had moved with her husband and two children to set up a general store in a small town in the vastnesses of upper Ontario. One winter's night, a fire broke out (arson was suspected) and in the ensuing holocaust, her husband was burned alive and she herself left in the hospital with two semi-orphaned young children. The details were gruesome. There had been time only for Aunt Rachel to leap out of a second story window, struggle erect, although she had sustained fractures in both legs, and catch the children thrown to her by my uncle before the floor caved in under him and he plunged to a fiery death.

The store was a soaked, smouldering shambles after the efforts of the local fire squad. My mother came to the rescue, leading a family group to the town, salvaging what was still salable, and giving Rachel an opportunity to reestablish herself once she had recovered. The gratitude of the family to my mother was, paradoxically, to lead to the tragedy that a few short years later brought about my final estrangement and the breaking of all ties with my upbringing.

In my senior year, I came under the influence of a young faculty assistant, David Gans, who introduced me to the mysteries of research, focusing my eager but blundering efforts on a problem involving the spectral analysis of emission lines excited in ammonia by electrodeless discharge. At the same time, I was completing my obeisance to the traditional curriculum by taking the chairman's year-long lecture course in advanced organic chemistry. Stieglitz had a fatherly interest in his students, and this course was his means of getting to know them. The lectures covered nineteenth-century organic chemistry historically, and even personally, as he regaled us with accounts of the various controversies in which he had taken part.

Stieglitz conducted his lectures with numerous asides to be filled by volunteers from the class. In this way, we saw that we would have opportunities to ingratiate ourselves by displays of brilliance in answering his usually obvious queries. All students aspiring to future advancement strove for seats in the first few rows. Over the years, Stieglitz had come to a resigned attitude about ever seeing a volunteer from the back rows. One day to his astonishment he saw a hand raised in the gallery in response to a query about the "vital" significance of lactic acid! "Yes!" he said in his slightly genteel German accent, "Mr. — —, and vot iss the vital importance of lactic acid?" "It's used in the manufacture of sauerkraut," came the response. Amid general laughter, in which Stieglitz joined, he remarked ruefully, "Val, Mr. — —, I vould imagine that is vitally important to you!" Then, he turned to his reliable front-row auditors and I supplied the expected answer about the role of lactic acid in muscle contraction.

I still have my written exam from the first quarter course, a prolix effort with a disclaimer stating "apologies are due for the illegibility in many sections of this paper—a condition which the writer could not remedy because of lack of time." Nevertheless, I managed to fill three exam books on the variety of subjects posed—mechanisms (as Stieglitz taught them) for oxidation of primary alcohols, hydrolysis of halothanes, Grignard reactions, hydrogenations of chlorinated aldehydes, reactions of the keto group, and so on. This earned an A. However, there was a C for my effort to describe the structural chemistry of amines. In the last of the three quarter courses, I earned an A with a learned dissertation on the photochemical transformation of 0-nitrobenzaldehyde to 0-nitrosobenzoic acid.

Meanwhile, at a somewhat diminishing rate, I continued my studies downtown at the Girvin Conservatory. I was playing viola almost exclusively, except for sessions with Lunn and with a young woman pianist, Jean Williams. I was smitten by her beauty as well as by her exceptional talent. She distinguished herself by winning a citywide contest to play a Stravinsky concerto with the Chicago Symphony. She was a rather tall, strikingly handsome girl with straight, jet black hair and prominent blue eyes. I met her because she accompanied the ballet at the university on the occasion of a performance of Jaromir Weinberger's comic opera *Schwanda the Bagpiper*. This work, enormously popular in Europe in the thirties, concerned the adventures of a rustic type whose playing of bagpipes cast a spell on all who heard it. For me, the major demand was a viola obligato in the second act, played while a sultry contralto—the "Queen of the Night"—belted out a passionate torch song. Incidental to this, there was a ballet featuring scantily clad coeds, which I was unable to watch because I had to pay complete attention to my accursed solo!

On another occasion, we presented Handel's *Xerxes* as produced by Thornton Wilder, then in residence at the University. The opening act aria, the famous "Largo," was a paean of praise to the shade of the plane tree, a simple stylized piece of branch set in a bracket. When the curtain swept back, it caught the branch and spun it around so that it was facing backstage, and the tenor found himself singing about shade that was nowhere to be seen. This production also featured ballet, as well as Wilder himself, who delivered a single line while sitting on the stage among a bevy of extras. This circumstance earned the production mention in *Time*.

Other ambitious projects promoted and performed by the Music Department owing to the enterprise of Carl Bricken included a Brahms Festival in the spring of 1933 to celebrate the hundredth anniversary of the composer's birth. Bricken also produced the final act of Wagner's *Die Meistersinger* in a concert version. The cast, including soloists, chorus, and orchestra, overflowed the stage and taxed the capacity of Mandel Hall sorely. I remember this event particularly because Bricken in his enthusiasm lost count of the final chords. We had finished the piece, when to our horror we saw him wind up for yet another blast. When he came down with a massive final beat, only a bassoon, oboe, a few violins, and I straggled in with a pitifully inadequate ending. Nevertheless, the effort was a great success, most of the unfortunate denouement having been lost owing to the applause that had begun even before the end of the act had been reached.

My association with the university choir had many perquisites. One was the chance to get into the Reynolds Club choir office, which directly overlooked Stagg Field. On Saturday afternoons, one could see most of the football games going on there. Another was entry to Dean Charles Gilkey's office in the inner sanctum of the chapel, where on any given night a small group of us would meet to lounge on the dean's luxurious furniture and drink in the elegant atmosphere while listening to the strains of the organ filtering in through the walls.

A meeting place of quite another character was the music repair shop of J. N. Copland downtown in an old office building at 63 E. Adams Street, just around the corner from Symphony Hall. Here the gang from the Girvin Conservatory, Chicago Symphony orchestra members, and many of the practising musicians in town would congregate, exchange notes, make contacts, and do business in general. Mr. Copland, or "Cope" as he was affectionately known by his devoted clientele, had formerly taught the violin, and had retired to make instruments and do repairs, usually at little or no expense to the needy. His "Greatest Violin Offer Ever Made" advertised a violin, a bow, a case, rosin, mute, tuning pipe, and music stand, as well as an extra set of good quality strings, for the sum of eight dollars. In those days one "Strad model" violin of adequate quality cost six dollars.

In the summer of 1933, Chicago celebrated its "Century of Progress" with a World Fair, an event eagerly anticipated by the burghers of Hyde Park, including the university community. The American Chemical Society organized a special national meeting for the occasion, bringing many famous chemists to the city. I was among a group of senior chemistry students drafted to serve as pages, registrars, runners, and what-not, helping keep traffic moving smoothly. In this way, my growing awareness that scientists were not bloodless automatons was dramatically enhanced, particularly during my meetings with various types who wandered into the area of my station to register. One day, a seedy-looking character wearing a shabby, shapeless suit and a tired old cap sidled into the hotel foyer where I prowled waiting for customers. I was sure he was a tramp who had stumbled into the hotel by mistake. I was startled, therefore, to have him approach me purposefully and ask how much it cost to register for the meeting. "Three bucks for members, eight bucks for non-members," I replied. "Three bucks," he cried, "that's going to break me!" Whereupon he reached into his pocket, dragged forth an object that might have been a wallet and extracted three limp dollar bills, which hung from his fingers like rags. "Okay, what's your name?" I asked. "Leo Baekeland—B-a-e-k-e-l-a-n-d," he replied, his eyes gleaming mischievously. The fabled inventor of the first photographic gaslight paper and the formulator of the first commercially successful polymer, Bakelite, this unprepossessing tramp could easily have bought the hotel!

Another time, the famous Swiss deep-sea explorer and inventor of the bathysphere, Dr. Jean Picard, arrived with his American wife to register. Ushering him to a seat at the desk, I gave him the form to fill out, while his wife stood behind him. He stared, in apparent agony, at the form, muttered confusedly, and, suddenly overcome by the complexity of it all, threw his arms back, and sank sprawling into what seemed to be a coma. Quite calmly, Mrs. Picard reached over his inert form, extracted the paper from his benumbed fingers and finished filling it out. Then, gently soothing, she revived him and led him away, while he complained piteously that it was all too much.

As a former professor of organic chemistry at the university, Picard was invited to lecture to his old colleagues and the students in the Kent Chemical Laboratory theater one afternoon. He arrived with his brother, Dr. Auguste Picard, a physicist who pursued cosmic ray studies in the stratosphere. One could call Jean the "down-going Picard" and Auguste the "up-rising Picard." Professor Arthur Compton sat with Auguste at the back of the room as Jean delivered a speech on one of his favorite topics—explosions and explosives. To ensure the audience knew what was meant by "explosion waves," he began with great energy mimicking the operation of digging a hole in which to plant a stick of dynamite. After a while, he paused, breathing heavily from his labors and appeared satisfied with the results. Then he went on to dig several more imaginary holes at appropriate intervals until he had traversed the whole width of the lecture platform. Finally, he said, "We put sticks of dynamite in each hole, set off the first one and then, vavoom! Explosion wave!" He accompanied this statement with a fast shuffle across the stage, flapping his arms to give the effect of a series of rapid detonations. Collecting himself, he returned to the blackboard and began a learned discourse on the mechanism of the explosion wave, based unfortunately on an irrelevant classical theory of sound wave propagation. Almost immediately from the back of the room there came Auguste's shout "But no!" Instantly, Jean wheeled about, pointed a finger in the direction of Auguste and shouted, "But yes!" There ensued a violent and incomprehensible argument in excited French between the two brothers, over the heads of the stunned and embarrassed audience. Professor Compton tried to intervene and lower the heat, but to no avail. The argument raged on for a few minutes, then was broken off suddenly by Jean who turned his back on the audience, and with folded arms stared darkly at his scratchings on the blackboard. Then, slowly letting his arms drop to his sides and turning to face the audience, he said, with a gesture to indicate how one must put up with all manner of idiots, "Let us proceed."

He talked in a most fascinating way about the manufacture of explosives, and in particular about how mercury fulminate was used as a filler in detonation caps. This highly hazardous material had formerly been stacked in piles on a table, he said, at which an operator sat and selected an empty cap which he or she filled by pushing a bit of the fulminate in with vigorous effort. "Thus," Jean remarked with infinite regret, "many were killed, but not all!" (I have often thought since that this statement could well serve to summarize the twentieth century.) He finished by assuring us that much more caution was now observed in these procedures, with operators moving about quietly and with great care. He illustrated this by tiptoeing off the stage with his finger to his lips, murmuring "Shhh."

The lectures, such as those by George Barger on thyroid hormone, Leopold S. Ruzicka on sex hormones, and Richard Willstätter on chlorophyll, were unforgettable. It was thrilling to see the chemists who had made all of these important discoveries and to hear their own accounts of their work. I still recall the excitement of seeing Ruzicka standing among a group of us students and displaying a small glass tube in which he had many milligrams of the male sex hormone, testosterone (crystallized as the propionate), which he and his associates had isolated and synthesized.

As a reward for our labors, the Chemical Society awarded us free student membership for the year. Of course, we were unaware that this honor required that we continue to pay dues every year thereafter if we wanted to remain members in good standing. This involuntary early induction made possible my celebration of half a century of membership in 1983.

Although the results of my first research were never clear to me, they helped earn me a B. S. degree with honors on graduation in the winter quarter of 1933. At this point I had an education that included a comprehensive major in physical chemistry, and minors in physics and mathematics, the latter including advanced studies in differential equations. I had also made a beginning in studying group theory and its applications in handling wave equations as applied in quantum physics and chemistry. Finally, I had earned election to the national honor society, Phi Beta Kappa.

As for the future, I had convinced myself that I should continue with graduate study. The previous summer spent fetching and carrying for David Gans and the resident graduate student, Henry Newson, had revealed to me some of the excitement and fascination of research, and I was eager to get more involved. Prospects for employment as a newly minted Bachelor of Science were minimal in any case, and for one with a major in physical chemistry they were nonexistent. Rumors had it that Ph.D.'s in chemistry were happy to find jobs at Macy's in New York for fifty dollars a month! At the university, students were eking out an existence on a kind of dole under the auspices of the National Youth Administration and the Works Progress Administration. Scholarship and fellowship awards had been sharply curtailed. Gans had told me, however, that Professor Harkins was willing to accept me as a replacement for Henry, who was on the verge of completing his research on interactions of fast neutrons with nuclei of low atomic number, and that I could count on some fellowship aid.

At home, prospects for some help had improved as the ring business downtown showed signs of prospering. I had been able to make enough working weekends in this business to supplement my campus earnings, even to the extent of paying off the balance owed on my viola. So, as I walked down the aisle at the chapel to receive my degree there was hope in the air and cautious optimism about the future.

3

Predoctoral Years

Tragedy and Transition (1933–1936)

THE GOTHIC aspect of the university quadrangle was mirrored in the ornate exterior of the George Herbert Jones Laboratory, where I was to spend most of my time for the next three years. One entered through a foyer graced by busts of the donor and the chairmen of the department, past and present. The homage to J. U. Nef, the deceased first chairman, made no waves, but busts of the living chairman, Julius Stieglitz, and the donor—a prominent steel mogul still very much alive—seemed an intolerable exhibition of hubris to certain graduate students in physics. One afternoon, Stieglitz, showing off the building to a Jones family group, was mortified to find that the busts of Mr. Jones and himself had been replaced by some repulsive likenesses of Neanderthal Man, transported there by parties unknown from the anthropology museum in Rosenwald Hall. The whereabouts of the missing sculptures was cleared up several days later when the university purportedly received a postcard from the Art Institute downtown stating that some objects, apparently belonging to the university but of no artistic merit, had been left (as a donation?) on the steps of the institute. The card went on to state that the objects under consideration were not acceptable as art. The university was requested to indicate what was to be done with them. They were duly returned to their proper pedestals and the Neanderthal Men to their proper roosts in Rosenwald Hall.

The same gang of graduate students, based in Ryerson Laboratory, was also credited with the invention of "Bastard Decelerators," dedicated to harassment of unfortunates apparently close to finishing their doctoral theses. Such types, known as "bastards" obviously required "decelerating." To this end, various devilish maneuvers were devised, such as tampering with the internal arrangements of meters and the drilling of tiny holes in barely accessible areas of vacuum line oil pumps. Carl Eckart subsequently told me that a leak in his vacuum line, which he failed to locate for two years, turned him into a theoretical physicist. While it may be a great experimenter was lost, it is certain a great theoretician was gained.

It should not be supposed that life was less interesting in the Jones Laboratory. The existence of an Emergency Room with a cot, presumably for the injured, but convenient for clandestine couplings, caused at least one administrative flurry when a graduate assistant and his girl friend were caught in flagrante delicto by the assistant's research supervisor. The faculty, meeting in emergency session, solved the problem by decreeing that henceforth there should always be a light burning in the Emergency Room. Still another crisis, caused by a janitor stumbling over a couple on the floor of a basement laboratory, was met by the ukase that all shades be removed from doors. With Prohibition in full swing, undenatured reagent grade ethyl alcohol from the store room was tapped frequently, despite heavy governmental restrictions and penalties. The inadequacy of bookkeeping then in accounting for alcohol calls to mind the futile efforts of the federal government to keep track of its plutonium and enriched uranium supplies some fifty years later.

Much of the extracurricular activity in these Depression years quite naturally centered on politics. The chemistry graduate students were split into vociferous and passionately involved groups on both the right and the left. The rightists were found for the most part among the organic and inorganic chemistry students, probably because they were still sought after by industry. The physical chemistry students had less obviously marketable skills, so none of them were likely to receive invitations for interviews by prospective employers. Their resultant feelings of alienation and disenchantment with the Establishment created strong sympathies for leftist causes. Charges alleging "Red influence" in the content of some courses in history and sociology were brought by the niece of a drugstore tycoon, and were taken up by the downtown press, eventually leading to one of many investigations of alleged communist and socialist influence at the university.

On May Days, there were always big parades organized by church, labor, and leftist groups, which set the stage for action by the political factions in the Chemistry Department. I recall that on the evening before one such event, signs proclaiming various slogans of support for left-wing causes were being nailed together in the basement of Jones Laboratory, where the physical chemists were concentrated, while up on the third floor in the organic wing, stink bombs were being fabricated to use in breaking up the same demonstration. Meanwhile, roving bands from the American Legion prowled outside looking for communists.

The antipathy between the two groups of graduate students was exacerbated further by the intellectual ferment attending the rise of quantum chemistry. Many of the physical chemists, particularly we "Lunnites," were passionately concerned, but the organic chemists were not in the least aware or interested. In the various courses taught in advanced organic chemistry, for example, isomerism and configuration effects were handled in the most naive fashion possible, no attempt being made to provide students with the general theory of permutation underlying the phenomenology of isomerism. The resistance of most practicing chemists to any attempt at mathematical sophistication in chemical theory was still pronounced in the early thirties. Willingness to accept the approach of quantum chemistry, with its formal invocation of wave and matrix mechanics, was still several decades away.

It is added irony that the one paper Lunn permitted himself to publish late in his career was again far ahead of its time. In 1929, he and James K. Senior, a member of the Chemistry faculty, who himself experienced little understanding on the part of colleagues in organic and physical chemistry, published a remarkably perceptive and original paper on isomerism and configuration[1] in which a general approach to systematics of isomer classification and prediction of isomer numbers was developed as an application of permutation group theory. Although it appeared in a reputable and widely read chemical journal, it had no better fate than Lunn's unpublished paper on wave mechanics of a decade earlier.

The itch to comprehend the new quantum knowledge drove me to sign up for more courses in physics and mathematics than I did in chemistry (behavior that obviously did not ingratiate me with the department administration). Our little band in the basement also organized our own seminars in group theory, using the classic texts of Hermann Weyl and Eugene Wigner. The physics graduate students—we might refer to them as the Ryerson Gang—noted with amused condescension our presence in an advanced Physics Department course given by Robert S. Mulliken. It was in this class that the intellectual pretensions of the Ryerson Gang were unmasked. One day, a ringleader of this group seated prominently in the front row, and one of the more outspoken deriders of the chemists (always self-banished to the back rows), raised his hand and asked Mulliken a question. Unaccustomed to interruption, and assuring himself that he had heard rightly, Mulliken looked up, blinked, and remarked with some asperity that if the questioner had been listening, he would have known that he had been talking about that particular topic for the previous two weeks. Suddenly, it was clear to us that our supposed intellectual superiors were in fact no better than we. We even began to question the professor's own basic understanding when we asked him to predict the magnetic properties of a compound of borane (diborane), which had just been measured by Simon Bauer, one of our chemistry group. It happened that this compound contained the same number of electrons as did molecular oxygen. The professor hazarded a guess that it would therefore be identical in its magnetic properties. In fact, it was devoid of any magnetism, whereas oxygen exhibited the opposite behavior, being quite magnetic. One could recoup by some ad hoc theorizing, but we felt that it was not much of a theory that needed to know the answers in advance.

I also attended lectures by Arthur Compton (electromagnetism) and Arthur Dempster (electron theory and thermodynamics), while crowding in both laboratory and lecture courses on all phases of advanced physical chemistry given by David Gans (experimental physical chemistry), William Harkins (surface chemistry and nuclear theory), Thomas Young and Thorfin Hogness (chemical thermodynamics and statistical mechanics), Simon Freed (solid and metallic states, symmetry and the physical sciences), and Warren Johnson (advanced inorganic chemistry, non-aqueous solutions). A formidable list of graduate courses in mathematics included introduction to theory of functions, higher algebra, partial differential equations, and tensor analysis.

The intellectual turbulence of the time was increased by a revelation from the Case Institute in Cleveland that Robert Shankland, a former student of Compton's, appeared to have disproved the observations made by Compton a decade earlier at Washington University. Shankland claimed that when photons were scattered by electrons there was no correlation between the directions taken by the two particles after collision, as expected from the "Compton Effect." As this early experiment of Compton's, which had earned him the Nobel Prize, was a bulwark of quantum theory, one can imagine the consternation created. In a course I took on quantum theory, given by Carl Eckart the spring of 1934, Eckart found himself in the dilemma of presenting two diametrically opposed versions of quantum mechanics, depending on whether one believed Shankland's or Compton's experiments. The resolution of the difficulty in favor of Compton came later, but not soon enough to avoid catastrophic confusion for the few months of the course. Eckart also performed nobly in teaching us the elements of Fourier Series in a special course of lectures, as well as providing our home-grown seminar on group theory with his classic paper on its application to quantum dynamics.[2] Through the next three years I busied myself reading publications that I thought might provide illumination, as prefigured in the works of Mach and Bridgman and other proponents of operationalism.[3]

Before beginning research for my doctorate, I had to pass the preliminary exams in chemistry. These were considerable sources of anxiety, requiring adequate performances in four written exams, each lasting three hours, covering the major areas of chemistry—inorganic, analytical, physical, and organic. If the candidate survived these, there remained the hurdle of an oral examination, at which any of the faculty might attend and question the candidate. My preparation in these areas was excellent, except in experimental organic chemistry, which I had found neither time nor the inclination to study in any depth, particularly at the bench. It seemed to me that the practice of organic chemistry was the rawest kind of empiricism, full of smells and devoid of intellectual content. I expended all of my intellectual energy in the intense pursuit of the new learning preached by Mulliken, Eckart, and others.

A nagging uncertainty existed for me as I approached the Ph.D. orals, having passed the written exams after a summer's intensive study. I had little bench knowledge of organic preparations. What I knew was largely "paper chemistry," learned in the courses given by

Stieglitz. The most prestigious professor on the chemistry faculty was the organic chemist Morris S. Kharasch, who was a terror at oral exams. He was known to have a dim view of physical chemistry candidates who presented themselves without having taken his courses in qualitative organic chemistry. He would sit at such exams unsettling the candidate by obvious shows of derision for the latter's efforts to answer his questions, usually devised to contain some trap into which the hapless victim was sure to fall. I was relieved to see that he was not on my oral committee, and that organic chemistry was represented by Professor Glattfeld, my first curriculum advisor. The others, all friendly to me, were Warren Johnson (inorganic), Dave Gans (my research mentor, representing physical and radio chemistry), and T. R. Hogness (advanced physical and statistical mechanics). Kharasch was stated to be out of town, although on a committee for Herman Ries, a fellow student and also a colleague of mine working under Harkins in surface chemistry.

My exam was scheduled for 1:30 P.M. When we had assembled, I remarked that there was a movie on surface tension showing at the Oriental Institute at 2:30 P.M. that might be of interest to the group to see. The committee agreed there was no real need to extend the exam beyond an hour. The exam began in this relaxed ambience with Hogness wanting an explanation of the importance for quantum theory of the occurrence of sharp line emission spectra of excited atoms. This evoked several minutes of high-flown mathematics from me, including an exposition of some statistical mechanics. Hogness broke in to say that he had not wanted such detail, only some intimation of how one might teach the principles of quantum mechanics to freshmen. I said I didn't know, whereupon Hogness grinned and observed that he didn't either. Johnson then took over to ask for a statement of the first law of thermodynamics, whereupon I launched into a discourse on exact and inexact differentials with relevant digressions on heat, work, and internal energy of closed systems. Johnson interrupted to say he had not meant me to get on so erudite a plane and asked if I would now tell him what possible reason there was to study solutions of isoamylammonium picrate in dianisole. (Surely no more remarkable a non sequitur to the first law could be imagined!) Remembering Johnson's career as a disciple of the famed father figure of American inorganic chemistry, Charles A. Kraus, who had pioneered studies on the relation of solvent dielectric to solution chemistry, I dredged up a reference to Kraus and his work.

Johnson was satisfied and indicated to Gans that he should take over, looking at his watch to see how we were doing on time. Gans, somewhat worried about the casual tone the session was assuming, asked some precise questions in my area of expertise—radiochemistry—and then nodded to Glattfeld to finish the questioning.

It was now a little after two o'clock. Hogness, Johnson, and Gans left, remarking that they would see me over at the institute soon. Glattfeld apologetically indicated that he would not keep me long. He asked if I might remember what "benzoin" might be. As he expected, I could not recall, so he supplied a hint by telling me that it was the product of condensing two molecules of benzaldehyde. I could draw the structure by running through the condensation reaction on the board and proceeded to do so.

Just at this point, the door was flung open and the devil himself in the person of Professor Kharasch burst into the room! As I learned later, he had hurried back to Chicago to be on time to spread shock and dismay. He had just come from laying poor Herman Ries low next door. Hearing I was being quizzed, he had cut off impaling Herman and hurried over to see what he could do with me. The atmosphere, so light and friendly before, now became fraught with apprehension for me. Glattfeld continued his line of questioning by reminding me that benzoin condensation went well in the presence of cyanide and asked if I could explain why. As Kharasch glowered and fidgeted, I went through a recital of a typical Stieglitz-type mechanism, invoking a bivalent carbon intermediate. Glattfeld then turned to Kharasch and indicated that he had finished with me.

Kharasch began by asking for the preparation of benzoic acid. I hazarded the suggestion that this could be done by oxidation of toluene with permanganate. Kharasch complained that this would be impractical because of expense, whereupon I bristled and retorted that I had no expectation of going into industrial chemistry. Karasch said that if I had taken his course I would know that benzoic acid could be made via phthalic anhydride, produced by passing naphthalene vapor over vanadium oxide as a catalyst in a hot iron tube. I said I had not taken his course. Kharasch brushed this aside, going on to ask how I would make normal propyl alcohol from normal propylamine.

Here my superficial encounter with amine chemistry in Stieglitz's course returned to haunt me. Sensing a trap, but not knowing where it might be, I said, "I don't know." Rolling his eyes upward and glancing at

Glattfeld to bear witness how little basis there could be for imagining I would ever become a chemist, Kharasch asked if I had ever heard of nitrous acid. When I nodded, he said, "What happens with methylamine?" (Methylamine is a one-carbon compound, an important point for what was to transpire.) I wrote a reaction showing the production of nitrogen gas and the one-carbon compound methyl alcohol. Unhappily, I used some speculative chemistry I had learned in Stieglitz's course that invoked the intermediate appearance of a compound called "methylnitrosoamine," which was supposed to be on the pathway to the final products. Feigning amazement, Kharasch turned to Glattfeld and asked if he had ever seen such a compound. I broke in to say that I had only shown some "paper" chemistry we had learned from Stieglitz and that in fact only nitrogen and methyl alcohol were products.

Now that we were back on the track, Kharasch went on to the case of two-carbon compounds, the next in the series being ethylamine. I told him the products were, accordingly, nitrogen gas and the two-carbon compound ethyl alcohol. Then, leaning back in happy anticipation, Kharasch asked what would happen with the three-carbon compound, "normal" propylamine. (This is where the trap was to be sprung.) The new twist was that while the one-carbon and two-carbon cases involved no possibility of producing anything but carbon compounds with "straight" chains, three-carbon compounds for the first time allowed the products to rearrange so that one got both "normal" straight chain three-carbon propyl alcohol and its branched isomer, "isopropyl" alcohol. Not having the wit to see this, I simply answered that I didn't know. Kharasch, enraged that I had not fallen into the trap, cried, "What's the matter with you. Can't you think? If methylamine and ethylamine give the corresponding alcohols, what do you expect from normal propylamine?" I retorted, "You'd expect to get normal propyl alcohol, but that's not saying what you'd get." Kharasch broke in with, "Don't you think organic chemistry is a science?" and I replied heatedly, "No! It's a Black Art!" Whereupon, Kharasch arose in great dudgeon and hurried from the room.

Down in the Jones basement a few moments later I broke into Sam Weissman's lab and yelled, "That bastard is back in town!" Then I noticed that the prim and proper T. F. Young was sitting there. Embarrassed, I withdrew. (Sam told me later that Young had smiled paternally and said he knew "how it was.") As I came out into the corridor, I encountered Hogness and Gans returning from the movie, which I had not gotten to see after all. They asked where I had been. When I described my encounter with Kharasch, Hogness's face turned dark and he stormed up to the conference room where the faculty members were meeting to discuss the fate of the Ph.D. candidates. I was told later by Gans that there had been a heated argument between Hogness and Kharasch, during which Kharasch had exclaimed, "I'll pass Kamen in organic only over my dead body!"—whereupon Hogness opined that that would be a good idea!

The upshot of this episode, which emphasized the basic schism in the department between the physical and organic wings, was that I would have to repeat the organic oral exam with Kharasch when I felt prepared to do so. I spent the next six months memorizing hundreds of commercial syntheses. Finally, I went to see Kharasch in his office. He received me cordially and said he hoped I had gotten over my prejudice about organic chemistry and that I really could have passed the first time but that he thought I needed a jolt. He then reminisced about an occasion when the great physical chemist John H. van Vleck had asked Kharasch for advice on what compound might have a certain symmetry needed for experiments he contemplated. Kharasch had suggested diazomethane. Van Vleck was most pleased because it turned out that diazomethane was just what was needed. Hence, Kharasch opined, a knowledge of organic chemistry could often be useful even to physical chemists. Then he told me to go to see another organic chemist on the faculty who would conduct the required repeat of my organic oral. Ironically, I was asked about the mechanism of the "aldol condensation," the kind of question I could have answered easily six months earlier, but had difficulty with at that moment because since my initial encounter with Kharasch I had concentrated so hard on learning syntheses rather than on remembering mechanisms. I did manage to resurrect an adequate mechanistic explanation, however, and so my prelim exam traumas were ended.

However, the faculty fight over me had worried Professor Schlessinger, and led to a bizarre interview with him shortly after the oral exams were over. Schlessinger, whom I had not seen since attending his famous freshman course in introductory chemistry some three years earlier, filled the post of departmental secretary and general worrier. He thought of himself as a pater familias to the graduate students, seeking to emulate the role filled by the chairman, Stieglitz. He began by asking why I did not like Professor Kharasch. I replied that Kharasch had shown little liking for me. Schlessinger said that it was his understanding that I was a very social type, a musician in fact, and he could not understand why I had behaved so belligerently at my exam. He thought that I might be more relaxed if I smoked, and inquired if I did so. I said I did not, whereupon he urged me to consider that possibility. Then he asked if the fact that I was Jewish could be a factor in the problems I seemed to be having. I denied this, saying that I had never thought of my origins as a problem until he had brought the matter up. Schlessinger, who was popularly known, or rumored, to be the son of a Milwaukee rabbi, pointed out to me that although he was "half Jewish," he had certainly experienced no difficulty in making his way to fame and fortune, and that I could take comfort in that fact. So saying, he dismissed me with the sincere hope that I would mend my ways.

In justice to Schlessinger, it is certainly true that riding herd on the faculty and students in the troubled Chemistry Department was not easy. The split between the organic and physical wings had its origins not only in disparities of economic expectation but also in the personalities of the principals. Moreover, there was the sniping and gossip about Stieglitz and Rising.

My original decision to work with Gans and Harkins while I was a senior in 1933 had not been easily maintained, because the arrival of Simon Freed around that time had given the department a new faculty member of unequalled imagination, generating new and exciting ideas where there had been so few before. Sam Weissman, one of my best friends, had opted to work with Freed, attracted by one of Freed's more daring proposals—to measure the gravitational analogue of the Zeeman and Stark Effects. The idea was that just as magnetic and electrical fields caused shifts and splittings in the spectral emission lines of excited atoms, so would a gravitational field. Unfortunately, the very tiny shifts to be expected were hopelessly beyond the means available for detecting them in the laboratory. The details of Sam's heartbreaking efforts over the next five years to observe them need not be recounted. The only point in mentioning his troubles is that Freed was long on ideas but not always sound in suggestions as to implementation. Sam's eventual doctoral dissertation was an analysis of the sharp line spectra emitted by rare earth ions.[4]

For me, it seemed safer to stick with the less exciting, if more tedious, prospect of working on some aspect of nuclear physics and chemistry with Harkins, Gans, and their graduate student, Henry W. Newson. They had built the first Wilson cloud chamber on the campus, considerably before one was seen in the Physics Department. Harkins's interest in nuclear structure received the same condescending and patronizing treatment from the physics faculty that we chemistry grad students had experienced from our physics colleagues when attending advanced predoctoral classes in the Ryerson Lab.

Harkins had reacted by becoming embittered and defensive in his efforts to claim some attention for his contributions in nuclear systematics. He came by his interest in this area quite naturally, having succeeded H. N. McCoy as the senior physical chemist on the faculty early in the century. (McCoy was one of the original workers in isotopy, and was held on our campus to be the discoverer of isotopes, but he lost out on priority to Frederick Soddy in England.) Indeed, both Robert S. Mulliken and Samuel K. Allison in Physics had done their doctoral dissertations under Harkins, the former on the theory of resolution of isotopic mixtures by diffusion and similar processes, as exemplified by separation of mercury isotopes using evaporation,[5] and the latter on an exhaustive disproof of claims, then popular, that nuclear disintegration could be achieved by high voltage discharges.[6] Harkins had gone on to publish an extensive series of articles on his empirical correlations of nuclear stability with nuclear constitution, first expressed in terms of the ratio of protons to electrons (presumed to exist in nuclei in the twenties) and later in terms of neutrons and protons, following Sir James Chadwick's discovery of the neutron in 1932. These correlations were stated in some fourteen "rules," given a final airing in 1933, just at the time I was starting my research under Harkins's sponsorship.[7]

A point of considerable frustration to Harkins was the failure of the physicists, in his view, to recognize that he had postulated the existence of a neutral nucleon—the neutron—prior to Rutherford. Harkins had published his prediction, based on his cogitations about nuclear systematics, in the *Physical Review* 15 (1920):72–94, the article being dated as received April 12. Rutherford had delivered his report in a Royal Society lecture, which was published in the *Proceedings of the Royal Society* 97A (1920):374–400, and dated as received June 3. Harkins even reproduced notes of his lectures taken by students to show that he had the idea of the neutron some time before his publication. All of this availed naught except to earn him the derision of faculty and students, who looked upon his claims in nuclear physics as merely senile outpourings. Some of this feeling rubbed off on me.

Harkins and Lunn were good friends, a fact I found hard to reconcile with the obvious disparity between them in intellectual stature. One reason they clung together, I supposed, was because they each perceived the other as unappreciated by the Establishment and a victim of calculated indifference.

Harkins's stature as a major figure in the history of surface chemistry remains unquestioned, but I read little of his writings in this area. Nor did I pay much attention to the considerable activity in the surface tension research laboratory next to the room housing the cloud chamber. I spent much time reading Harkins's papers in nuclear physics and could not see them as other than collections of data with endless variations on one idea—that the greatest stability resided in those nuclei for which the numbers of protons and neutrons were even. While the germ of the concept of "magic numbers"—so basic to the eventually successful shell theory of nuclei—could be said to exist in these empirical studies, there was no theory as such. Many years later I made a point of examining the writings of the physicists mainly responsible for the formulation and development of nuclear shell theory, such as Maria G. Mayer and others, and found practically no mention at all of the voluminous empirical contributions made by Harkins.[8] Nevertheless, his labors over two decades to bring some system into the inchoate mass of data on elemental abundances were worthy of more attention than they received, as Maria Mayer agreed in a conversation I had with her a few years ago. The main result of my poring over the Harkins papers and related material was, however, that I began to develop some specific notions about what I might do for a doctoral dissertation.

I had assimilated the prevalent idea that the statistics of nuclei were accounted for adequately by assuming that protons and neutrons were the ultimate particles ("nucleons") making up nuclei. By analogy with a successful treatment of chemical binding forces in the hydrogen molecule (H_2), nuclear forces could be explained as attractive because of the existence of an "exchange" force between them. On this basis, it could be predicted that, if one measured the distribution of angle between protons colliding with sufficiently fast neutrons and the recoiling neutrons, there would be a preferential scattering of the protons in a forward direction.

Attempts to establish this fact and to prove or disprove calculations that indicated no such forward scattering until one reached energies of some twenty million volts were indecisive at the time for many reasons. First, the neutron sources available were too weak to get enough proton recoils for a satisfactory statistical sampling to be made. Second, there was always residual contamination in the experimental setup giving spurious alpha particle tracks from radioactive materials present in the materials used. Third, some authentic recoils could be lost at very narrow or very wide angles. Fourth, the energies of the incident neutrons were not all the same. Fifth, the neutron source was not a point in space, so that one could not measure the angle of recoil very precisely, using a line drawn to the origin (source) as the reference. Finally, there was always the possible occurrence of multiple scattering to worry about—that is, one could not assume with certainty that the invisible neutron path was a straight line from source to the starting point of the visible proton track. Neutrons, being uncharged, left no wake of charged water particles to reveal their tracks in cloud chambers.

I decided that it would be worthwhile to try improving the methods of observation in our Wilson cloud chamber apparatus to obtain data that might settle this crucial aspect of observations on the nature of nuclear forces. But first I had to acquire more laboratory skills. I had made a start of sorts on machine shop practice during my senior honors project, but had no skills in glassblowing.

The glassblowing shop was a research support area of basic importance in a Department of Chemistry. Often, in science departments, the character presiding over its operation was an individualist, full of temperament and sometimes even of genius. George Reppert, the departmental glassblower, lacked none of the former, whatever question there might be about the latter. An obese and garrulous man, careless in dress and language, he exhibited alternating periods of moodiness and joviality. For those graduate students whose problems required complex glass setups (and this was most frequently the case), it was essential to stay on the good side of Reppert. It helped to take his course, which I did in the summer of 1933. We were given a start in the art of blowing beads and seals, grinding joints and stoppers, and making all kinds of lab items of which they formed integral parts—condensers, thistle tubes, T-tubes, capillary discharge tubes, vapor traps, and so on. As it turned out, however, I never had to acquire real expertise because I had no complex glass lines to fabricate.

Much of 1933 was spent learning to cope with the massive machine and accessory equipment comprising the Wilson cloud chamber that Gans and Newson had built to forward Harkins's developing interest in studying nuclear reactions. They had described its construction and operation in the literature that year.[9] Newson was finishing his thesis work doing studies on the energetics of disintegrations induced by fast neutrons on nitrogen and other light elements. He was lanky, sanguine, and dryly humorous, presenting a great contrast to the nervous, fast-moving, and impatient Gans. When I came upon the scene, they were in the midst of an effort to check out a claim that beryllium was naturally radioactive. They had been unable to detect any alpha particles, but the possibility that there might be beta ray emission remained, and I was assigned the task of checking this out.

The business of getting conditions just right to see thin, difficult-to-visualize beta tracks as compared to easily visible thick alpha tracks, baffled me for weeks as I crouched in the darkened lab bent over the cloud chamber, straining to see some indication of tracks from the pieces of beryl mineral embedded in the floor of the chamber and strewn about its top and sides. The room was unventilated and it was a hot, humid Chicago summer. The lab was saturated with the odor of ozone produced by the discharges of the capillary flash tubes illuminating the chamber during photography synchronized with the rapid expansions producing the vapor condensates around which tracks would form, if present. The cracks of the discharges and the bangs of the expansions helped create headaches already well on the way from the heat, humidity, and ozone. I had a vivid impression of research at its worst! Of course, I found no evidence of beta ray activity in beryllium. The plausibility that there might be such activity had arisen as the result of an error in estimating the binding energy of beryllium, which was later corrected.

In an attempt to put a more positive aspect on the work, I tried to see evidence of beta activity in a few elements such as phosphorus, sodium, and chlorine after slow neutron irradiation, as reported[10] by the Fermi group later in 1934, but found none because our neutron source was apparently too weak. It consisted of mesothorium oxide, and salts of radium D, polonium, and thorium X, equivalent in radioactivity to twenty-eight milligrams of radium, mixed with beryllium powder. Neutrons were released by the action of alpha particles from the radio-elements impinging on beryllium. Fermi's work had been done with sources up to twenty times more active.

The preparation of our neutron source provided the only instance in which I got to see Harkins in the laboratory. He wanted to take charge of the preparations, because Gans was not available at the time and he did not like the idea of entrusting the expensive new shipment of thorium salts to me for handling. So, while I watched and handed him reagents, he proceeded with the preparation. This involved carefully grinding the beryllium powder and mesothorium and thorium X salts together in a small mortar with a specially finely polished pestle, then transferring the mixture to long, thin silver cups, which were closed with a solder bead and packed into a platinum cylinder, about a centimeter in diameter and with a wall about three millimeters thick, to absorb out as much gamma radiation as possible without making the source too large for the subsequent measurement of scattering angles. Unfortunately, the old gentleman had a bad case of the shakes and spilled some of the mixture on the paper placed below the cups standing upright in clamps. Seeing that neither his vision nor his hands were steady enough to carry on the operation, he turned to me and suggested I clean up the mess and finish the procedure.

The Wilson cloud chamber setup featured two rigidly positioned cameras for photography, and later stereoscopic reproduction, of tracks in the cloud chamber. They were supported on steel tracks attached to concrete pillars embedded in the floor. In the center stood the cloud chamber assembly. A piston, the top of which was the floor of the chamber, could slide up and down and be adjusted to give a controlled increase in volume when it fell suddenly, thereby producing a quick expansion of the chamber gas and supersaturating it with water vapor emanating from a wet film of gelatin painted on the chamber floor. As the piston fell, it released a trip mechanism that hurled a shorting wire through a spark gap, causing a high voltage discharge through capillary tubes to light up the chamber. The high voltage current was supplied by a battery of large condensers. The water vapor settled after expansion and whatever charged particles were present were cleared away by an electrostatic field from a ring of foil placed between the glass side and top plate of the chamber. The whole series of events—expansion, exposure, field-clearance, and photography—was timed and synchronized by a system of gears and cams connected mechanically and restoring the apparatus for a new cycle every ten or fifteen seconds. Film was fed to the cameras for photographic recording by the same system of gears and cams.

Needless to say, there was plenty that could go wrong. The lights could fail, the cameras could jam, the chamber could leak—any and all of these things could happen, singly or in combination. When the machine was running well, I could sit outside, insulated from the ozone and noise, and listen apprehensively for whatever trouble might betide.

The one bit of experience I brought to this research was the knowledge I had acquired as a photographer's assistant, helping my parents in the studio. I was able to improve the quality of the photographic record by modifying the developing and fixing solutions used to process the film record. This was read after each run by threading the two films through two projectors and examining the tracks in a darkroom. Whenever usable tracks were found, they could be marked (assuming both films showed clear views) and then replaced in the cameras, and projected back on a white matte plate mounted in place of the cloud chamber so as to show the source center (indicated by a calibrated pointer). The reconstructed tracks, now reproduced precisely in space, could then be measured for length and orientation to the source.

It was not until the summer of 1934, a year after I began work, that I started on my own thesis problem. Henry Newson had left to work at the Radiation Laboratory of the University of California in Berkeley under Ernest O. Lawrence. This laboratory had become a Mecca for all aspiring young nuclear scientists in the country because of the success of the cyclotron, as developed by Lawrence and Stanley Livingston, in producing beams of high-energy projectiles far surpassing in intensity those available elsewhere.

By the end of that summer, I had helped finish off some work with Newson and Gans on neutron disintegration of neon and fluorine (as CF_4), looked at several hundred feet of film of heavy hydrogen recoils produced by fast neutrons, installed a cylindrical lens on the chamber to improve the lighting system, removed and rewound a transformer for the high voltage system, repaired the gear system by putting in collars to prevent slipping, installed a new trip mechanism for the lights, devised and put into use a better system for the 400-volt clearing field, started making a new steel chamber, and "improved photography." I also fabricated a new neutron source, designed by Gans, in which radiothorium salt was mixed with beryllium powder in a small, spherical soft glass bulb housed in a steel vessel about a millimeter thick. For the fall, I proposed to finish work on heavy hydrogen, install the new chamber, and make another attempt to see beta rays. Also, I had to take the German and French reading exams required for the doctorate in chemistry. Finally, I needed to make a new photographic developing outfit and blacken the camera mount. In the meantime, I continued to fill my notebooks with summaries of current work in all areas of nuclear research.

Times were still hard, but back at home we were managing to keep afloat. We had kept our apartment on Cornell Avenue on a rent-free basis because the mortgage holder had consented to this arrangement as long as rentals on the eight other apartments kept some returns coming in on the investment. The studio, under my mother's management, was meeting expenses. The ring business downtown was even making some money. Once or twice a week I went there to help with orders and make a little money to put aside as savings. In addition, I had been receiving scholarships that covered tuition and expenses at the university. Most fortunately, in my second graduate year I received one of the Reynolds Fellowships, which paid tuition and an additional pathetic two hundred dollars per annum for me to squander on myself.

One might wonder if I had any social life in view of the heavy demands at the lab, but in fact there was no slackening in the social whirl. I had long since given up the idea of joining the feeder orchestra of the Chicago Symphony, but still maintained a professional level, playing the viola in the University Symphony and in innumerable chamber groups.

Meanwhile, I was making headway in taming the cloud chamber and accumulating significant data on neutron-proton scattering. By late 1934, I had found that background "contamination," as evidenced by the occasional appearance of alpha particle tracks from ambient trace amounts of natural radioactivity, was quite negligible, amounting to less than one such track every forty expansions. Nor was there any visible cosmic ray background, as the beta and gamma rays produced no background fog under the conditions we had established were best for visualizing heavy particle tracks. I also made a number of mechanical improvements in the construction and operation of the apparatus and perfected procedures for obtaining pure heavy hydrogen, argon, and oxygen. I had obtained the argon from the Compton group in the basement of Eckhart Hall and intended to use it for neutron-induced disintegration experiments after finishing the neutron-proton problem.

By the beginning of 1935, we had obtained enough usable proton recoils to publish a short note[11] corroborating and extending the work of Franz N. D. Kurie at Yale, which showed a marked preferential forward scattering of the protons at neutron energies much less than expected theoretically.[12] Kurie's results, like ours, were in disagreement with some preliminary findings by earlier investigators.

I had heard about the successful use of specially treated photographic emulsions made by the Ilford Company in England for visualizing heavy particle tracks from cosmic rays. This inspired me to try some experiments to see if our neutron source could produce protons in them. If I could see them, it might be the basis for an alternative approach that would be superior to utilizing the cloud chamber, because there would be many more protons produced in the solid material of the emulsion than in the gas phase of the cloud chamber. (Indeed, a few years later in England, Cecil Powell used just this method to demonstrate the existence of mesons in cosmic rays, thereby winning a Nobel prize.) However, nothing came of these efforts.

The award of the Reynolds Fellowship brought with it the dubious honor of running the graduate seminar in physical chemistry. This chore required me to dragoon speakers, procure food and drink, and be responsible for all housekeeping details. It was not an easy assignment, particularly because of the idiosyncratic demands made by faculty tastes. For example, Hogness required coffee thick enough to satisfy a Turk, while Young blanched at the sight of a coffee bean. Nor did anyone care for the institutional food the local caterers provided.

The general drumbeats of complaint were muffled one wonderful Monday morning just before a scheduled noon meeting when Hogness informed me that the speaker for the day would have to yield, as Linus Pauling was in town and was willing to appear. Pauling, still only in his thirties, was the acknowledged leader in probing the nature of the chemical bond. His classic text on the subject was to appear shortly. The approach he and others were using involved so-called "atomic orbitals," in contrast to the molecular orbitals we had been trying to learn about from Mulliken. This approach gave more attention to the intuitive desire of chemists to invoke localized bonds between atoms as a starting point for quantum dynamics than did that of Mulliken and others, which started with the molecule as a whole. Pauling used the concept of "resonance"—an empirical blending of all the electronic structures one could reasonably write for any given molecule.[13]

I do not recall the occasion for Pauling's appearance in Chicago except that it had something to do with an award from the local section of the American Chemical Society, which he was scheduled to address that evening. Pauling arrived just before noon. We students were charmed, if slightly surprised, to see a bouncy young extrovert, wholly informal in dress and appearance. He bounded into the room, already crowded with students eager to see and hear the Great Man, spread himself over the seminar table next to the blackboard and, running his hand through an unruly shock of hair, gestured to the students to come closer. He noted that there were still some seats vacant at the table and cheerfully invited students pressing in at the door to come forward and occupy them. As these were seats reserved for faculty, the students hung back, but Pauling would have none of that. He insisted, and they nervously edged in, taking the seats.

The talk started with Pauling leaping off the table and rapidly writing a list of five topics on which he could speak singly or all together. He described each in a few pithy sentences, including racy impressions of the workers involved. For instance, he mentioned that Oliver Wulf, the investigator whose research on infrared spectra he had listed, was working at the Fixed Nitrogen, or fertilizer, laboratory of the Department of Agriculture. He grinned as he said this, pointing out that Wulf was often referred to as a "man full of his subject." Another topic inspired him to note that the subject "harkened" back to 1915. To make the allusion to Dr. Harkins and his notorious preoccupation with priority crystal clear, Pauling then went on to ask, "Is 'harkins' a word?"

The seminar he gave was a brilliant tour de force and made a never-to-be-forgotten impression on all us students, although I do not suppose any of us could recall it in detail now. Later we were happy to hear he had taken special pains at the Chemical Society dinner to needle Kharasch and the organic chemists as a whole for their naive empiricism in correlating heats of combustion with bond energies.

Progress toward my Ph.D. got a boost when I passed the language reading exams. The department required that all candidates show proficiency in translating scientific reports in German and French. I had no worries about German, or so I thought. I was not aware that my old friend, Professor Kurath, was lying in wait. When I arrived to take the exam, he smilingly handed me the bulky physical chemistry monograph by Jellinek and indicated that I was to translate two sections, each of about five printed pages. The text was a particularly murky example of German scientific prose. I found that Kurath had picked an especially dense portion. The first section dealt with the theory of perfectly reflecting surfaces. The subject of the first sentence was nowhere to be found. All I could see was a remark translated as "let us consider numerous," followed by no object. Much panicky floundering finally revealed it buried two paragraphs back, where it appeared as "elements of surface." Having sweated through this, I ran into the next thicket—the second selection, which turned out to be a philosophic discourse on the significance of the second law of thermodynamics for living systems. Nothing could bring out the lurking mysticism of a German pedant more than this subject. Jellinek's text gave the impression of a scholar awed by the subject and its implications flat on his back pawing the air. It took me another hour to decode the message. I passed, but Kurath had had his revenge for my forsaking German back in my freshman days.

As for French, I knew nothing about it whatsoever. Considering that I had barely managed to squeeze past in German despite the considerable fluency I had in it at the time, I despaired of getting through the French exam. I made the half-hearted gesture of reading a little scientific French, provided in advance by the examiner as an example. It so happened that he was also the second bassoon in the university orchestra and held me in some awe because I was first viola. The text he provided for practice was simple enough. One hardly needed to know any French to read it, because it consisted of simple sentences such as, "One passes two amperes of current through five ohms of resistance for three hours." Arriving at the exam, I found I had been given precisely the same text to translate that I had plowed through in preparation. I passed easily, while a colleague, who had studied at the Sorbonne for several years and was as fluent in French as in English failed. In fact he failed several times before he succeeded in getting over this hurdle. His exams were thorny masses of highly idiomatic French that would have puzzled an Academician. The examiners were antagonistic because he had patronized them, boasting about his great experience and expertise, a mistake for which he paid dearly.

Meanwhile, the research project was not proceeding with uniform success through 1935. The new radon-beryllium source proved disappointingly weak. Argon was not disintegrated by our neutrons. Work on the tetrafluormethane showed no results that could be interpreted as the consequence of carbon or fluorine disintegration. All that was left was the neutron-proton project, and here useful data accumulated very slowly. I suggested we try some of the stable boron hydride compounds Anton Burg was synthesizing in Schlessinger's laboratory, but now I find no mention of work on them in my notes. It was clear that we needed a fresh source, and I discontinued work for a few months until a new mesothorium preparation could be procured.

Meanwhile, life off campus was taking on new and more exciting aspects. Around the corner from our apartment on Cornell, Gavin Williamson and Phillip Manuel, an internationally renowned harpsichord duo, had taken an old house. I went there often and was introduced to the great traditions of French and Italian baroque music.

Of more significance for me, however, was another house, next to that of Williamson and Manuel, in which three young ladies were living on the top floor, sharing space and expenses. One of these, Marjorie M., had been married for a short time to my old friend, the violinist Ben Senescu, and had since become involved in a torrid affair with one of my colleagues. I thus came to meet Virginia A., one of her roommates. Virginia was a pert, handsome blonde for whom I generated an immediate infatuation. I was ripe for this initial entry into the Rock Garden of Love, considering that I was twenty and still a virgin. Soon we had a passionate affair going, which raged on for the next five years before running its course.

Nor was my emancipation wholly in matters of the heart. I even started to smoke—although I should hasten to add that Dr. Schlesinger's admonition had nothing to do with this development. Rather, the milieu in which I was living encouraged it. Everyone smoked. It was a social gesture symbolic of the ersatz hedonism we were cultivating. I even had my first (and fortunately my last) hangover.

The pall of the Depression still pervaded our lives. If life took on a happy cast at times, we nevertheless had a certain foreboding that all was not well. The one unfailing source of strength was my mother. I could count on her support and love, no matter what. While I was home but little, I knew she was there, and this knowledge seemed to me a guarantee that all would work out well in the end. Knowing that she looked forward to my obtaining my doctorate and launching out on my own, I redoubled my efforts to bring the neutron-proton project to a successful conclusion. By the end of the summer of 1935, the new source had been put in place and I was grinding out data that Gans and I could see as sufficient to support the preparation of an adequate thesis for a doctorate.

Meanwhile, my Aunt Rachel, who some years previously had been through the terrible trauma that I have already described, had shown the indomitable courage so characteristic of my mother and her other sisters. She had made a successful comeback, establishing a new business and raising her children wholly on her own. By summer 1935, she thought it appropriate to celebrate and in particular to express her family's gratitude to my mother, who had done so much to put them back on their feet. So, late in August, it was arranged that I stay at home to take care of the studio and monitor the operation of the ring business downtown, while my mother, father, sister, and aunt drove off to Canada for the Labor Day holidays to join in Aunt Rachel's celebration. Our Buick held these four comfortably, but my mother, always generous and giving, also invited two of our cousins from the far North Side who had asked to come along. My father objected strongly, because this would result in too much crowding, but my mother was adamant. The six of them jammed into the two seats, three in front and three in the back.

I heard nothing more until the terrible Tuesday (September 3, 1935) after Labor Day when my father phoned and said in a strained tone that there had been an accident near a place called Hillsdale in Michigan on their return trip. He said no more, except that everyone was well but my mother. I spent a bad night, but it was as nothing to what came next. The family came dragging in with the news that my mother was dead. In a daze, I heard my father tearfully explaining that there had been a wet road, a hill, and a truck coming from the opposite direction that had sideswiped our car. I think he mentioned that our car had skidded into the path of the truck; but in any case, the car had spun around and the door on the right front side had flown open, throwing my mother to the pavement. She had sustained a fractured skull, which proved fatal by the time they reached the hospital. No one else had been hurt, except that my father had a bruised chest where he had been thrown against the steering wheel. I realized bitterly that my mother happened to be sitting in the fatal position because of the crowding caused by her agreeing to bring the cousins along.

The news got about fast, and my friends rallied to give comfort, but it availed little. I went about in a daze, hardly knowing where I was or with whom. I recall being driven to the funeral parlor where I took a last look at a body I could not believe was my mother's, so deathly white and expressionless did she appear and so unlike her real self. At the funeral I went through all the rituals, numbed by the calamity and only beginning to sense a little of the catastrophe that had befallen us.

A spiritual malaise, the legacy of a childhood beset by pressures of being regarded as some kind of scholarly and musical prodigy, had deepened. My experiences in academia further reinforced this feeling. As might be expected, I had become strongly attached to Dave Gans, my mentor and guide in graduate research. Dave was caught in the unenviable position of being an assistant professor striving for tenure, while essentially committed to Professor Harkins as his assistant. This meant that he had to take responsibility for the actual day-to-day instruction of Harkins's students and act as their hand-holder, while the professor remained aloof and free of the pressures graduate students can generate. Moreover, Gans had to do much of the work Harkins contracted for in his extensive consultation practice.

I recall one terribly hot afternoon when Gans was sweating away, running an electrodeless discharge apparatus that Harkins had asked be set up for experiments under contract with the General Mills Company. The objective was to see if irradiation of foodstuffs by this apparatus could provide a basis for challenging the Steenbock process, involving production of ergosterol by ultraviolet radiation. It seemed to me highly unethical for an academic to be engaged in research that might undermine a source of revenue for the University of Wisconsin. To make matters worse, Gans was suffering from a serious eye infection incurred during these experiments, which he had caught from a towel that the professor, who had a similar affliction, had contaminated. I thought such academic slavery outrageous and resolved that if ever I became a professor or research supervisor, I would never so exploit my assistants and students. I myself did not see Harkins at any time during my research. Gans and I were wholly responsible for both the idea behind my thesis work and its implementation. Yet when I wrote the paper, Harkins finally appeared and signed it as senior author.

In retrospect, the behavior of the professor was not particularly atypical for the time. Examples still abound in the United States, having been imported largely from the Continent, particularly from Germany, where the tradition of the *Geheimrat* was developed (the term, originally meaning "privy councillor," came to apply to anyone with autocratic tendencies). The feeling that I must at all costs avoid such behavior added to my burden of spiritual detritus.

There was still the business of finishing my doctorate, however, and to this objective I bent every effort now that I realized I was entirely on my own. The need for work as an antidote to grief that would not abate was overpowering. I felt that I had to get away from the world I had known and that the key to a new life would be the doctorate. The anomalous results I was getting were a worry, and one day in the early spring of 1936, after I had resumed work at the lab, I obtained an invitation to visit the renowned Eugene Wigner and the young Eugene Feenberg at Madison, Wisconsin, where they, as outstanding experts on nuclear theory, were giving seminars and workshops at the university. I spent an afternoon with them going over my data on angular distributions of protons scattered by high-energy neutrons. They confirmed that there was no obvious fallacy in the way I treated the data, but held out no hope that these data could be reconciled with theoretical expectations. Nevertheless, they felt this was no reason not to publish the data, and so encouraged I returned to begin putting finishing touches on a dissertation, with the prospect of more data to be added, perhaps in the next few months. We expected we could increase our recoil yields using the new source and thereby provide a sufficiently large number of events to establish the reality of our results.

The new source performed well. I was getting a good yield of one track in eleven expansions, with a small contamination background, in a series of runs beginning in May and extending through the summer. There was the usual crop of mechanical failures, but by the end of June, I had measured a total of 700 proton tracks in hydrogen, ethylene, and hydrogen sulphide. The observed maximum in the angular distribution at 25° rather than 45° remained, as did the discrepancy with theoretical expectations. I busied myself with the writing of a paper for publication in the *Physical Review* in which I described the data, the apparatus, and the precautions taken to exclude errors arising. There was also a high-flown discussion of the results, in which I flexed some mathematical muscles. Realizing that the results were in gross disagreement with the then current ideas on nuclear forces, we promised to reexamine the data for possible errors not yet uncovered, although actually we never did. I say "we" because the paper appeared with Harkins named as senior author and Gans, Newson, and myself bringing up the rear. The paper, which appeared in volume fifty of the *Physical Review* (pp. 980–91), was accepted as a doctoral dissertation after the formality of printing a new title page naming me as sole author and binding it as a reprint. I received a Ph.D. degree in Chemistry at the winter 1936 convocation, just three years after admission to graduate school and six after entering the university as a freshman with a major in English. In the end, the correctness of the theoretical predictions that no marked dissymmetry in angular distribution should be observed when neutrons hit protons at energies below 20 million electron-volt equivalents was upheld when stronger neutron sources of homogeneous energy became available during the forties and as detection techniques improved.

While my doctoral work was winding down, so was the world as I had known it. My circle of friends was dwindling as they began to disperse to the far corners of the country. Virginia had left for San Francisco. The family was disintegrating, although we still lived for a while at the Cornell apartment. My father abandoned the old studio and opened a small business in a loft at a nearby bank building. Eventually, my sister and aunt moved. No family ties remained to provide psychological security for me, even if my tendency to flee family involvement had not existed.

Finally, I had a feeling of inadequacy about how much I really knew of science and my ability to make a career of it. After all, I had been working on it a very short time. The fact that I had had such meteoric success in obtaining a Ph.D. in chemistry only six years after starting was not really much help. Gans indicated that as far as my future was concerned, it might not be a bad idea for me to spend a little time at the Radiation Laboratory in Berkeley with E. O. Lawrence, as Harkins was planning to build a cyclotron at Chicago, stealing a march on the Physics Department as usual, as he had done with the cloud chamber. I had saved a few hundred dollars working after hours downtown while the ring business still flourished and by playing odd musical jobs in and around the South Side. So I decided to risk all, using most of my savings to buy a one-way fare to San Francisco, hoping I could exist on the rest (a few hundred dollars) until something showed up. Moreover, by going to the Radiation Laboratory I would be near Virginia, whom I was still pursuing. Late in December 1936, eager but scared about the uncertain future, I boarded a luxury "streamliner," The Challenger, for the Coast, my youth ended and my ties to the past cut.

4

A Time and Place for Euphoria

Berkeley

(1937–1938)

IN THE middle thirties, railway passenger travel in the United States had not yet deteriorated. There were still great crack trains, especially those leaving Chicago for the West. One of these was the Union Pacific's Challenger, exemplifying the wonderful service everyone then associated with cross-continental travel. The coach seats were large and luxuriously appointed. With luck one could have all four seats of a section to oneself, as I did. And then there was the diner, with its immaculate table settings, real linen, and mind-boggling details such as salt shakers filled with vitamin B enriched, orange-colored salt! Meals were priced so modestly that even my meager budget suffered negligible attrition. For three nights and two days I rode on this modern magic carpet. As the vastness of the plains and the majesty of the Rockies unfolded, the drabness of the life I had left behind was quickly forgotten. Euphoria built up mile by mile. I could hardly contain my excitement when the train finally rolled onto the Oakland Mole and I took the ferry ride across the Bay to the Embarcadero.

My first view of the San Francisco skyline evoked such a feeling of exhilaration that I hardly knew how I eventually reached my room in the Cartwright Hotel on Sutter Street, where Virginia had made the reservation for me. Nor was I affected much when she arrived with a bad cold and the news that she had a new boyfriend.

The next day I had a strep throat, which likewise did little to affect my euphoria. I went to a doctor who swabbed my throat with an evil-tasting concoction and urged me to stay in bed for a day. Instead, I bethought myself that an old friend from University of Chicago days, Alfred Frankenstein, was the new music and art critic of the *Chronicle,* and so I called on him, a hopeful first step in establishing a musical beachhead in the East Bay. Alfred had been the lucky beneficiary of a slot opened at the *Chronicle* when its previous music critic left in the wake of a Newspaper Guild strike. He received me cordially and gave me a complimentary ticket to a quartet concert scheduled in the city that evening.

Meanwhile, I had been wondering how I might find my high school friend Adolph Brez, all 6 feet 8 inches and 280 pounds of him. That night I ran into him at the concert, wholly by chance, and we had a riotous reunion. Adolph was working as a reporter for the Hearst afternoon newspaper. He had wanted to join the Chinese in their effort to oust the Japanese from China but had been frustrated in his efforts to get to the Far East by the general harbor strike. We agreed that as soon as I got settled in Berkeley, he would move over so that we could share expenses. (Virginia had found a room I could use temporarily in a kind of slum apartment building on Center Street, near the University of California campus.)

It was a gloomy, rainy December day and I still had my cold when I set out, eager and apprehensive, to find the Radiation Laboratory. It turned out to be housed in an old frame building at the eastern end of the campus. Entering, I was hospitably welcomed by the group gathered around Ernest Lawrence, who was conducting a kind of pep session. The cyclotron happened to be undergoing repairs and renovation, and he was using the time to organize the lab staff for a systematic attack with high-energy deuterons on the elements of the periodic system. Franz Kurie, for whom I was looking, was there, and we immediately began to discuss how to proceed to work on our mutual interests—the angular distribution of fast neutrons scattered by protons, and neutron interactions with the light elements in general. He had constructed an elegant cloud chamber, ingeniously adaptable to work at low as well as normal pressures, which he invited me to share with him as we began our collaboration. At the same time, Lawrence suggested I try some disintegration experiments on bismuth and thallium—these being the only elements still left unassigned. I was told the cyclotron would soon produce deuterons traveling at speeds equivalent to their being accelerated by six million volts or more.

However, there were some tedious, unexciting repair jobs to do. The one that fell to my lot was to help the future editor of *Science,* Phil Abelson, then a young graduate student in physics, to replace a broken glass plate in a condenser used in the circuitry to provide the high voltage needed for the deflector that pulled the high-speed ions away from the circular orbits maintained in the cyclotron and into the periphery where they could strike the targets. The job involved cutting an inch-thick plate of glass to the right size. We scratched a deep groove straight across the glass with a metal scribe and then, placing a steel cylindrical rod under the glass, jumped on the projecting piece in an effort to crack off the desired length. We worked at this most of the afternoon.

There were other hazards in the laboratory besides high voltage. The operation of the cyclotron, by 1937 in its 37-inch phase (this refers to the diameter of the acceleration chamber), produced a veritable sea of neutrons and high-energy electromagnetic radiations, through which the lab staff waded quite without concern. By the time I arrived, perhaps owing to the expressed fears of John Lawrence, Ernest's brother, who was an M.D. and in charge of the newly organized program on the therapeutic applications of cyclotron-produced high-energy particles (particularly neutrons), some massive steel tanks filled with water had been constructed as radiation shields. The control desk had been moved to the adjoining room and a remote communication system (a so-called Voycall) installed. The operator sat at this desk pushing appropriate buttons and watching an array of meters telling him about the performance of the machine hidden from sight in the next room.

Occasionally, it was necessary for one of the two-member "crew" operating the cyclotron to adjust portions of the machine while the other monitored the results at the controls. Once I was making such manipulations while Lawrence (E. O. L.) presided. The particular maneuver at that time required me to push and pull a coil coupling the primitive "self-excited" oscillator providing the alternating accelerating voltage to the cyclotron electrodes (called "dees" because of their shape resembling a letter *D*). E. O. L. would push a button, note that the cyclotron was not in resonance and yell through the Voycall to me to pull or push the coils. I would put a grounding hook on the coil to discharge the voltage, then massage the coil, take off the grounding hook, and yell "On!" At one point, we got out of synchrony, with E. O. L. pushing the button while I was still grasping the coil. Through the Voycall, he complained he was getting no response, not realizing that I was still holding the coil. He was getting no reaction from the machine, I assumed, because my body capacitance was throwing the dees out of resonance. Sweating with fear that if I let go, the current from the oscillator would surge through me, I gingerly withdrew my hand. Nothing happened, however, as I weakly cried "Off!"

On yet another occasion, Jack Livingood and I were engaged in making some radioactive sodium (^{24}Na) for use at the university hospital. This involved bombarding a target of sodium metal distilled onto the target plate and maintained in the cyclotron vacuum. Several microamperes of six million electron-volt (6 MEV) deuterons impinged on the highly reactive sodium, which had to be cooled by a continuous supply of liquid air, fed into the target backing through a funnel and tube at the edge of the cyclotron. The deuteron activation of the sodium produced copious amounts of gamma rays (high-energy x-rays) and neutrons, so that we were exposed to enormous radiation dosages in a few minutes of operation. Jack and I took turns pouring in the liquid air. We were carrying radiation dosage meters (pocket "Victoreen" ionization chambers). After the run of about twenty minutes was over, we found that each of us had accumulated several hundred daily doses. Such experience led to much coarse humor about production of monsters among our progeny in years to come, but we all thought from what little we knew of elementary genetics that no great consequences would ensue if our children did not intermarry.

A continually recurring piece of cyclotron troubleshooting was hunting for leaks in the vacuum chamber. This was done by "gassing"—that is, a torch attached to a gas line was opened and the gas allowed to flow out of the nozzle on to the surface suspected of leakage. If a hole existed, the gas entered. The consequent increase in pressure could be noted by the operator at the control desk. The whole practice was a race between asphyxiation and location of the leak. When one was detected, it was fixed by liberal application of a glyceryl phthalate paint called "Glyptal." Sometimes, plasticine was applied. In time the outside of the vacuum chamber came to be a mass of glyptal and plasticine.

In the basement below the cyclotron, various electrical connections existed that occasionally required inspection. The operator had to crawl into a murky blackness, often the favorite abode of black widow spiders and other bad-tempered creatures, feel around and hope he found the wires before dangerous ones found him.

Within a few days of my arrival at Berkeley, Virginia reported that she had found a room available in a house on Benvenue Avenue, just south of the campus, where she and some of her literary and artist friends had taken up residence. This house was an antiquated red brick structure with a turretlike bay. The bottom floor was occupied by the landlady and her daughter. On the upper floor there were six rooms and a bathroom for rent. I had a large rambling room in the bay section, with the one sink on the floor. There was no heat other than that provided by a portable contraption in which oil was ignited at the base, hot air being supposed to emerge at the top. Actually, what came out was mostly oil fumes. The heat vanished a few inches into the room. I was charged twelve dollars rent a month. In addition, a "breakage fee" of five dollars was extorted as a deposit to cover whatever repairs the landlady deemed necessary when the roomer left. It was accepted that she would always find excuses not to refund it.

I phoned Adolph about my good fortune and he moved over to share the room with me, thereby cutting my rent in half. I still had about two hundred dollars left from the savings I had brought with me from Chicago. I calculated that with extreme frugality, I could last about six months before some new source of funds would have to be found. However, Adolph was making the fabulous sum of something like 135 dollars a month and was a possible source of financial help if I needed it.

The sleeping arrangements were less than ideal. The cot provided for Adolph was at least a foot too short. He slept with either his head or his feet hanging over its edge. Somehow we could find no place for the cot except in a corner near the door opening into the corridor. In the early morning, other roomers would dash in to use our sink, usually banging against Adolph's feet or head, whichever happened to be exposed. I often awakened to a flow of his inspired invective. Thus unceremoniously aroused, Adolph would stagger to the sink, where the discovery that there was no hot water would bring forth fresh streams of profanity.

It mattered little what my prospects might be. I was riding a wave of euphoria that submerged all doubts about the future. I could see, even in the short time I had been in the East Bay, that its cultural and social ambience accorded with my notions of paradise on earth. Virginia had introduced me to a remarkable group of creative writers and artists, among them the fledgling poet Robert Duncan, soon to embark on an illustrious career. Hardly a week passed when I did not play at least two or three times. (These sessions, ranging over the whole region, including Berkeley, Oakland, San Francisco, and Marin County, often involved trips on the ferry, the galley of which featured a corned beef hash I still remember fondly.) Through David Blumenstock, who had been a graduate student at the University of Chicago and had played in the university orchestra, I got to know Leila Hassid, an enthusiastic and quite competent amateur violinist who organized regular quartet sessions at her home. There I met her husband, W. Zev Hassid, the first of those scientists on the campus who were to prove decisive influences in the development of my scientific career. Zev was a struggling young plant physiologist at the start of what was to be a brilliant career in carbohydrate chemistry. Through him I met two other scientists, association with whom proved singularly fruitful for my productivity in biochemistry in later years. One was Michael Doudoroff, a wonderfully extroverted expatriate Russian, whose father had been an admiral in the czarist navy. Mike was to make seminal contributions in microbial chemistry. The other was H. Albert Barker, already a distinguished soil microbiologist and biochemist. Through them I was introduced to others in the Biochemistry Department and in the Life Sciences Building.

Not long after my arrival in Berkeley, I was drawn into the orbit of J. Robert Oppenheimer, who introduced me to yet another lively crowd in San Francisco. "Oppie," as he was affectionately known, once took me to a New Year's party in the city given by Estelle Caen, a pianist and the sister of the popular newspaper columnist Herb Caen. On our way there, Oppie remarked that he was not sure of the address, but that he knew her apartment was on Clay Street and that the number was made up of double digits divisible by seven—1428, 2128, 2821, and so forth. So we tooled up Clay Street peering at houses on the way until we found Estelle's apartment at 3528.

After only a few months, the arrangement with Adolph blew up when he suddenly found himself out of a job. After a gloomy review of our new financial situation, I went back to the lab. Returning late in the afternoon, I was amazed to find a note from Adolph, wrapped around two dollars and a few cents he owed me. It informed me that he had come into some money and invited me to join him over in San Francisco that evening. All was explained when I met him a few hours later at the YMCA where he had rented us two rooms for the night. Adolph had solved his immediate problem by an unheard of feat—extraction of the breakage fee from the landlady. Exerting all his considerable charm, he had drawn her into reminiscences about her youth in Berlin, where she had been a young aspirant to an opera career. He had urged her to sing for him, and, by enduring half an hour of excruciating torture listening to her, and even jovially bouncing her on his lap, he had persuaded the old lady to refund his five-dollar deposit. (As far as known, this was the one and only time she ever did this.) Leaving the small amount he owed me with the note, Adolph then spent a quarter to get back across the Bay and bet the remaining two dollars on a long shot in the Santa Anita handicap. The odds were thirty to one, but he maintained he had gotten wind of a fix. In any event, the long shot paid off and Adolph suddenly had sixty dollars. He went on a spree, buying tickets to the movie premiere of *The Good Earth,* and arranging dates with two handsome girls he knew. He met me at the YMCA smoking an expensive Havana, and with both girls on his arms. Waking the next morning, we were totally broke again.

It is impossible to describe the enthusiasm and zeal for accomplishment that pervaded the Radiation Laboratory in those magical years. A common determination to make the cyclotron bigger and better actuated all of us, producing the warmest camaraderie. The contrast with conditions as they had existed at Chicago was too painful to contemplate. My hope was that somehow I could earn the right to stay on at Berkeley working with E. O. L. into the indefinite future.

Lawrence had a genius for organization and for getting the maximum effort out of his troops. His drive and forcefulness were complemented by the gentle Don Cooksey, who kept a fatherly eye on us and provided the temperate consideration needed in moments of crisis, which were not infrequent. The work of maintaining the machine and keeping the laboratory operating around the clock fell to "crews" of two, with hours divided between morning, afternoon, and night shifts. At the beginning of each week, E. O. L. would appoint a crew captain responsible for operations for the week. This assignment was a high honor, and the recipient considered it his responsibility to keep the machine running at maximum capacity. If a breakdown that necessitated disassembling the cyclotron occurred, everyone within reach rallied to help, whether on the crew or not. The heady ambience of work in the lab was accentuated by daily lunches in the Faculty Club where we had a special table at which E. O. L. presided. Almost always he ordered corn flakes, strawberries, and cream for his lunch.

There was no lack of excitement on crew duty, even in routine operations. One night, after a particularly uneventful and tedious evening shift, I went to the Chemistry building basement to replenish our liquid air supply. The liquid air reserve was maintained in a large Dewar can there, supplied with a transfer tube connected to an air pressure line to push the liquid air over and into a receiver Dewar. This setup was immediately adjacent to a large rack holding the circuit breakers and relays controlling the operation of the liquefaction plant that produced all the low temperature liquid oxygen, nitrogen, and helium for cryogenic research. Mindful of stories about what might happen if things got out of hand—an explosion that would blow up the whole Chemistry complex and wreak devastation several blocks beyond Sather Gate—one entered the room a bit apprehensively, there to be greeted by the banging of circuit breakers opening and closing. As I stood sleepily holding the transfer tube, with one hand on the air pressure hose, there was a sudden jolt and the whole relay rack and wall seemed simultaneously to rise and move toward me with a hideous clanking. Riveted to the spot, I wildly imagined myself to be living my last moments on earth. Gradually the shaking ceased, however, the wall settled back, and I recovered volition. Still trembling, I crawled back to the lab to be met by an excited Paul Aebersold calling, "Hey, did you feel the earthquake?"

Despite the high morale generated by E. O. L.'s inspired leadership, it was inevitable that in so high spirited and talented a group some tensions would occasionally develop. Welcome and eagerly anticipated relaxation was afforded by our annual party at Di Biasi's—an unpretentious Italian restaurant in Albany, near the Bay. The arrangements were made by Paul Aebersold. Paul, happy and extroverted, took his responsibilities as the impresario in charge of the Di Biasi parties most seriously, creating the program and dragooning staff to produce poetry and appropriate acts for general amusement. Damage to the walls and fittings was often considerable, so that Paul had to give the management a breakage deposit to cover whatever repairs had to be made.

The most enjoyable release from the lab routines was provided by the ever generous Don Cooksey, who made his ranch available in the summers. Situated in Trinity County on the south fork of the Trinity River (a tributary of the Klamath), it was reached by an arduous drive west from Redding, the nearest inhabited site being the hamlet of Forest Glen, not far from the small towns of Weaverville and Hayfork. Cooksey's camp consisted of several log cabins presided over by a genial Lithuanian, Bob Sihlis.

There was a stream at the headwaters of the Trinity a short distance from camp in which salmon could be found in their spawning cycle. On one memorable occasion, I was in camp with two lab associates, Bob Wilson and Bill Farley. From the small footbridge we saw a gigantic salmon lurking under a rock a few feet upstream. A plan was devised to capture it. Wilson found some piano wire and fashioned a noose. The rock lair could only be reached by scrambling down a deep declivity and afforded space for only one person lying prone. The wire was not long enough to reach the fish unless one lay on the rock and dangled it in front of the prey, which Wilson did while Farley held his legs to keep him from falling in. The fish obligingly stuck its head through the noose, whereupon Wilson jerked up, attempting to hand the wire to Farley. Reacting, the fish jerked forward so strongly that Wilson's hand was pulled down so that he could not hand the wire up to Farley. He held on grimly as the fish pulled forward. Farley just as stubbornly held on to Bob's legs, so that Bob slowly lifted off the rock, forming a bridge between Bill and the salmon. This tableau lasted a few seconds, then the wire broke and everyone fell into the river.

Bob, angered, returned to camp, got his .45 revolver and went back to the rock. While Bill and I watched, he waited for the fish to emerge. When it did, he took careful aim straight downward to avoid scatter that would occur if the bullet entered the water at an angle. He fired at a distance of only a foot from the salmon's head. The massive impact of the bullet created a great turbulence, which on subsidence showed the fish back under the rock somehow not incommoded in the least. No one yet knows how Bob, a dead shot with his cannon, could have missed.

It was Bob who prevented a sudden end to my career. A 25-mile hike had been scheduled to put variety into our camp program. The party including myself, Bob Wilson, Robert Cornog (a genial giant and graduate student at the lab, later the co-discoverer of tritium with Luis Alvarez), and a few others started out early in the morning to negotiate the rough terrain. A few miles later, I was picking my way along a ridge down which there was a steep declivity plunging directly a hundred feet into a mass of rocks in the river rapids. Suddenly, I lost my footing on the slippery grass and began falling. Bob, who was directly behind me and had on proper mountain-climbing boots, quickly reached over and caught me by the neck. I was saved but, in twisting about to regain my footing, I pulled some muscle in my back and gradually became completely immobilized as the hike proceeded. Many hours later we were met by a courier from the nearby ranch, run by a Mr. Ostrat, responding to our call for help with "Old Silver," a decrepit nag that broke wind at every step. I was hoisted onto its back and hauled back to Ostrat's ranch.

For years afterward, I had recurrent low back pain. But at least I did have those years afterward, for which Bob Wilson can be thanked! As everyone should know, he went on to a great career in nuclear physics, his most notable achievement being the creation of the Fermi Laboratory and the successful design and construction of its high-energy accelerator.

At the lab, conditions for creativity were ideal, even though the emphasis on keeping the machine running seemed exaggerated at times. Occasionally there were complaints that we were spending too much time on plumbing and not enough on physics. E. O. L. met these outbreaks with the confident assertion that if we got the cyclotron working at full volume the discoveries would come automatically.

A major source of dissatisfaction was the continual need to provide regular therapeutic sessions for John Lawrence's patients, part of an attempt to prove the potential of neutron irradiation in the treatment of cancer. For each session, scheduled whenever patients were available, a treatment cubicle, hiding the cyclotron, had to be assembled and installed by Paul Aebersold, who hoped the machine would operate once the patients arrived. The tensions created by these sessions plagued the crews badly, and they made a pill-popper out of Aebersold.

The therapy program was based on two hopes. First, E. O. L. genuinely felt that the neutron rays might prove more effective than x-rays in cancer treatment and at the least would be a useful extension of radiotherapy. Lawrence's mother was apparently cured of seemingly lethal cancer by such therapy.[1] Secondly, the major sources of funds for support of research were in the hands of medical foundations. While the time would come when nuclear physics could command enormous subsidies, it was still a shoestring operation in 1937. So E. O. L. hoped that the medical applications in prospect could provide badly needed funding for cyclotron development. The same argument was used to obtain support for the use of cyclotron-produced radioisotopes.

During those early years, E. O. L. managed to do some experimental work, and haunted the laboratory around the clock inspiring and urging on his staff. As administrative demands increased, it became more and more difficult for him to assume an active role in research, a fact he found extremely frustrating. If a particularly long time elapsed, say several weeks or a month, when he had been away from the bench, his frustration would mount until he could contain it no longer. Tossing on a ragged old lab smock, he would descend from his office on an upper floor in Le Conte Hall, the physics building, and rummage through the lab, snatching as much as he could carry of what equipment was not nailed down. Triumphantly, he would then disappear in the direction of his office, his arms full of meters, with resistors and assorted lab items such as C-clamps dangling from his belt. The lab gang, fully aware a raid might be imminent, would have mounted a watch beforehand to alert everyone that he was coming, so that one could hide whatever might be needed for continuation of one's own research. It was great sport to see how well we could anticipate these depredations and minimize our losses.

Art Snell, a quiet redhead from McGill University whose phlegmatic exterior hid a uniquely dry wit, quickly established himself as one of the top experimenters in the lab. It was he who was entrusted with the important task of diverting the beam away from the cyclotron vacuum tank, past the deflector and down a long tube, where it could more usefully be exploited for many experiments in prospect. (One of these was the series of neutron scattering experiments Franz Kurie and I wanted to run in extension of our previous research.) This operation, called "snouting," required much artistry and intuition in fabricating flat pieces of iron ("shims") of varying sizes and shapes, placed over the tube and around the acceleration chamber to produce a magnetic field that would lead the beam through and focus it adequately on a target. The neutrons so produced were supposed to pour through a hole in the water tanks toward which the snout was aimed. Kurie and I were delighted to observe that when we had the cloud chamber running opposite the hole we saw hundreds of recoil tracks, whereas there were hardly any when the chamber was moved to one side. For me, the sight of so many recoils each expansion was absolutely paralyzing. I had been accustomed to see only one recoil every ten or so expansions back in Chicago with the weak neutron source we had available. The Berkeley cyclotron could do thousands of times better!

Unfortunately, the profusion of neutrons was as fatal to further progress as the lack of neutrons had been before. Kurie and I found that the neutrons were emerging not as a well-collimated beam but rather as a melange, coming not only directly from the target but from many other directions, owing to multiple scattering. We thus could not assume a fixed direction for the neutrons impinging on the target nuclei in the cloud chamber. We nevertheless did some studies on neutron disintegration and scattering, which we reported at a Physical Society meeting later that year.[2] This was my first such communication.

A phenomenon later to prove of considerable importance for me was the appearance in our cloud chamber photographs of many short stubby tracks when we filled the chamber with nitrogen gas. These recoils, being of constant and short range, were the result of "thermal" neutrons. These were present as a minor component in the neutron flux through the water tank hole. They arose because the fast neutrons initially made in the cyclotron collided often with the water hydrogen in the shielding water-filled tanks. These slow neutrons, it was subsequently discovered, caused a disintegration of the nitrogen nuclei of mass fourteen (^{14}N) to make ^{14}C and a proton (hydrogen nucleus of mass one). ^{14}C, a long-lived radioactive isotope of carbon, became the most important single radioactive tracer, inasmuch as carbon is central to living systems.[3]

Those first few months at the Radiation Lab were a tremendous learning experience for me. As I worked on the cyclotron crews, I was filling big gaps in my knowledge of electronic gadgetry, electrical circuits, and practical physics. There were weekly lab seminars when exciting new findings at Berkeley or elsewhere would often be described and discussed, invariably with Oppenheimer and his students in attendance to clarify relevant points. There were lectures on the theory of nuclear structure at which I took voluminous notes. Working with Kurie, I improved my general expertise in identifying the different kinds of nuclear radiations and their interactions.

Early in 1937, an event occurred that was to have a profound effect on the course of my research career—my meeting with Samuel Ruben, then a graduate student in chemistry. It happened that E. O. L. and James Cork, a visiting professor of physics from the University of Michigan and an authority on x-rays, had collaborated on a study of disintegrations of platinum by 5 MEV deuterons the previous year. The technique they had used was simple—thin platinum foils were stacked in a pile and placed in the path of the deuteron beam. As the deuterons passed through, they lost energy, gradually becoming attenuated to lower and lower energies until finally they were stopped. Each foil represented the energy residual left after passage through the preceding foils in the stack. Cork measured the radioactivities produced in the platinum foils, giving the average disintegration yields for each corresponding deuteron energy. Plotting these yields against energy, Cork and E. O. L. expected to see a simple monotonic curve showing a smooth lowering of yield with energy. Instead, they were surprised to observe that there were "bumps" in the yield curves they obtained. These bumps seemed to establish unequivocably that there were energy levels in the platinum-deuteron system at energies of many MEV.

Oppie developed an elegant theory to rationalize these results, although they were certainly preliminary in that they were dependent on hurried, qualitative chemical separations to establish that the radioactivities actually were due to platinum and nearby heavy elements in the Periodic Table, such as osmium and iridium. These radiochemical characterizations had been done by Henry Newson practically on his way to the train to return to Chicago, where construction of a cyclotron was in prospect. The work was duly reported[4] and it was proudly pointed to as a major contribution of the laboratory when Niels Bohr came on a visit later in the year.

A special seminar was arranged and the lecture room was filled to overflowing to see Bohr and hear E. O. L. and Oppenheimer give their report. After a short presentation of the data by E. O. L., Oppie gave a typically stupefyingly brilliant exposition of its theoretical consequences. We all sat in dumb admiration as he filled the blackboards around the room with arcane symbols and expressions in a seemingly effortless and self-deprecating fashion. Through it all, Bohr sat impassively until Oppie came to a pause. Bohr thereupon raised his hand diffidently, almost apologetically, to remark in his guttural and almost incoherent manner that he had much difficulty in believing that the data could be valid and therefore that the theory Oppie was expounding had any real basis. He had just finished evaluating some as yet unreported results obtained by Otto Frisch in Denmark and had developed his liquid-drop model of nuclei, which quite positively excluded the results Cork and E. O. L. were claiming. As can be imagined, these remarks produced shock and dismay. E. O. L. was especially disturbed because he was just recovering from a bad episode in connection with some erroneous claims he had made about the neutron mass and deuteron instability. He was anxious not to be involved in yet another such imbroglio. We were all aware that such repeated mistakes would do his cause no good with the Nobel Prize Committee, which, as everyone had surmised, was considering him as a possible candidate. To have Bohr, whose influence was considerable, unfavorably impressed was not at all a happy circumstance.[5]

Understandably upset, E. O. L. called in Ed McMillan, who had already in a very few years in Berkeley established himself as a brilliant young investigator, to straighten this matter out. Ed looked over Cork's data, and realized that to really make sense of it, adequate chemistry was needed to sort out the activities Cork had observed. Perhaps aware of my radiochemical background, he suggested I join him. E. O. L. had in the meantime sought help from Wendell Latimer, the dean of the College of Chemistry, who recommended young Samuel Ruben, then beginning his graduate studies in chemistry. So the three of us joined forces to reexamine the so-called "resonance transmutation" of platinum by deuterons.

It soon became clear that the results reported could not be verified. First, Ed found that the resonances did not occur in the same foils from run to run. Moreover, some yields of positron emitters formed were much larger than was reasonable. There was the ever-present bugaboo of contamination by impurities, especially those of low atomic number, which could be preferentially activated by deuterons, and thus give misleading radioactivities in the absence of adequate chemical separations. I carried out an experiment using Cork's reported procedure for pretreating the foils, which he had assumed would remove trace contaminants, and found, sadly enough, that it actually made the foils much dirtier than they were in their pristine state. Cork had washed the foils in dilute nitric acid, rinsed and dried them, and then flamed them with a torch to volatilize whatever contaminants remained from the acid pretreatment. Actually what he was apparently doing was precipitating and baking lab dust into the foil. The material thus introduced accounted for all the anomalous yields, particularly of positron emitters, one being ^{13}N, a 10-minute positron producer. The experiment I did was to rub a foil in transformer dust, representative of lab air-born contaminants, and put it through Cork's procedure. Ed then bombarded it and found it was at least an order of magnitude more active than the same foil not so treated.

We decided to start over, rejecting all the results previously obtained and using neutrons to establish the authentic activities of the platinum metal group. The chemistry required was extremely difficult and time-consuming, involving dissolving the foils in boiling aqua regia, followed by an exhaustive series of repetitive precipitations that were not wholly effective in separation of some radioactivities. A complete analysis took eighteen hours of continuous effort. We found that the previous anomalies were replaced by new ones—namely, an overproduction of negative electron emitters, the assignments for which were too many for the isotopes available in the platinum group. We concluded that we had produced a large number of nuclear isomers—that is, nuclei that had equal mass numbers and charges but differed in radioactive properties. This finding was quite surprising because until then nuclear isomerism had been confined to a relatively few examples. Only one was known among those radioactive isotopes produced artificially. Nevertheless, this result was confirmed many times later. Publication established nuclear isomerism as a phenomenon of widespread occurrence.[6]

There was little appreciation of what chemistry was all about among my "Rad Lab" confreres. This was apparent in many ways. There was a tendency to disparage the contribution radiochemistry might make in the studies underway at the lab. Chemistry was considered a kind of service subordinate to nuclear physics. On one occasion Ed McMillan discovered at the last minute that he could have a bombardment at two o'clock one afternoon (the cyclotron worked erratically, and when the opportunity for a bombardment arose, it was seized). I protested that our chemistry took eighteen hours straight without a break, so that Ruben and I would get no sleep. Ed, understandably impatient, retorted, "We'll make this run if it takes you guys all night!" In most collaborations with the chemists, the physicists early claimed the senior role. For example, Jack Livingood was the senior author on the first papers he published working with Glenn Seaborg, who was just beginning his distinguished career as a nuclear chemist and was a graduate student in chemistry.

The traumata Ruben and I had experienced in working out the platinum chemistry had drawn us together in a relationship that quickly developed into a strong friendship. We decided to form a partnership in which the facilities of both the Rad Lab and the Chemistry Department could be exploited to the maximum. We reached an understanding that we would split senior authorship on any results we obtained, in the sense that whatever concerned production and characterization of cyclotron-produced isotopes would be my responsibility, while their application in chemical research would be Ruben's.

I was simultaneously involved in a number of other collaborations, one in particular with Phil Abelson in which we were familiarizing ourselves with an ionization chamber setup E. O. L. had become enthusiastic about as an assay procedure alternative to the use of Geiger counters. By this time, I had been doing enough isotope separations as the lab chemist to become so contaminated that I could not enter a room in which a counter was running without raising the background to prohibitively high levels.

One morning, I stopped at Abelson's room to pick him up for a session on our ionization chamber setup. I noted that his electric clock was running backward and asked if he was aware of this. Yawning, Phil said that he was, and in fact had the clock running backward on purpose because trying to figure out what time it was woke him up. This remark set the tone for the day's activities. We had a particularly hard time with an erratic drift in the background rate of the apparatus. After much futile effort to find out what was causing it, Phil noticed that the drift was associated in some way with my movements around the room. Ordering me to remove my pants and stand in the corner, he took them (the only pair I had) and gingerly approached the apparatus. Sure enough, radiation was coming from my pants! We localized it on the fly, where apparently a large amount of radiophosphorus had accumulated.

The cavalier attitude to chemistry was well exemplified by the casual manner in which samples of cyclotron-produced radioactive materials were sent through the mails. Because there was a large demand for them, such traffic was considerable. As an example, red phosphorus, after bombardment, was simply placed in a waxed envelope or whatever was handy and mailed without further ado. Needless to say, such samples often arrived in a state of some deterioration—having turned into phosphorous and phosphoric acids en route. After I arrived and became the lab chemist, I had the responsibility of seeing that samples were properly processed before mailing. One such shipment, made while war was raging in Europe, consisted of some radioactive phosphate I prepared for the grand old man of isotope tracer methodology, Georg von Hevesy, then in Denmark. I sent the samples to him divided into three batches so that if any of the ships carrying them were torpedoed, the chances of at least one getting through would be greater. Hevesy received them all and wrote a charming acknowledgment.

While E.O.L had the bad luck to be on a campus that did not house a faculty of biologists by and large ready to seize the opportunities presented by the cyclotron's production of radioisotopes imaginatively, the demand for them on the Berkeley campus and elsewhere was growing.[7] It was clear that there was need for a resident radiochemist to monitor their production and deal with requests for them. E. O. L. had learned that I was primarily a radiochemist (I believe Luis Alvarez had helped to bring this to his attention), and one wonderful day he proposed that I join the staff as a Research Fellow at a salary of $1,200 annually. Just prior to this, Don Cooksey had heard that Harkins back in Chicago was interested in my returning there to help build his cyclotron. Don was concerned that I might accept and dropped a hint that E.O.L might make me an offer. I was only too happy to assure him I would stay no matter what salary E. O. L. was able to raise. (As it turned out, the Chicago proposal was several hundred dollars less.) With my new-found wealth, I was able to move in with Paul Aebersold, who had found a room in a house on College Avenue, a great step up from the Benvenue tenement.

Thus, only a half year after my arrival that dismal December day in 1936, I was on the payroll with an important role in the lab program. However, I soon discovered that handling the biochemical and medical needs of John Lawrence's group alone was a full-time occupation. First, there was an almost insatiable need for radiophosphorus (^{32}P) to implement experimentation on therapy of cancer and various blood diseases, particularly polycythemia vera. The ^{32}P was prepared by bombarding red phosphorus pressed as a paste into a water-cooled knurled copper plate and covered with thin gold foil to minimize mechanical loss in the initial phases of bombardment by the 6 MEV deuteron beam. The target was external to the vacuum chamber, from which it was separated by a thin aluminum window, through which the beam passed after being guided out of the interior of the cyclotron by the deflector. The design for this external target had been worked out by Kurie,[8] and was later improved by William Brobeck, whose arrival brought real engineering expertise to the laboratory. The need for water cooling is readily appreciated when one realizes that a beam of 6 MEV deuterons of some hundreds of microamperes requires dissipation of up to a kilowatt or so of power—roughly equivalent to the heat generated by an oxyacetylene torch.

After a bombardment of many hundreds of microampere hours, the "hot" target was removed, with its millicurie equivalents of a fearful mixture of copper, zinc, and other contaminating radioactivities in addition to ^{32}P activity, which itself amounted to many hundreds of millicuries. The mass of elemental phosphorus, copper, and other metals had to be processed and presented to the "medics" as an isotonic sodium phosphate solution, free of contaminating radioactivities and "pyrogens" (substances that could induce fevers). In other words the preparation had to be safe to drink. Similar demands were made for the use of calcium, sodium, strontium, and iron isotopes. In one instance, I was included as a co-author in such a study.[9] The correspondence initiated and maintained by me with cyclotron customers from 1938 onward was to prove important in building my reputation as an authority on tracer applications in biology and medicine. I began to accumulate large masses of data on cyclotron yields. In the same period I did a few radiochemical assays on radioactivities observed by various staff members, who found them using bombardments with high-energy deuterons and alpha particles.[10]

Meanwhile, in my collaboration with Ruben, studies were begun on neutron activation of rare earth metals, beginning with cerium and europium. We had devised gravimetric procedures for the difficult separations required and had made some progress in sorting out some radioactivities when Sam had a brilliant idea that completely changed the course of both our research careers. One day late in 1937, when I was at the 37-inch cyclotron control desk on a crew assignment, E. O. L. excitedly dashed in with a young fellow in attendance, whom he introduced as I. L. Chaikoff, an assistant professor in physiology. E. O. L. informed me that Chaikoff had a proposal to use short-lived carbon 11 to study carbohydrate metabolism. (This radioisotope had a half-life of twenty-one minutes.) When I inquired how this was to be done, particularly how the ^{11}C could be incorporated quickly into the starting material, such as D-glucose, E. O. L. said the isotope would be given to green plants as CO_2. By photosynthesis the plants would in short order synthesize the radioactive glucose, which could then be fed to the rats used in Chaikoff's studies. It had been calculated that with the ^{11}C yields then available, there would be plenty of time to administer the labeled glucose by feeding or injection and study its fate in the rat. I promised to help in any way I could to provide adequate amounts of ^{11}C-labeled CO_2.

Later that day I saw Sam and told him about the Chaikoff proposal. With some heat, Sam informed me that the original idea had been his. He had mentioned it to Chaikoff and was incensed that the latter seemed to be passing it off as his own. Sam had, in fact, urged Chaikoff to approach Lawrence, and showed me the calculations referred to by E. O. L. I was intrigued by the boldness and ingenuity of the proposal—so typical of what one might expect from Sam—and began to schedule cyclotron runs to produce the ^{11}C by deuteron bombardment of boron.[11]

I should emphasize that the Department of Chemistry held the university biologists and their work in low esteem and it took extraordinary courage on the part of a young instructor such as Sam to commit himself to a program in biological research. The prospects of such research providing a basis for promotion in the department were exceedingly dim. Nevertheless, Sam persisted, and his doing so eventually aided the creation of many great research laboratories on the Berkeley campus dedicated to tracer research in basic biology and medicine.

Sam had read the standard textbook exposition of green plant photosynthesis as a process in which CO_2 and water combined to produce glucose and oxygen. He reasoned that if radioactive CO_2 was administered to a green plant in the light, it would incorporate the radioactive "label" in all the six carbons of glucose. The use of such labeled glucose would make it possible to determine for the first time how animals utilized sugars in their metabolism. The relatively short half-life of the ^{11}C available was a drawback, but not necessarily fatal to some success in achieving insights into the intermediary metabolism of sugars. Already, R. L. Schoenheimer and his associates at Columbia University, using heavy hydrogen (deuterium) and the rare stable isotope of nitrogen (^{15}N) had revolutionized studies on dynamic turnover of organic compounds in the metabolism of fats and proteins.[12]

Sam and Chaikoff recruited Zev Hassid, already a friend of mine through my participation in the chamber music sessions organized by his wife. Zev, as an accomplished young plant physiologist with expertise in handling carbohydrate chemistry, was obviously an ideal collaborator for the initial isolation of the labeled glucose. Chaikoff had a large group of graduate students, so the enterprise was well-staffed. Sam brought over the boron oxide as target material, pressed into a copper target plate I supplied, and waited while I bombarded it. Then, quite heedless of the massive radioactivity to which he was exposed by the hot target, he would scrape the radioactive material into a ceramic boat, place it in a preheated combustion tube he had installed on a bench near the cyclotron, add a piece of filter paper to assure the presence of carbon for the $^{11}CO_2$ driven off the target material, and carry through a combustion, catching the $^{11}CO_2$ in a U-tube immersed in liquid air. He would then dash to the antiquated old shack where he had his laboratory on the ground floor and where Hassid waited with reagents and plants. This building, known as the "Rat House," was a short distance away from the cyclotron building and no more than a few minutes was needed to carry the $^{11}CO_2$ there, allow it to diffuse into the desiccator used to hold the plants, and illuminate them. After a total of perhaps ten minutes of elapsed time from preparation of the $^{11}CO_2$, the experiments could commence. The plant leaves were chopped up and dropped in boiling dilute ethanol.

The usual chemistry to find in which molecules the radioactive carbon resided required that the procedures be modified to circumvent a difficulty that arose because the actual number of molecules labeled was too small to find with the ordinary analytical methods. For instance, in a given experiment the actual concentration of labeled glucose might be no more than a millionth of a milligram. This would be insufficient to ensure that, when the glucose was treated with the special reagent to bring down a precipitate that could be counted, any such precipitate would appear. To ensure its appearance, it was necessary to add relatively large quantities (several milligrams) of unlabeled glucose with which the presumed labeled glucose could mix completely. Thus, the total concentration of glucose would be high enough to give the desired precipitate. One could then measure it, and determine how much labeled carbon was in it. The unlabeled glucose was termed "carrier," and in any experiment we undertook, our problem was to make sure we added all the different kinds of compounds in carrier amounts so that whatever molecules became labeled would coprecipitate with the carrier added.

To find carbohydrates, such as sugars, we used glucose as a carrier, and treated the radioactive mixtures made in our tracer experiments with the reagent phenyl hydrazine. The resultant "osazone" precipitate, prepared by Hassid, was collected on a filter paper, dried and covered with cellophane. These samples could then be wrapped around the cylinder of a Geiger counter to assay for radioactivity.

Sam had set up a primitive but effective counting circuit. It was spread out "breadboard" style so that Sam could troubleshoot as needed, which was often. A radio blaring loudly through the experiment was a monitor of unwanted electrical disturbances, which occurred, particularly if someone on the two upper floors started using a spark coil tester on a vacuum line. When this happened, as evidenced by static from the radio, Sam would utter an oath and dash upstairs. There would be shouts, threats, and recriminations easily heard downstairs, then an ominous quiet.

After a few weeks, I saw that Sam was looking depressed. It developed that the yields of $^{11}CO_2$ were quite erratic. Worse than that, the plants were sometimes exhibiting adequate CO_2 absorption, sometimes not. Most puzzling and frustrating of all, when there was a successful run insignificant radioactivities appeared in the osazones, showing that of the initial fixation observed, hardly any was in glucose or other related carbohydrates. In the meanwhile, Chaikoff and his expectant students were becoming restive.

During a recital of these troubles, Sam suddenly stopped, his eyes widened, and he blurted, "Why are we bothering with the rats at all? Hell, with you and me together we could solve photosynthesis in no time!" From that moment, we were out of everything but the photosynthesis business. I was just as excited as Sam at the prospect of solving the Big Problem—identifying at long last the initial product of CO_2 fixation in green plant photosynthesis—a mystery that had plagued chemists and biologists ever since the original discoveries of Joseph Priestley, Horace de Saussure, Jan Ingen-Housz, Jean Senebier, and others in the eighteenth century and thereafter.

Our strategy for the solution of the Big Problem was simple in prospect, if complex in implementation: feed the plants $^{11}CO_2$, wait for predetermined lengths of time, from a few minutes up to an hour, add carrier (measurable amounts of whatever compound we guessed might be labeled in the $^{11}CO_2$ exposure period), extract, isolate, and see if any or all of the radioactivity appeared in the carrier. If we had guessed right, most, if not all, of the radioactivity would be localized in the compound defined by the carrier.

It became evident very soon that implementation would be much more difficult than we had supposed. Experiments came to consume all our waking hours not otherwise required for our routine duties. There was hardly time for sleep. The first difficulty was to devise procedures that would assure adequate and reproducible supplies of $^{11}CO_2$ as needed. The cyclotron was not always available on schedule. Runs had to be made when it was not preempted by other demands, which meant squeezing them in at odd times. We were helped by the ^{11}C's short half-life of twenty-one minutes, inasmuch as yields not far from maximal ("saturation") could be gotten with bombardment times of only twenty or thirty minutes.

Meanwhile, the counters and associated circuits had to be kept in working order and be ready at a moment's notice. Sam had this responsibility and guarded his counting setup fiercely, forbidding anyone else to touch anything. A sign particularly aimed at the biochemists collaborating with him, such as students from Chaikoff's laboratory, stated in very direct language what would happen to any interloper or "button pusher." Sam had rigged up a signal light to indicate when the cyclotron was operating several hundred feet away in the Rad Lab, so that we would not be trying to use the counter when the background was increased in some unpredictable way by the stray radiation reaching the Rat House. It was also helpful in telling him when a run I was doing had ended, warning of my impending arrival with the radioactive CO_2 sample.

Then there was the troublesome business of the actual sample production and delivery. Sam's awkward early technique of pressing the boron oxide powder into a water-cooled knurled copper plate had to be scrapped because yields were small and erratic, not to mention the radiation exposure I had to undergo when removing and processing the bombarded target material. I developed a simple and highly effective technique based on our discovery that during bombardment most of the ^{11}C activity appeared in the gas filling the space between the target plate and the cyclotron window through which the deuteron beam emerged into the external target chamber. By connecting an evacuated brass cylindrical vessel some ten times greater in volume than that in the target chamber, the radioactive gases, mostly present as carbon monoxide, could be sucked out of the target gas space and removed for production of the labeled CO_2. In this way, radiation exposure was also minimized.

At the Rat House, Sam and Zev would be waiting for me like sprinters at the starting gate. Beakers would be filled with boiling water or other solvents and pipettes ready to suck up measured volumes of radioactive solutions onto absorbent blotters, which would be held by tongs over hot plates and dried. All the necessary reagents and apparatus would be in place. The counter would be ticking away establishing the background activity. Each experiment had to be planned ahead in every detail so that no time was lost in confusion or delay in deciding what procedure to follow. Anyone looking in on the Rat House when an experiment was in progress would have had the impression of three madmen hopping about in an insane asylum, what with the frenzied activity punctuated by loud classical music from the radio monitor, and Sam's yells to get on with it and hand him samples while he sat at the counter table, feverishly taking background and sample counts. We had no idea of what had happened until hours later when, with all samples assayed, we sat in exhausted consultation, calculating and evaluating the results.

A typical experiment required construction of protocols based on procedures devised after hours of library work, looking up the analytical procedures for an interminable list of possible compounds in which carbon derived from CO_2 might appear.[13] Often we had to devise wholly new procedures because those available did not produce samples convenient for our blotting-paper technique. They had to be absorbable or applicable to the paper and amenable to quick drying over a hot plate before being wrapped in thin cellophane and slipped over the counter for assay. We developed group tests to test whole classes of compounds with molecular weights of less than 500 or so, and then focused on individual compounds if indicated by the group tests. As time went by, both Sam and I acquired a considerable expertise in qualitative and quantitative organic analysis, a fact that would have greatly surprised Professor Kharasch in my case. Some years later, I had the opportunity to write a book in which I described some of our special preparations, as well as providing reproductions of original protocols and the time course of a typical experiment with $^{11}CO_2$.[14]

Dean Lipmann had claimed that nitrogen fixation occurred in germinating seedlings of non-leguminous plants, and we plunged into a series of experiments using the ten-minute ^{13}N isotope. Our results pleased the dean in that while we could not state positively that such fixation occurred, the data indicated further work with short-lived ^{13}N might be definitive.[15]

The extremely demanding procedures involved in our experiments with the short-lived carbon isotope, the constant pressures to get on with the work, and the realization that the answer to the problem of fixation did not seem to involve carbohydrates at all forced Zev to discontinue his collaboration. He suffered from high blood pressure, which was not helped by the hectic nature of our work, and his physician strongly advised him to desist. Sam, who had no physical disabilities and thought anyone who worked less than eighteen hours a day was a malingerer, had not been too sympathetic, especially as we had to slow down at times to keep Zev from collapsing. Zev had no option but to withdraw, which explains why his name appears only on the first few papers we published.

As 1938 drew to a close, I could take great satisfaction in the dramatic improvement in my prospects. I was a regular member of the leading nuclear physics laboratory in the world, filling a position of responsibility vital to its present and future. I had the confidence and friendship of E. O. L. and warm, even affectionate, relations with the associate director and my colleagues. I had a host of new friends and acquaintances in Berkeley and the East Bay. I had met and married Esther Hudson, an attractive brunette, with whom I found great compatibility, a shared love of music, and every prospect of happiness. The marriage took place the night of the Big Game in 1938 so that it easily escaped notice. We celebrated with a spectacular feast at a Basque restaurant in San Francisco.

Another very significant event occurred a few nights later. Frances Wiener, the concert mistress at NBC, invited us to go with her to a chamber music party at the house of the budding young violin virtuoso Isaac Stern. She assured us that we would find him a most remarkable young man. A lifelong friendship resulted.

The East Bay was at that time producing an amazing array of young prodigies—Yehudi Menuhin, Giulia Bustabo, and Ruggiero Ricci, to name a few—mostly students of Louis Persinger's. Isaac was not of this group, but eventually turned out to be the most successful of them all.

I should not end this chapter without attempting to impart more of the flavor of these exciting times. Three "Tales from the Berkeley Woods" may suffice.

One of the many distinguished scientists to visit the Berkeley campus was James Franck, a Nobel laureate in physics, and a truly noble individual. Franck had left Germany in protest of the Hitler terror. The Fels Foundation in America agreed to support him provided he work in photosynthesis. He was not happy at the prospect, but adapted. I came to know him through the years and developed a warm relationship with him. Franck had been invited to give the prestigious Hitchcock lectures in Berkeley. As a gesture of gratitude to all involved in arranging the invitation and lectures, he gave a banquet at the then luxurious Hotel Leamington in Oakland. Attending were numerous distinguished luminaries of the university, notably, the chairman of the College of Chemistry, G. N. Lewis, and many other members of the Chemistry faculty. Also present was an assortment of professors from other departments, including Oppenheimer from Physics and—possibly as an afterthought—the famous biologist Richard Goldschmidt. Sam Ruben and I were admitted to this august assemblage, being the only faculty types directly concerned with photosynthesis. Goldschmidt, who as a biologist ranked low in the pecking order, sat more or less isolated at the foot of the table.

By the time we reached the brandy and cigars, everyone was feeling mellow and garrulous, particularly Franck, who leaned back in his chair, smoking a Havana, and remarked casually, "Well, gentlemen, let us face it. There are some problems science cannot solve." He voiced this sentiment with no intention of starting a serious philosophical discussion. However, G. N. Lewis, ever combative, snapped "What problem, for instance?" Franck, taken aback, reacted testily after a moment's hesitation by firing back, "Well, what is life?" Lewis thereupon launched into a learned diatribe against biology and biologists in general, who in his view were devoid of ideas of their own but leeched off the physical sciences, using concepts not intended for application to living systems. He maintained that biologists needed to get their own ideas if science were to make inroads on a problem such as the nature of life. Franck argued that Lewis was being too narrow in his critique. The discussion was beginning to wax hot when Oppenheimer, seeing his role as that of moderator, suggested that it was really a matter of semantics. This made no impact on the disputants, who continued to wrangle. Suddenly noticing Goldschmidt, who was calmly eating away at the far end of the table, and had heard nothing of the dispute raging at its head, Oppenheimer cried, "Ah, we have a biologist here. Let's see what he thinks about this question." Goldschmidt's attention had to be aroused by a jab in his side from the person sitting next to him. Looking up with his mouth full of food, he was startled to hear Oppenheimer repeat the query about what he thought of the proposition by Lewis that biology was in need of its own ideas. Slowly, he marshaled his thoughts, methodically finished chewing, and said something like, "Well, gentlemen, I really have little time for questions like this. I work at the bench all day and find no energy to dwell on such matters except perhaps when I am relaxed in the toilet!" Needless to say, this ended further discussion on the relevance of science to the unraveling of life's mysteries.

Later in Franck's visit, I had the special privilege of accompanying him on numerous walks around the campus while he reminisced about the great days back in Göttingen, where he had been involved in the development of quantum mechanics during the twenties. He recalled the discussions that arose when Heisenberg first enunciated his famous principle precisely stating the ultimate limits of error in physical measurements. The need to find a proper name for this principle prompted Max Born, one of the famed physicists at Göttingen, to suggest the rather profound appellation "Uncertainty." Franck recalled how he had vigorously fought this notion, asserting that Heisenberg's proposition had nothing to do with uncertainty in its broad philosophic sense, being only a specification of the precision possible in strictly physical measurement. Franck suggested that a term like "incommensurability" would be more appropriate, but he failed to move Born, and so the Principle of Uncertainty was promulgated, bringing with it, as Franck predicted, all manner of miasmic nonsense purporting to show that science had found a place for God. As Franck wryly pointed out, the operative measure in Heisenberg's principle was Planck's Constant, which required God to be squeezed into a space on the order of 10^{-26} centimeters.

One of my favorite Berkeley stories is the tale of the Rope in the Clarinet. This has become veritable legend through the years and has been so refined in the telling by the various participants that it deserves, so to speak, to be a footnote in the Bible. Its overtones are profound enough, providing an allegory of the way science gets done. There is the Deductive Type (read: theoretical physicist) who is confronted with the description of an apparently unacceptable phenomenon by a Pragmatist (read: experimental physicist) whose knowledge of theory is slim. At first, the former scornfully dismisses the observation by the latter as obviously a figment of the latter's imagination, but on being confronted with it in reality quickly finds an explanation, thereby modifying previous theory or adding an important new principle where none had been suspected before. Finally, the phenomenon is taken over completely by the Deductive Type, leaving the Pragmatist lost in the background.

One day I was visited by Julian Schwinger, a member of the brilliant coterie around Oppenheimer and destined to be a Nobel laureate in physics some years later. He informed me that a very promising young theoretician named Leonard Schiff had just arrived on a National Research Council Fellowship to work with Oppenheimer. In those days, these fellowships were scarce and difficult to get, so the award to Schiff, who was hardly twenty and had received his doctorate only a short time before, indicated the esteem in which he was already held. Schiff was to become a leading theorist and developer of quantum mechanics, while holding chairmanships of physics departments at Pennsylvania and Stanford. What concerned him at the moment, Schwinger explained, was his understanding that I was involved with chamber music groups in Berkeley. An enthusiastic clarinetist with limited experience playing in chamber groups, he wanted me to include him in these activities, hoping to improve his knowledge of the literature for the clarinet. This alarmed me somewhat, as it indicated that Schiff was probably not sufficiently skilled to join the highly proficient groups into which I had managed to work my way, starting from the bottom of the scale on first arriving in Berkeley. I was not anxious to create embarrassment by bringing someone not up to the level of performance at our chamber music sessions.

Soon Schiff came to see me and told me how he had come to play the clarinet. His mother, an excellent pianist, had avoided teaching him on the grounds that this might cause tension between mother and son, so he had grown up hearing music and loving it but frustrated in playing it. In the high school he attended, there was the opportunity to enter the band and play a band instrument, so he had taken up the clarinet, the one that happened to be available. Soon, he conceived the notion that there must be something for the clarinet other than Sousa marches, admirable as they might be. A search in music stores and libraries revealed that there was indeed a great chamber music literature for the instrument, with contributions from such as Mozart, Brahms, Reger, Weber, and Prokofiev, to mention but a few. He obtained the scores to quartets, quintets, and other combinations written by these composers, and methodically went through the clarinet parts, learning each by rote. Now he wanted to play them and hoped he could do so at the sessions he had heard I attended at least once weekly.

This story did nothing to allay my fears. Obviously, Schiff had no experience playing with others. I could imagine a session with him getting lost over and over again, and the evening being spent in fruitless stops and starts. So I did nothing to bring him into chamber music activities for almost a month. Then, as Leonard kept the pressure on, I gave in one day. Waldo Cohn, an enthusiastic and highly competent amateur cellist whom I knew as a biochemical colleague at the university, had a session scheduled. I phoned him and explained the situation. Waldo, quite as expected, offered his hospitality and the suggestion that I bring Leonard that evening. He added that we could perhaps try one piece with Leonard to see how it went. We would have our usual quartet, consisting, in addition to Waldo and myself, of Frank Houser, a rising young professional violinist, who eventually became concertmaster of the San Francisco Symphony and Opera orchestras, and Edward Becker, a fledgling lawyer and excellent semiprofessional violinist.

When Leonard and I arrived at Waldo's house, he was cordially welcomed and invited to take a seat while the quartet warmed up with a few compositions by Haydn and Beethoven. These took almost an hour, while Leonard sat impatiently clutching his clarinet. Finally, I turned to Frank and said, "How about playing something with Leonard?" Frank replied, "Sure, what's he want to play?"

Leonard had thought of trying the Brahms Clarinet Quintet, but as the moment of truth approached, he lost confidence and decided to fall back on the Mozart. He quickly set up his stand and passed out the parts, but just as we were about to begin, he stopped and looked worried. He had suddenly realized that the Mozart Clarinet Quintet was written for an A clarinet. His was a modern B-flat instrument. It was customary for clarinetists to make the necessary transposition, playing a half tone low. Leonard confessed he had not learned how to transpose, as the need had never arisen when he was playing by himself. Everyone favored me with a sour look for making such an impasse possible. Leonard suddenly brightened and said, "Why don't you tune your strings up a half tone?" We thought this might work, but quickly found how little tolerance there is in stretching strings. They produced a characterless sound and kept slipping down. Drawing on his knowledge of the theory of the fundamental characteristics of pipes, Leonard then tried pulling the mouthpiece of the clarinet up as far as it would go without falling off completely, in the hope that by increasing the length of the clarinet tube he could lower its fundamental pitch at least a quarter tone. Then, he explained, we would need to tune up only a quarter tone. This procedure only produced a mournful cacophony.

Leonard now made the amazing suggestion that we string players transpose! The notion outraged us, as violinists never were called upon to do such a thing unless we happened to be of a musicological bent and read ancient scores in exotic clefs. Occasionally violists and often cellists were required to read in a clef other than their accustomed one, but never the lordly violins. Nevertheless, we gamely agreed to try. The Mozart Quintet now exhibited a feature of which we had until then been unaware. The frequency of notes per bar, low at first with two half notes to a measure, rapidly increased in powers of two until very shortly there came a passage in fast sixteenths. I can still see Frank's face contorted with effort as he slogged along note by note with the first violin part, thinking for each note "half tone high, half tone high." When he climbed to the top of the plateau from which a rapid passage down to Avernus began, he froze in anguished confusion.

At this juncture, Frank suddenly remembered something he had seen at a party he had attended a short time before that might solve the impasse. The following exchange ensued:

FRANK: This guy I saw at the party had a clarinet and he couldn't transpose either. He hung a sock in it and it came out A.

LEONARD: You can't be serious. What would hanging a sock in there have to do with it?

FRANK: But I saw it!

Turning imploringly to me as a colleague, Leonard asked what I thought. "Well, I can't see how it could work but why not try it?" I said. Leonard eyed me scornfully indicating he had expected a more rational opinion and support of his position. Meanwhile, Waldo, who had found the fast passage in his part a little farther along, had been practising it fortissimo, a half tone high, and had heard none of this conversation. Calling him to halt, we ordered him to produce a sock. Completely taken aback and confused, he stammered, "Wh-what color?" With more than a hint of fine sarcasm, Leonard turned to Frank and asked, "Makes a difference what color?" Frank doggedly replied, "But I saw it." He added with a touch of asperity that he was no Great Brain, like Leonard and me, but he knew what he saw.

While we sat in impatient silence, Waldo went out and, after what seemed an eternity, reappeared waving a length of clothes line. "I couldn't find a clean sock," he explained. "Will this do?" Leonard, shrugging his shoulders, but resigned to the trial and sure he had fallen among madmen, muttered "Oh, well, all right!" Unscrewing the mouthpiece to expose the threaded portion of the clarinet pipe, he unraveled the end of the rope so that it could engage the threads and let it hang down the pipe, protruding untidily at the bottom. He then replaced the mouthpiece and, with the rope now firmly in place, puckered his lips and blew, his expression one of wary apprehension, mixed with resigned disbelief.

At first, no sound except escaping air emerged, but as Leonard blew harder, A, or something close to it, came out, a bit furry but undeniably A. When Leonard heard the note, his facial muscles betrayed shock and not a little dismay. Frank gleefully broke in with, "See, I told you it came out A!"

The Deductive Type had been confronted by the unarguable existence of a phenomenon. Reacting according to type, Leonard began to cerebrate even as he continued to blow. Almost immediately, he put his instrument down. Favoring us with a confident smile, he said, "Of course, it's obvious!" The rope being flexible, he went on to explain, could vibrate along with the wind column, thereby adding weight to the vibrating system's air mass. As everyone ought to know, the pitch depended inversely on the mass. The rope happened to add just enough mass to lower the natural vibration frequency from B-flat to A.

The work of activating the vibrating column was considerably greater than usual, because Leonard had to move the rope as well as the air, so he was exhausted by the time we finished the Mozart. We also noticed that he was a bit out of tune in the upper register, as well as exhibiting a rather fuzzy quality of sound in the middle and low registers. These effects were all consistent with his theory, Leonard explained, being "second-order perturbations on the initial coupling." Leonard, as typical of Deductive Types, seized the opportunity to show he was not helpless with his hands. He went to the machine shop the next day, filed a small groove in the top of the threaded portion of the pipe, and installed a small dowel from which the rope could hang as a permanent addition when he needed an A clarinet. He cut off the extra piece of rope dangling from the bell and thus completed the transformation.

Through the years, I told this story to other scientific colleagues who happened to be clarinetists and the same reaction of incredulity was invariably the result. One of these was Theodore Rosebury, a microbiologist of note, who had the resources to check it. One evening during a party at Ted's house, I regaled the group present with an elaborate version of the story, whereupon Ted, showing disbelief, produced a length of clothes line. When he repeated Schiff's procedure, the note came out G, one full tone below A. Apparently, Ted's rope was thicker. The original rope had chanced to be the right thickness to produce A.

Years later, there was a reunion with Leonard at the University of Pennsylvania, where he had assumed the chairmanship of the Department of Physics. Pressed to confirm the rope story, he flashed a typical shy, infectious grin and replied, "Well, it's a bit exaggerated, and anyway, I've learned to transpose!"

5

New Vistas in Photosynthesis

Excitement about imminent breakthroughs in understanding photosynthesis was not confined to our trio in Berkeley. A few hours' ride away was the Carnegie Institution at Stanford, a major center of photosynthesis research. Soon we were journeying there to exchange ideas and information at numerous seminars and other sessions where we could be enlightened by the local experts—Robert Emerson, William Arnold, Stacy French, C. B. van Niel, P. A. Leighton, and J. H. C. Smith, to mention a few—and authorities such as Hans Gaffron and James Franck, who were occasional visitors. This was a uniquely effective way to acquire a proper sense of the grandeur and sweep of the global process whereby green plants and certain bacteria grew at the expense of light.[1]

The modern study of photosynthesis may be said to have commenced when the Flemish natural philosopher, J. B. van Helmont (1574–1644) placed a twig in a pot of dirt and grew it to a sizable tree in the course of several years, using no nutrient other than rainwater. He assumed, however, that all the plant material made somehow came from the water. Analytical chemical methods needed to establish CO_2 as the source of all the carbon were still to be developed, beginning with the investigations of such as Robert Boyle (1627–91). Others, among whom may be mentioned John Mayow (1643–79), noted the presence in air of what we now know as molecular oxygen (O_2), but it was Joseph Priestley (1733–1804) who made the great breakthrough in the late eighteenth century and showed it to be the characteristic product of the only kind of photosynthesis then known, that in green plants. Priestley also established that yeast fermentation produced a gas (which we know to be CO_2) identical with that given off by plants respiring in the dark, a discovery that resulted from his observation that plants could "purify" night air.

Chemistry came of age in the nineteenth century with the development of quantitative methods based on the chemical balance and gas analysis, but it still took many decades before the overall steady state reactions of green plant photosynthesis could be shown to be represented by the simple bookkeeping statement that for every molecule of carbon dioxide absorbed, a molecule of oxygen was liberated.

How carbon dioxide (CO_2) could supply all the carbon needed for growth of photosynthetic organisms was an unsolvable riddle until tracer methods became available. The carbon in the CO_2 could then be "labeled" by adding one of the carbon isotopes—either radioactive carbon 11 (^{11}C) or the rare stable isotope carbon 13 (^{13}C)—as a "tracer." Such labeled carbon could be told apart from the carbon already present in the cells because the latter was "unlabeled"—that is, did not contain ^{13}C, except in small amounts. At last we had a method of tracing the carbon through all its intermediate forms, from the time it started as CO_2 to when it appeared as the carbon of cell materials such as fats, proteins, carbohydrates, and so on. Likewise, we could use the rare stable isotope of oxygen, oxygen 18 (^{18}O), to trace the origin of molecular oxygen by incorporating it as a label in the oxygen of water (H_2O) or in CO_2. We could even try using the radioactive oxygen isotope, oxygen 15 (^{15}O), even though it decayed almost completely in fifteen minutes. Such experiments with labeled oxygen might tell us if Helmont's guess was at least partially true.

For want of a definitive technique such as tracer methodology to locate, isolate, and characterize the few molecules of the first products of CO_2 assimilation among the thousands of other compounds—fats, carbohydrates, lipids, proteins, nucleic acids, and so on—manufactured over the life of the cell, and constituting the whole plant, hundreds of molecules had been proposed in the literature as candidates as time went on. Sam and I felt certain that with "labeled" CO_2 we had the means to pick out the molecule(s) in which the carbon from CO_2 found its first lodging.

Other developments, some certainly momentous, were occurring at this period in the history of photosynthetic research. Physico-chemical measurements of reaction rates in chemical systems were being refined so that the laws that quantitatively described the course of any given chemical reaction (the "kinetics") could be determined. From these, clues as to mechanisms could be deduced. In photosynthesis, they showed that at sufficiently low intensities of light, the photoevolution of O_2, or equivalent absorption of CO_2, was strictly proportional to light intensity. Although it was clear that the process of assimilation had to include "dark" reactions (that is, reactions not dependent on light absorption) coupled to the initial photoreactions, these were not limiting until rather high rates were reached, because the dark reactions were dependent mainly on temperature, whereas the light reactions were not. Eventually, the light reactions became too fast for the dark processes, so that the overall reaction reached a "saturating," or maximal, yield determined by temperature. The fact that the rate of oxygen production was greatest at low light intensities meant that light absorption was most efficient at low levels of illumination. Unfortunately, under such conditions the light reactions were slow, so that the "background" uptake of oxygen by the dark respiration process, going on simultaneously in the leaf system or algae used as test objects, was an appreciable fraction of the total gas exchange. The uncertainty of the correction to be applied to the observed rate of oxygen production in the light arose mainly from the possibility, difficult or impossible to exclude, that light itself might change the rate of respiration, which had to be determined separately in dark "control" periods set up to alternate with exposures to light.

Other troubles arose in these efforts to determine light efficiency. There was evidence, found by Robert Emerson and others, that in the transition from light to dark periods in the manometric assay procedure, devised to determine the "dark correction," the theoretical and experimentally established steady state ratio of one molecule of CO_2 absorbed to one molecule oxygen evolved did not hold. In fact, depending in a very complicated manner on the state of the cells and experimental conditions, there could be a "burst" of CO_2 released in the first few seconds or minutes of illumination—an effect precisely opposite to the usual absorption of CO_2 induced by photosynthesis. Conversely, an anomalous uptake of CO_2 could be observed to occur on transition back to dark. Despite these difficulties, Emerson and Arnold and many others pressed on, and by the middle thirties they, and others, had appeared to demonstrate that the efficiency of storage of light energy as chemical energy in simple green plants (algae) was not greater than about 10 percent. This was a very high value even so, as no photochemical systems in the laboratory were known with efficiencies of more than a few percent. Nevertheless, other investigators, led by the great biochemist and Nobel Prize winner Otto Warburg (with whom Emerson and others in the field had studied), held the view that nature was as close to perfect as allowed by the laws of thermodynamics and that energy efficiencies should approach 100 percent.

In the parlance of quantum mechanics, then being established, light rays of the visible spectrum, and electromagnetic radiation in general, consist of bundles of energy, termed quanta. In the case of green plants, red light, which is specifically absorbed by the special photosynthetic photosensitiser chlorophyll and known to support maximal rates of photosynthesis, had an energy equivalent for each quantum of about 40 kilocalories, whereas the total energy required to run the photosynthetic process producing cell material and oxygen from CO_2 and water was about 112 kilocalories. This meant that at least 112/40, or 3, or at most 4, quanta per molecule of O_2 would be needed. Such was the "right" answer. Any lesser efficiency (higher quantum number) meant something was wrong with the experiment.

Emerson just as stubbornly adhered to the view that only values from experiments done properly should be believed. Warburg refused to budge and set out to refute Emerson. The history of the resultant controversy is a tragic instance of how the time of two great investigators can be wasted. Warburg devised one ingenious procedure after another to refute Emerson. He eventually thought he had crawled out of this dilemma by suggesting that extra energy came from endogenous (internal) respiration, which he proposed was coupled to the primary photochemistry.

Emerson was one of the most remarkable men I have ever met. A descendant of the famed New England family of Emersons, he embodied all the virtues, as well as the rigidities, of his heritage. His fierce desire for independence and individual self-reliance was manifest in his living arrangements, which required no dependence on outside sources for energy and food: he raised vegetables and other foodstuffs in his backyard and had a generator for supplying electricity. He displayed intellectual courage and integrity of the highest order in all his research and inspired unqualified respect, and even love, not only among associates, but among those who knew him only as an acquaintance. During World War II, indignant at the brutal treatment of the Japanese-Americans in California, he suspended his career and devoted the war years to resettlement projects to rehabilitate these victims of war hysteria. (Needless to say, these efforts did not meet with popular approbation, but Emerson was not deterred.) His great abilities as an investigator were never fully realized as he lost much time in the quantum yield controversy. Also, there was family tragedy—a brilliant son afflicted with epilepsy who died an untimely accidental death. Emerson himself died prematurely in the first accident of an Electra; ironically, it was his first airplane flight.

Another scientist who experienced anguish in the quantum number controversy was the Nobel Prize-winning physicist James Franck. His expertise dictated that he give attention to what might be deduced from data on fluorescence emission and kinetic analyses, efforts that were to take up the rest of his career, first at Johns Hopkins and later at the University of Chicago. Franck thought that Warburg's dominant position as perhaps the leading biochemist of the time—he had won the Nobel Prize in medicine and physiology in 1931 for his work on respiratory enzymes—gave good reason to believe his claims. Regrettably, Warburg turned out to be wrong and Franck's deductions based on Warburg's assertions became irrelevant.

Ironically, the efforts Emerson made to catch up with Warburg's maneuvers were fruitful in the end, because one of Warburg's findings—that a little blue light raised the efficiency of photosynthesis in cells simultaneously irradiated with the red light absorbed by chlorophyll—set Emerson on a new path of research in which he discovered that two light systems were active in photosynthesis, not just one (see below).

My one meeting with Warburg was unfortunately under conditions that did not present him in the best light. It was early in the fifties, after Emerson had established his research group in photosynthesis at the University of Illinois in Urbana. The Russian occupation of Berlin was beginning to bode ill for Warburg who, though partly Jewish, had received special treatment as an "Honorary Aryan" in Hitler's Germany. Emerson generously decided to bring Warburg to America, where he might find refuge and possibly an appropriate appointment at some university. Warburg was established as a guest of the University of Illinois, living in the Faculty Club and working with Emerson, who thought that this would also offer an excellent opportunity for resolving the differences in quantum yield measurements. He was naive in this expectation, as Warburg was not about to admit that there was anything but error in Emerson's approach.

Arrangements were made to introduce Warburg to all the workers in photosynthesis who could conveniently journey to Urbana, thus forwarding the campaign to get Warburg located in a proper faculty slot. I was among those who received an invitation to meet him. I was excited at the prospect of seeing and talking to this seminal figure in twentieth-century biochemistry.

I was ushered into the Presence by Emerson with the astonishing announcement, "Er ist ein Physiker." Emerson explained later that had he identified me as a biochemist, I would have been dismissed curtly after a brief exchange of pleasantries. Warburg insisted on talking in German, although he spoke English fluently. He was sharp and perceptive in his comments about current research, but exuded intellectual arrogance based on his awareness of the superior position he held among contemporary biochemists.

He demonstrated to me his newest scheme to overcome uncertainties in the manometric procedures for determination of photo-evolved oxygen, using a background of white light to cancel out the possible fluctuations in respiration caused by illumination. He then irradiated the vessel with a bit of red light in the visible region of best yield for photosynthesis. He reasoned that the additional oxygen so produced could be used to calculate quantum yields because the uncertainty in oxygen production—owing to the effect of light on respiration—was automatically cancelled by the simultaneous illumination with white light.

After this laboratory diversion, we talked for some time about current progress in biochemistry, especially interesting to Warburg, who had been somewhat isolated from developments that had taken place during the war years. I mentioned Fritz Lipmann's introduction of the "energy-rich" phosphate bond, whereupon Warburg sharply inquired what that might be. I explained in my halting German that this concept was exemplified in the Warburg laboratory's work on the production of enol phosphopyruvate in glycolysis. (I had realized that it was best to slant the discussion toward Warburg's own accomplishments.) He reacted by asking why something he had already done was considered so novel. I responded that Lipmann had built a general theory of bioenergetics based on the kinds of reactions with phosphate exemplified by the glycolytic formation of enol phosphopyruvate. (But of this more later.) Warburg then inquired who Lipmann might be. I was on the verge of telling him Lipmann had worked with Otto Meyerhof when I remembered that Warburg had just a few moments before voiced some uncomplimentary opinions about both Meyerhof and Franck, whom he had thought quite unworthy of the Nobel Prizes they had received. So, I mumbled some innocuous story about Lipmann's having been a refugee from the war situation and went on to another subject.

Emerson's scheme to help Warburg came to naught. Warburg had arrived in an extremely wary frame of mind. Not surprisingly, he felt very much on the defensive, essentially a stranger removed from his native heath and sources of power. His aide-de-camp, one Heiss (sic), who had come with him, was a sort of gray eminence. He had the responsibility of handling all of Warburg's housekeeping needs. Emerson had arranged a room for Heiss in a fraternity house. Only one room was available for Warburg in the Faculty Center and he had not imagined that Warburg would require Heiss to be living with him. Warburg visited the fraternity room and was repelled by the primitive state of the accommodation—not too surprising to those who knew how fraternities and other student living quarters look even at the best of times. The situation had to be resolved by installing a cot for Heiss in Warburg's room. I mention this seemingly trivial circumstance because it emphasizes that Warburg trusted no one so much as Heiss, who was intensely loyal to him.

In his efforts to find Warburg a haven, Emerson arranged a symposium to bring together most of the academic researchers in the field, to be chaired by Farrington Daniels, a prominent physical chemist and leader in the area of quantum efficiency measurements, who had helped establish the units of photometry for the American Bureau of Standards. Warburg embarrassed Emerson and irritated Daniels at this meeting by maintaining stoutly that the Americans were in error and that the German standard unit of light intensity, the "candle"—as determined by his father Emil Warburg, a famous physicist—was correct.

In Warburg's defense, it should be said that he was a proud man, unhappy at the thought of receiving charity, and quite rightly pessimistic about the prospects for obtaining a position comparable to that he had enjoyed in Germany. He expected and deserved an appointment that would take into account his stature as the foremost enzymologist of his time. Eventually, he returned to Germany, where he was installed as the head of a special laboratory in Dahlem, a suburb of West Berlin, in the system of Max Planck Institutes. I doubt that Emerson, with a sense of values based on self-denial and wholly at odds with that of Warburg, ever really understood that his well-meaning efforts to help Warburg were foredoomed to failure.[2]

I came away with the feeling that I had met an authentic genius, however self-centered he might be. I would very much have liked to work with him. There is no doubt that he had a great gift for training experimenters, as witness the list of remarkable biochemists who emerged from his laboratory. Of course, there are others, Lipmann for one, who achieved similar results through mutual respect, rather than unquestioning deference on the part of students and associates. I am reminded of a like situation in music, where Arturo Toscanini, by terror, and Bruno Walter, by gentle persuasion, both produced great orchestral performances.

A kinetic finding of real relevance at the time I was beginning to read about photosynthesis in 1937 was that the actual efficiencies at low light intensities required that only one chlorophyll molecule in every hundred or so present be active in conversion of light energy. Emerson and his associates at Stanford's Carnegie Institution had shown by the middle thirties that the chlorophylls in the photosynthetic apparatus—the chloroplasts of algae in these experiments—acted as though several hundred chlorophyll molecules cooperated in light absorption by any one of them. This notion of a "photosynthetic unit" had been presaged by earlier work of K. Wohl and H. Gaffron in Germany. A realization of the complexity of the light absorption system was thus beginning to emerge.

At about the same time, a momentous development at Cambridge University in England was the first successful application of the biochemical approach to the analysis of photosynthetic reactions—that is, the separation of the complex process of photosynthesis into component chemical systems. In 1937 Robin Hill demonstrated that chloroplasts, the site of photosynthesis in green plants and algae, could produce oxygen photochemically without the need to be coupled to the CO_2 fixation systems. This chloroplast reaction, or Hill Reaction, as it came to be called, occurred if one supplied an oxidant—that is, a hydrogen acceptor—to take up the reducing power normally produced in photosynthesis.[3] There had been prior claims in the literature that such reactions were possible, but not until Hill's work did it become clear that they actually occurred in isolated chloroplasts with yields of oxygen as high as, or higher than, the intact leaf or algae.

By the time Sam and I ventured into photosynthesis research, the field was thus beginning to open up. Encrusted tradition was crumbling. The primitive state of knowledge about chloroplast enzymology was soon to be apparent.[4] Daniel Arnon and his associates at Berkeley were to show in the decade ahead that the chloroplast was a complex living organelle with a full complement of enzymes to make the whole array of photosynthetic products. For example, there were plenty of indications that at least the biosynthetic apparatus for starch formation from sugars was present in chloroplasts. There were recorded observations that starch was deposited in isolated chloroplasts even in the dark when leaves were floated in solutions of soluble sugars. We immediately thought of an experiment to use sugar labeled with ^{11}C and isolate the insoluble labeled starch that should form if these claims were correct. Our readings suggested numerous experiments other than those directed to isolation of fixation products.

About this time, Kenneth Thimann published a note in *Science* in which he suggested that the CO_2 production from enzyme breakdown of organic acids—pyruvate, formate, oxalacetate, and so on—might be the reverse of the process whereby CO_2 was absorbed in photosynthesis. He made the point that this suggestion was at least a working hypothesis consistent with a number of facts, whereas the earlier assumption that CO_2 combined with chlorophyll, sanctified in the literature and espoused by Warburg, had no experimental basis at all. Thimann emphasized, as had others, that there was even less evidence for an "activated formaldehyde" as the first reduction product of CO_2 and remarked that "the persistence of these unsupported theories must be ascribed to the absence of any plausible substitute."[5]

Thimann cited work by Hans Gaffron, who had noted that certain photosynthetic bacteria could absorb and incorporate carbon dioxide, like green plants, if they were fed organic acids.[6] There were a number of observations recorded in the literature that showed definitely that carbon dioxide was involved in the metabolism of even those bacteria that did not photosynthesize. Some microorganisms made propionic acid from glycerol[7] by fermentation, and others hydrogen from formic acid.[8] More important, it was beginning to be realized that "cyclic" mechanisms occurred in biochemical systems—the most famous example being the early demonstration that urea was made from carbon dioxide and ammonia in liver slices through the participation of certain intermediate molecules that did not appear in the overall equation.[9]

It was clear to us that we had the means to investigate these exciting suggestions definitively. It was not surprising that Emerson and Arnold both reacted so enthusiastically when they were made aware of our prospects, going so far as to declare that all work should be held in abeyance until we had the chance to fully exploit our labeling techniques.

Of course, this was not taken seriously by anyone, least of all by Emerson, who continued to work night and day in his efforts to clear up the Warburg matter. The history of this research, which led to the discovery of the existence of anomalous bursts of gas in the transitions from light to dark and, more importantly, the seminal demonstration that two separate photosystems existed, is a fascinating example of the rewards in hewing for long periods of time to a line of work that seems routine. In Emerson's case, he had been continually frustrated in his careful examinations of Warburg's manometric results by the latter's remarkable ability to come up with new sets of conditions refuting the conclusion that his previous experiments had been erroneously interpreted. It was known, mostly from Emerson's work, that the yield of oxygen did not parallel the absorption of chlorophyll in the red. Going toward lower energies, the yield fell off faster than did the amount of energy absorbed. This discrepancy was resolved when Emerson investigated Warburg's discovery that addition of higher energy light, as in the blue part of the absorption spectrum, gave better yields in the low energy region of the chlorophyll absorption band. From this it was a short step to the elaboration of the basic mechanism that light utilization in green plants requires the interplay of two separate systems. In turn, the realization that such a mechanism is operative inspired a whole new field of modern research on photosynthetic mechanisms for energy storage.

6

Happy Years

Berkeley

(1938–1940)

BY EARLY 1938, our work with short-lived carbon had begun to achieve visibility abroad as well as at home. Our Rat House operation was a focus for increasing activity in a variety of research areas, including studies not only in photosynthesis but in bacterial metabolism. Young investigators already making their mark as researchers, such as Stanley Carson, Jackson Foster, and H. A. Barker joined us as collaborators, bringing deeper appreciation of the important problems in biochemistry that could be approached using tracer methods. Emerson, an acknowledged world leader in photosynthesis research, was enthusiastic about the potential of our methods. C. B. van Niel likewise saw in our technology a key to the establishment of his cherished hypothesis that CO_2 was a universal metabolite.

Thus encouraged—although we hardly needed it—we pressed on, averaging at least three runs weekly for the next three years. This despite the formidable obstacles posed by the vagaries of cyclotron performance, never-ending demands for invention of new analytical procedures to fit our short-term experiments, and the competing requirements of our duties in the Chemistry Department and Radiation Laboratory. Each run usually generated masses of data that required long hours of reduction to results before they were assimilated in some sufficiently rational manner to indicate the next experiment. Great piles of experimental protocols accumulated, all meticulously organized by Sam, who acted as scribe. The bulk of this paper, which easily fills a large filing cabinet, documents the effort we and our colleagues put forth in those exciting years.[1]

Requests for appearances at seminars and symposia began to increase as our papers appeared. We were established leaders in tracer methodology, a new and fascinating science, a fact gratifying to me and especially to Sam, who while still a graduate student, had provided the initial impulse. He had come far in the few short years from the days when, as a beginner fresh from industry and working with Willard ("Bill") Libby, then an instructor, he had made a determination of neutron absorption in an iodine resonance level.[2] Regrettably, we had not reached our goal of establishing the identity of the CO_2 fixation product, despite the many hundreds of experiments we had performed. Hindsight shows that we lacked the necessary mix of technologies. The long-lived carbon isotope (^{14}C) had yet to be discovered and paper chromatography yet to be developed. We had nevertheless made some solid contributions to understanding the general nature of the photosynthetic process. Our experiments clearly established the existence of two systems, one a complex of dark reactions for CO_2 uptake with production of reduced cell material, and the other a light dependent process for the simultaneous evolution of molecular oxygen. These two systems in the plant had to be closely coupled so that one did not get ahead of the other, but the means for accomplishing this still remained unknown.

An aside about the chemistry of carbon should aid in understanding our research. Carbon has a dense central nucleus with a positive electrical charge of six, and six orbiting electrons, each with a single negative charge. Two electrons fill the orbit closest to the nucleus. The next shell, or orbital level, needs eight electrons to be a stable configuration, but carbon only has four electrons in this shell. To attain stability carbon either gives up these four electrons or acquires four others through interactions with other atoms. Because carbon has this versatility built in, it can make chemical bonds with itself to form a vast multiplicity of compounds to an extent not exhibited by most other elements. This is the basis for its central role in the chemistry of living tissues.

Oxygen, with a nuclear charge of eight, has two more outer electrons than carbon (making it the eighth element in the Periodic Table). Thus it is two electrons closer to forming the stable arrangement of eight in its unfilled shell. Each oxygen atom can accept two electrons from carbon to form a "double" bond, so that two atoms can produce a stable compound with one atom of carbon. The result is carbon dioxide, CO_2. As this process removes all the valence electrons of carbon, it is said to be "fully oxidized"—that is, it cannot donate or share any more electrons. Conversely, if the reacting element is hydrogen (element number one) with only one electron in orbit around its nucleus with positive charge one, the carbon nucleus can take up and share its four valence electrons with four hydrogen atoms to form CH_4, or methane. It is then said to be "fully reduced," in that it cannot take up any more electrons. All carbon compounds consist of carbon atoms in combinations that exhibit sharing or donation of electrons with other elements to extents intermediate between the extremes of CH_4 and CO_2.

In so-called organic acids, the determinative moiety is a carbon combined respectively with one oxygen atom, and another oxygen atom to which is also attached a hydrogen atom providing one of the electrons this oxygen atom wants (a combination, symbolized OH and called "hydroxyl"). This leaves one electron of the four in carbon to form a bond with another element, or with another carbon in the case of an organic acid. To show its composition the combination is designated COOH, called carboxyl. Its acidic nature arises from its tendency to lose its hydrogen as a proton, a process called ionization, in which the hydrogen electron remains with the oxygen atom and the hydrogen comes off as a bare nucleus.

The addition of one electron to the carbon of CO_2 is the first step in the reduction of CO_2 and the basis for the formation of the first fixation product in photosynthesis. We showed in our tracer studies that this was a compound formed by a readily reversible reaction in the dark whereby the CO_2 could react with the primary acceptor or acceptors to form carboxyl. Further, we showed that as the time of exposure of such labeled products to light was increased, less activity appeared in the carboxyl and more in carbon reduced further in an irreversible manner.[3]

We concluded that the primary fixation product was most probably a charged molecule with carboxyl and hydroxyl groups.[4] It was a year before we thought to ask how big it might be. No methods existed for determinations of molecular weight that did not involve solutions with concentrations of at least several millimolar. We had only enough of the labeled product to make solutions perhaps a million times less concentrated. We could not add "carrier" because we did not know the precise nature of the labeled molecule(s). We thought we had found a way to circumvent this difficulty by combining measurements of the rate at which the labeled product diffused through porous glass discs and determinations of the rate at which it could be spun down in a high-speed ultracentrifuge. For the former measurement, I spent several months running potassium chloride, sucrose, and other compounds in a simple diffusion apparatus we had built, using "sintered" glass discs of varying thicknesses and porosities. We measured the appearance of labeled material in the solution in which the vessel containing the labeled material was confined behind the glass disc. The rates we found seemed to indicate a size between that of potassium chloride and sucrose (approximate molecular weight 100 to 400).

To proceed further, we needed an ultracentrifuge that could be adapted especially to determine changes in concentrations in almost vanishing amounts. Our optimism about eventually finding such a machine was based on reports from Stanford where J. W. McBain and his associates had described a rotor, or "top," that could be spun and then stopped for sampling the concentration gradients established during centrifugation. This apparatus was ideally suited to our purpose, as all we had to do was fill the rotor with a little of our labeled sample solution, spin, stop the top, and withdraw a portion from the periphery. By comparing its radioactivity to that of the initial sample, we could calculate a velocity constant for the rate of sedimentation. This, combined with our diffusion measurements, we reasoned would give us an accurate value for the molecular weight of our labeled product.

It proved difficult to get reproducible results for the diffusion determinations. We found that our measured values for diffusion rates varied as much as 25 percent from run to run. I suspect we were not sufficiently precise in controlling times of exposure to our labeled CO_2 or conditions of algal growth. However, we could still specify a size range adequate for our purpose if we could do as well with the ultracentrifugations.

McBain was gracious enough to offer his help and the use of his ultracentrifuge if we wanted to spin our tops in it. He suggested we work with a former graduate student, L. H. Perry, who would provide needed expertise in handling the apparatus. Calculations showed that we could make enough labeled product on exposures to cyclotron-produced $^{11}CO_2$ to last not only for the two-hour auto trip to Palo Alto but for many hours afterward, in which we could carry out the ultra-centrifugation and sedimentation measurements. We spent a little thought on how to cut the driving time to Palo Alto. (Sam first suggested a police escort; then at 2 A.M. one morning he woke me by phone to announce excitedly he had the solution—carrier pigeons!)

We proceeded to set up a Geiger counter apparatus in McBain's laboratory. Our plan was to have Sam stationed there to assure the counter was working. I would carry out the bombardment, process the labeled carbon to carbon dioxide, administer it to the plants, prepare the fixation product solution, and bring it to Stanford.

A few runs were made without either police escorts or carrier pigeons, as we found that we had plenty of time to do the experiment with the samples we made. Then Perry informed us that a McBain apparatus had been built and was working at the Shell Development Company laboratories in Emeryville, only a few miles from the Berkeley campus. Permission to use their ultracentrifuge was obtained with no difficulty, thereby allowing us to do our experimental procedures in the Rat House, except for the actual centrifugations.

The values we obtained for sedimentation rates were roughly reproducible but could only be regarded as preliminary. When taken together with the diffusion results, the molecular weights indicated might be anywhere in the 500–1000 range. There is where the matter had to rest until the actual isolation and identification of the fixation products some fifteen years later when ^{14}C became available together with chromatographic procedures adequate for reliable identification and isolation of the fixation products.

I ruefully recall how we convinced ourselves that the molecules reacting with carbon dioxide must be big, with molecular weights in the range indicated by our experiments with the ultracentrifuge. We knew from reading the literature that the reactions of carbon dioxide to form a carboxyl product were not favored unless the molecules reacting were complex, such as in certain polyphenols. In these cases, the reactions to form carboxyls might occur significantly at high temperatures, such as 200° C. We assumed that in the algae special conditions existed that made possible such reactions at room temperatures. Of course we did not know then that such rationalizations were irrelevant because the "special conditions" involved systems in which there was no requirement for complex large molecules reacting in a simple chemical equilibrium.

A major finding, quite unanticipated, was that all the living systems examined, from bacteria and yeasts to rat liver and muscle extracts, could fix CO_2 in the dark, as did photosynthetic organisms. Yeast showed a marked retention of radioactivity when exposed to labeled CO_2, extracted, and the resultant solutions acidified to remove unwanted residual labeled CO_2.[5] The phenomenon in a nonphotosynthetic organism had already been found a year or two before our work began. Harland G. Wood, a student in the laboratory of C. H. Werkman at Iowa State University, had shown that in a bacterium that lived in the absence of air by fermenting a three-carbon alcohol, glycerol, to its corresponding oxidation product, the three-carbon propionic acid, there was production of the four-carbon succinic acid. He established that the extra carbon came from CO_2, one mole of CO_2 being taken up for each mole of glycerol disappearing.[6]

Hardly anyone believed Wood at the time. Eventually, the reaction—a carboxylation of the three-carbon product of glycerol, pyruvic acid, to form a four-carbon compound, oxaloacetic acid, containing two carboxyl groups—was established. Called the "Wood-Werkman Reaction," it was a forerunner of numerous others later discovered, supporting the general occurrence of CO_2 fixation in heterotrophes.[7]

However, the idea that CO_2 could be fixed in animal and plant tissues ("heterotrophes") without involvement of light was not accepted by most biologists, especially animal physiologists trained in the classical manner to regard growth with CO_2 as sole carbon source ("autotrophes") as a unique property of photosynthetic organisms. The realization was slow in coming that in respiration or fermentation, uptake of CO_2 was masked by the much greater production of CO_2 resulting from oxidation of organic nutrients. Tracer carbon was needed to establish the generality of the phenomenon.

Wood, who is certainly to be credited with the discovery of heterotrophic fixation, criticized our claim to have established it as a general process in living tissues, because we failed to show the specific compounds in which the fixed carbon appeared.[8] However, the techniques we used in showing CO_2 fixation did not require such demonstrations. Our evidence, while indirect, was strong that the carbon-hydrogen bond in organic compounds tested had been replaced by a carbon-carbon bond, thereby establishing that fixation of carbon from CO_2 had taken place.[9]

Wood's experience at Ames was very much like ours at Berkeley when we asked our biochemistry colleagues across the campus in the Life Sciences Building for advice on the significance of our finding of dark CO_2 fixation in green plants and algae. We were met by grins and/or scowls of disbelief and the remark, "you can't have photosynthesis in the dark!" Of course, a few of the more enlightened types, microbiologists familiar with the so-called chemosynthetic bacteria that fixed CO_2 autotrophically by using energy from inorganic oxidations of sulfur or nitrogen compounds, were ready and even eager to accept our results. I learned later that such ideas were already in the air, having been sensed by farsighted and creative biochemists like A. Lebedev and Hans Krebs.[10] By 1940, Earl Evans and Louis Slotin, using labeled CO_2 produced by the Chicago cyclotron, proved its incorporation into oxaloacetate to produce α-oxoglutaric acid in pigeon muscle preparations.[11] Such experiments had become a bit distasteful to us after a few experiences trying to make animal tissue preparations. I recall one particularly unhappy occasion when Sam and I tried to probe the possibility of reversing a particular reaction in pigeon breast muscle. While I was over at the cyclotron making the radioactive CO_2, Sam was trying to cope with a frightened pigeon from which we hoped to dissect out some muscle for a fresh preparation of the enzyme. Escaping from Sam's clutches, the bird flapped around the room destroying glassware and Warburg manometers and creating general havoc. Enraged, Sam finally caught the bird and tore its head off with a pair of pliers. When I arrived, he was sitting shakily in a chair in the midst of the carnage, wholly unable to continue. We had two other birds, obtained from Oakland's Chinatown. These we eventually liberated the next night because the cyclotron was not working. The pigeons, raised for the table, and unable to fly, waddled off into the bushes, where they probably helped solve the food problem for some fortunate feline. Our inexperience was further impressed on us when we realized that such birds, whose breast muscles were hardly functional, might have yielded poor preparations. We concluded that it was better to work with bacteria and algae, which presented no such complications.

Also quite unappreciated by us was the apprehension with which our efforts were regarded elsewhere. For instance, Wood quickly realized that we could do expeditiously and simply what he was trying to do with much more cumbersome methods employing mass spectrometry using the rare stable carbon isotope, ^{13}C. He wrote Ruben asking to come and collaborate with us. We were delighted to have him bring his bugs, as we were with all our microbiological colleagues. There was an unfortunate hitch—Wood had not first asked his professor, Dr. Werkman, for permission. Werkman took a dim view of having this work moved out of his laboratory and categorically refused to let Wood visit us. The result was that Wood, joined by Lester Krampitz, had to struggle to continue using ^{13}C. They spent a miserable Christmas away from family and friends, coaxing their homemade mass spectrometer to function. At Harvard, Birgit Vennesland, who had been a classmate of mine in Chicago, had joined A. Baird Hastings's group, which proposed to study the metabolism of ^{11}C-labeled lactate in its transformations and storage as glycogen in rat liver. Vennesland had guessed CO_2 might be fixed and proposed that an experiment using labeled CO_2 and unlabeled lactate as a control be included in the protocol. The control experiment, run somewhat grudgingly, showed to Hastings's surprise that there was CO_2 fixation in glycogen in the presence of lactate. Vennesland wrote later that "we had been thinking and discussing and looking at the (so-called) 'Krebs Cycle acids,' in order to explain how CO_2 got into glycogen and it seemed clear that Rubin [*sic*] and Kamen were thinking about that, too. . . . I was sure we were going to outrun Rubin and Kamen. In our youth we regarded scientific activity as an athletic competition."[12] If there was a race, we were certainly unaware of it back in Berkeley. That our entry into biochemistry was creating such waves was news to me even when I heard of it some thirty years later.

A kind of turning point in my own enlightenment about biochemistry came late in 1938 when E. O. L. sent me to a symposium at Ohio State University honoring Harold Urey. I went as a representative of the tracer work being done at the University of California with cyclotron-produced isotopes. E. O. L. was anxious to promote an appreciation of what radioactive tracers could do in solving previously intractable problems in basic medicine, physiology, and biochemistry, and seized every opportunity to spread the message. This is not surprising when one remembers that practically all funding for research in those days came from medically oriented foundations. Big Science had not yet arrived.

At this meeting I learned about the work R. L. Schoenheimer and his colleagues were doing at Columbia University, using samples enriched in the heavy stable isotopes of hydrogen and nitrogen (^{2}H and

^{15}N). Schoenheimer had begun his career in Berlin, later working at Leipzig and Freiburg. Hitler's rise to power had forced him to leave Germany. It was Urey's good fortune and E. O. L.'s bad luck that Schoenheimer landed not in Berkeley but at Columbia. He laid the basis there for a new science of intermediary metabolism, using tracer methodology to extend his previous observations on cholesterol, fat, and protein biosyntheses. This development was a key factor in the rise of molecular biology and the start of a golden age in biology in the second half of this century.[13]

The kind of experiment Schoenheimer and his associates did can be illustrated by the one they reported on feeding labeled leucine to adult rats kept on a standard stock diet.[14] The leucine was labeled in its carbon skeleton by replacement in its C-H bonds of the normal hydrogen isotope (^{1}H) with heavy hydrogen (^{2}H), or deuterium. The amino group ($-NH_2$) was labeled with the rare stable isotope ^{15}N. The leucine was thus doubly labeled so that the fate of the leucine carbon and nitrogen could be followed simultaneously. According to the old ideas of an internal, or "endogenous" metabolism separated from that of "exogenous" material ingested as nutrient, the administered leucine carbon and nitrogen should have been eliminated as excreted material, or have appeared only in the endogenous leucine of the animal's protein. Instead, the amino nitrogen was found in many other of the protein amino acids, as was the leucine carbon. It was clear that protein was being degraded and resynthesized even when the animals needed no new protein. Schoenheimer masterfully demonstrated that in all biochemical systems matter was in a state of flux, dynamically regulated so as to keep overall quantities and distribution constant in the adult steady state. Each constituent had its own characteristic rate of turnover, depending on its location and function.

I had the privilege of meeting Schoenheimer when he made a trip west in 1940, driving all the way from New York and back. He was in high spirits and enjoyed an evening of chamber music I arranged at home. From this meeting, I could see what a warm and impressive personality he possessed. Although he was a highly creative and sensitive individual, with great knowledge in his area of expertise, as well as a broadly cultivated intellectual, he carried his learning lightly. He was one of a few seminal figures in the new biology, and his premature death by suicide in 1941 was an incalculable loss to science. Apparently, the weight of domestic upheavals and loss of family and friends in the Nazi holocaust in addition to the difficulty he was experiencing in achieving the recognition that was his due proved too heavy a burden. Occasionally he would suffer severe depressive periods, from the last of which he did not recover. I count myself fortunate to have had at least this one opportunity to meet him.

Although biochemistry in Berkeley was not marching in the forefront of contemporary developments there were some talented young investigators working in the Life Sciences complex, particularly those in the groups headed by David Greenberg in biochemistry, Dennis Hoagland in soil science, and Israel Chaikoff in physiology. From these there came increasing requests for radioisotopes. My good friend and fellow chamber musician, Waldo Cohn, was working for a doctorate with Greenberg. So were Nathan Kaplan, Harold Tarver and Harold Copp, all of whom were to be heard from later. The samples of radiophosphorus I prepared for them were a fair proportion of the total I had to make for John Lawrence and his clinical associates.

There was a continuous flow of scientists from foreign countries through the laboratory. Most visible was Emilio Segrè, who had left the Fermi group and thrown in his lot with us. He was to become a permanent member and one of the Nobelists coming out of the Radiation Laboratory in later years. The spectroscopist S. Mrozowski came from Poland, experiencing the trauma of losing his country while he was still in midocean on the way to America. Perhaps the most glamorous of the foreign scientists was Count Lorenzo Emo. A warmhearted, easygoing aristocrat, Lorenzo had teamed up with Reginald Richardson who had the major cloud chamber rig in the laboratory.

Sam Ruben was beginning to achieve some visibility in the department, so that some graduate students were attracted by problems he described in strictly physicochemical areas. These were centered mainly around correlations that might exist between metal bond type and ability to exchange with free metal. Others were studies of mechanisms involved in organic oxidations, one in particular being the differences in products obtained under acid, as compared to alkaline, conditions during permanganate oxidation of propionate—a problem stemming directly from our research in CO_2 fixation. These were reactions used to establish distribution of fixed carbon in the products of fixation. The fact that Sam had students working with him on nonbiological problems was helpful in that the department could perceive he was not wholly committed to biochemical research. Students also came from other departments. A notable example was Albert Frenkel, from Plant Sciences, who joined vigorously in research on chlorophyll chemistry and studies of exchange reactions involving iron and magnesium chelates.

An outstanding collaborator early in our work was Don DeVault, a graduate student who was doing thesis work on recoil chemistry of bromine nuclear isomers in Bill Libby's lab. He had inherited a Geiger counter setup of Libby's from Donald Lee, the graduate student who had originally built it. DeVault improved it in various ingenious ways, including installation of a decade scaler—one of the first, I believe, to be available in the Chemistry Department. His analytical ability and acuity in perceiving solutions to questions as they arose was very welcome. In problems of relevance such as those dealing with chlorophyll chemistry and bond type, he provided detailed summaries of structural features, such as the various resonance forms possible for chlorophylls and porphyrins. We acknowledged his help in one of our first reports.[15]

Meanwhile, our visibility in the field of tracer applications to metabolism was increasing. We were visited by highly distinguished groups from Harvard and the Massachusetts Institute of Technology, led by Robley D. Evans from the latter institution. They had been organized by James B. Conant, the president of Harvard, who felt that there was inadequate appreciation of tracer possibilities among the faculty of his institution and persuaded them that they should be talking with us. One day in 1940, none other than A. Baird Hastings, whom I had last seen as the rising young star of physiology back in Chicago in my undergraduate days, dropped in on us and showed us data he had some difficulty believing, because they seemed to indicate unqualifiedly that during metabolism of labeled lactate to glycogen in rat liver there was incorporation of CO_2. The idea that CO_2 could be a metabolite was wholly foreign to Hastings, who had been raised as a classical physiologist. We assured him his data were old hat to us and that he could believe them.

There were other visitors, some of whom were to foreshadow later phases of my career, although of course I could not be aware of this at the time. Charles Coryell came from Cal Tech, where he had already won much acclaim for his work with Linus Pauling on the magnetic properties of iron compounds, particularly the blood pigments, myoglobin and hemoglobin. Meeting him for the first time in the Rat

House, I was immediately drawn to him as a kindred spirit, thus beginning a lasting friendship.

On quite another level, we were visited by Dr. O. L. Inman, director of the recently established Kettering Laboratory at Yellow Springs, Ohio, the campus of Antioch College. Inman had interested Charles Kettering, inventor of the Delco-Remy self-starter and numerous other important innovations, and a vice-president of General Motors, in setting up a laboratory for the study of photosynthesis. He brought a large case, which he opened with a flourish, revealing an enormous array of little glass tubes, each containing a differently colored compound. These, Inman informed us, were all derivatives of chlorophyll related to the breakdown products produced during digestion of grass by cows. He then closed the case and bade us farewell. I certainly did not think at the time that in another decade and another place I would be enjoying an intimate association as collaborator and consultant with the Kettering Laboratory.

Another happening was the prelude to an enduring friendship with L. P. "Larry" Bachmann, a script writer from the M.G.M. studios, who came to San Francisco to write a story on the "hospital ship" then anchored in the Bay. His decision to make maximal use of some extra time to interview John Lawrence, showed that the new clinical program at the Rad Lab was attracting wide attention. At Larry's invitation, Sam and I visited Hollywood where we had many exciting encounters with the movie greats, and also visited Charley Coryell at Cal Tech. The oily condition of the beaches at Malibu was just as evident then as now.

My duties at the Rad Lab had escalated as the demand for radioisotopes in John Lawrence's laboratory, around the campus, and from all corners of the world increased. John had assembled a large group of associates working on basic and clinical aspects of neutron therapy as well as in uses of radioisotopes in medicine. Most notably, Alfred Marshak had come from the East to investigate genetics and the radiobiological effects of neutron beams. The need to keep all these clients happy and provided with radioisotopes was a full-time job I was finding more and more difficult to manage. Nor was I the only one pressured. The demands on cyclotron time and the never-ending push to improve cyclotron performance all but interdicted any consistent research efforts on the part of the lab staff. This is not to say no research was done, but that there was a strong feeling that less was being accomplished than we might expect from the unique and unmatched facilities available to us. Important discoveries, such as induced artificial radioactivity and the neutron, had been missed.

We realized we were wading through a sea of neutrons much more intense than existed anywhere else, and the lab itself was alive with radioactivity induced by cyclotron radiations. Before the water tank shields were installed, one could walk through the laboratory when the cyclotron was running and activate silver and copper coins in one's pockets, not to mention gold and silver fillings in teeth. The metallic innards of the cyclotron itself were a Golconda of yet undiscovered radioisotopes created by deuteron bombardment. Emilio Segrè, who had visited the lab in 1937, obtained a piece of the discarded deflector plate from the old 27-inch cyclotron and took it back to Italy, where he earlier had been an associate of Enrico Fermi. There he extracted one of the missing elements in the Periodic Table, No. 43, which he called "technetium." Jack Livingood and Glenn Seaborg were just at the beginning of their productive research on deuteron-produced radioisotopes and eventually identified many of the radioisotopes then lurking in the cyclotron parts.

E. O. L. maintained that if beam intensities and energies could be increased, the discoveries would follow. As Luis Alvarez has noted in a moving tribute, "He often spoke with satisfaction of the proven value of his long campaign to increase beam current (often in the face of opposition from younger colleagues who wanted to 'use what we have' rather than shut down for improvements)."[16] It is amusing to recall that one of these "younger colleagues" was Alvarez himself. At one point, he even mounted a public protest to sway E. O. L., using a chart to show how publications from the laboratory had fallen off. E. O. L. was not moved. He reminded everyone that Luis was having no difficulty in maintaining a productive career, as was certainly true.[17]

The lag between demand for radioisotopes and their production grew still greater and some new approaches were urgently called for. Out of this pressure there came a development that was to be crucial for my own subsequent career—the "probe," or "internal" target. As E. O. L. had been noting for some time, the meters of the oscillator power supply always showed a loss of power much greater than expected when the beam came on. It seemed reasonable to suppose that the extra load came from a large circulating current inside the cyclotron that was not getting out through the deflector region to the target chamber. He set Bob Wilson to explore this possibility. I was interested because the postulated current might be useful for radioisotope production. As Bob and I both liked working at night, we arranged to be the "owl crew," working on the problem from midnight to early morning—a time when there was no interference with routine operations of the laboratory.

Wilson had already demonstrated that he could use a copper tube, bent and fitted with a flat copper jacket to expose a surface to oncoming ions and water-cooled to dissipate the resultant heat, as a "probe" to measure the ion currents it could collect as it was pushed back and forth inside the acceleration chamber. Later, he invented a vacuum-tight sliding seal to facilitate its placement, as well as its withdrawal through a vacuum lock.[18] This "Wilson seal" came into general use as a brilliantly simple and convenient means of obtaining rotational and translational motion of internal probes. In short order, Wilson showed that the expected ion currents did exist and were of magnitudes one or two times greater than that of the external beam, as E. O. L. had suggested. They were of lower energies because they came from orbits further inside the vacuum chamber than those of the external beam drawn from the peripheral regions.

The problem now was to exploit these probe targets to produce useful radioisotopes. Wilson found that a goodly portion of the internal ion currents could be intercepted without affecting the amount of external beam produced. We were able to estimate that milliamperes of several MEV deuterons could be collected by a probe inserted in the 37-inch cyclotron, although the external beam currents were some ten or twenty times weaker in intensity. This meant that we would have to solve the major problem of how to dissipate many kilowatts of power to utilize these "internal beams."

I started experimenting with a large number of refractile substances made up of useful elements such as phosphorus, sulfur, iron, and so on, and discovered that a layer of iron phosphide could be bonded to the copper surface of the probe with a "hard" solder called "phoscopper." This phosphide withstood the extremes of heating in the cyclotron vacuum for days on end without producing serious emissions of gas or disrupting cyclotron operation in other ways. Probes containing iron phosphide were installed as routine targets, which could easily be inserted through vacuum locks for bombardment, simultaneously with the utilization of the external beam. I worked out chemical procedures for isolation of the "hot" isotopes, using a new chemical laboratory setup built to my specifications in the Crocker Laboratory.

In this way, we triumphantly solved the problem of radioisotope production and raised the morale of those who had despaired of ever satisfying the insatiate clinicians. The probe targets extended the capacity of the cyclotron by whole orders of magnitude. As I wrote in our published report, "obviously the method of internal targets should find its most important application in the preparation of radioisotopes which are long-lived and difficult of activation, as well as in the demonstration of the existence of many radioisotopes as yet undiscovered."[19] This prophecy was to be fulfilled in especially dramatic fashion a little less than two years later with the discovery of long-lived carbon 14 (^{14}C). The probe target was the key factor in the first successful production of this radioactive isotope of carbon.

Our exploits earned us a glowing commendation from E. O. L. at a crew meeting. Bob and I were the heroes of the moment. I was taken aback a few weeks later, therefore, to find that Bob had been banished from the lab. This astonishing turn of events came about after a frustrating week spent by E. O. L. and the crew in trying to get the 37-inch cyclotron to work for a distinguished group of visitors whose support was needed to further the search for funds to build the 60-inch machine. The cyclotron kept exhibiting gassiness and inability to hold a vacuum when the power was turned on. Despite all of E. O. L.'s skill and persistence, there was no beam. The visitors finally left without seeing the cyclotron operate successfully. E. O. L., deeply disappointed, ordered the "can" to be pulled and examined. Immediately the culprit causing the trouble was found. One of the probe gaskets in a Wilson seal had lodged inside a "dee" and had sat there producing gas when the power was turned on. From its appearance, it could have continued to produce gas forever. It was clear that the gasket rubber had made its way there through sloppiness in our owl crew operations. Bob was called in and grilled. He admitted he had withdrawn a probe through the vacuum lock hastily at the end of a long night session, and that one of the rubber gaskets had apparently been sucked inside the can. He had been unable to find it by feeling around, and had concluded that he probably had used only one gasket. He closed the lock and let the machine pump down for the next morning's operations. It was his bad luck that the rubber had lodged up inside the dee instead of on the floor of the can, where he might have found it. On hearing this, E. O. L. flew into one of the few rages anyone can remember and ordered Bob out of the lab, excoriating him for his carelessness and irresponsibility. Actually, it was remarkable how few times Lawrence did lose his temper, when one considers how often there was ample provocation.

Away from campus, family and musical activities were not only satisfying and exciting, but constantly expanding. Waldo Cohn had married an associate in the Biochemistry Department, Elma Tufts, a uniquely gifted postgraduate fellow, and they had moved into an apartment where regular soirées provided opportunities for gaiety and the production of music far into the night, as did meetings at the Wolskis' across the Bay. An important new friend was Henri Temianka, the famed violin virtuoso, whom I met at one of the parties Isaac Stern gave in San Francisco during his customary spring recess from tours. Henri arranged some outstanding parties, many at the home of his fiancee, Emma Mae Cowden, who was fashioning a career as a sculptress. The friendships with Stern and Temianka were two major benefits I was to derive from my years at Berkeley.

I was enjoying a happy family life. Esther had the same thirst for musical expression I did. We entertained often, our parties being attended by lab associates and visitors. On one occasion, Felix Bloch and Merle Tuve, two of the ranking physicists of the day, were seen to crawl into our living room through a window at street level while a party was in progress. Music was usually produced by a group including Frank Houser, Eddie Becker, and Waldo Cohn as regulars. Occasionally Detlev Olshausen, a San Francisco Symphony violinist and violist with a scholarly bent, would participate. One evening, Detlev offered me the chance to examine his father's library, which included early editions of the works of Helmholtz, Hertz, Kirchhoff, and Maxwell, and take what I wanted. In this way, I came into possession of a remarkable collection of books by the greats of nineteenth-century physics.

Social events on the campus included many wild beer parties at Tilden Park, given by the Chemistry graduate students and presided over by one Dan Luten who could recite endless verses of Robert Service poetry about the Yukon, balancing precariously on a bench while it was being rocked violently by his manic colleagues intent on plunging him down the hillside. At a cultural extreme were the appearances of noted musical groups on campus. One such event was the eagerly anticipated appearance of the Kolisch Quartet in five concerts devoted to the music of Schubert and Bartók, given in Wheeler Hall in the summer of 1938. A member, Felix Khuner, later settled in Berkeley where he became a valuable addition to our chamber groups.

Toward the end of 1939, Esther and I were lucky enough to rent architect Bernard Maybeck's former studio on Buena Vista Way, a cottage nestled in a wooded area on a hill overlooking the Bay in north Berkeley. Into this idyllic abode we settled, deliriously happy, as the end of 1940 approached. Although Esther was no longer working, the income I was earning at the Rad Lab, while not munificent, was ample. We had all we needed for the Good Life—wonderful friends, cultural outlets, the eucalyptus-scented ambience of the East Bay, and, as I will describe in the next chapter, bright prospects because of a lucky break in my research.

I was wholly unaware of impending cataclysm, even when, early in January 1939, news reached us that nuclear fission had been discovered in Otto Hahn's laboratory in Berlin. This astounding finding had been missed at the Rad Lab—one more missed discovery to add to the others that might have been made there.[20] The phenomenon was quickly confirmed and set a number of our research groups on a new and exciting course of investigation—particularly Abelson and McMillan in the lab and a group consisting of Glenn Seaborg, Emilio Segrè, Art Wahl, and Joe Kennedy in the Chemistry and Physics Departments. The news from Berlin was particularly disquieting to Abelson, who had been finding that in his new and elegant bent crystal x-ray spectrometer no characteristic x-ray emissions from the "transuranic" elements supposed to be produced by neutron activation of uranium could be seen. Instead, he saw only very intense lines apparently from lighter elements. These, he now realized, could have been produced only by splitting uranium, instead of by simply adding a neutron to it. As for me, the impact of the great new discovery was largely peripheral. I found myself designing and arranging massive neutron bombardments to produce quantities of the new fission products, as well as of the new elements produced not by fission but by neutron capture—a process established by McMillan and Abelson. These new elements were true transuranics, called neptunium and plutonium after the planets next to Uranus in the solar system.

7

The Carbon 14 Story

Bright Clouds with Dark Linings (1935–1941)

MY SEARCH for carbon 14 (^{14}C), which was to end successfully on February 27, 1940, had begun some five years before, without my realizing it, when I was helping Gans and Newson do experiments on collisions of fast neutrons with nitrogen nuclei. In the cloud chamber I could see the tracks of the charged particles made when the neutrons hit the nitrogen nuclei. The way these tracks looked depended on what made them. If the particle involved was a positive singly charged hydrogen nucleus, or proton, the track was long and thin; if it was a positive doubly charged helium nucleus, or alpha particle, it was thicker and shorter. In the cloud chamber, the gas space was saturated with water vapor. When this space experienced a rapid expansion, the gas in it was cooled suddenly, so that the chamber volume became supersaturated with water vapor. Under these conditions, any small particles, especially if electrically charged, could act as centers for condensation of water vapor, just as in cloud formation. In the passage of the protons or alpha particles through the chamber at the instant of expansion, any electrically charged fragments of molecules present thus would become visible as tiny droplets of water. One would be looking at the vapor trails left in the wake of the passage of the protons or alpha particles. The interaction of the swiftly moving nuclear particles tore electrons out of the gas molecules they met, leaving positively charged fragments. For a given energy, the alpha particle, with four times the mass of the proton, moved much more sluggishly. This, combined with its double positive charge, compared to the single charge of the proton, produced much thicker and shorter tracks for alpha particles. With experience, it was relatively easy to tell whether a track was made by a proton or by an alpha particle.

Always associated with either the long, thin tracks of the protons or the shorter, thicker tracks of alpha particles were very short, dense tracks of the recoiling nucleus created by the disintegration of the target nitrogen nucleus by the neutron. Franz N. D. Kurie at Yale had reported that occasionally he also saw long, thin proton tracks in experiments like ours.[1] He realized, as we did, that the loss of a proton from the nitrogen nucleus required that the residual nucleus be an isotope of carbon. (Carbon of nuclear charge six has one positive charge less than nitrogen with its nuclear positive charge of seven.) Because it was not certain that the long, thin tracks were not made by hydrogen nuclei other than the proton, but still singly charged, as from isotope hydrogen nuclei of mass two (deuterons) or three (tritons), the recoil nuclei produced could be carbon with masses twelve, thirteen, or fourteen.[2]

T. W. Bonner and W. M. Brubaker, working in Pasadena at Caltech with a cloud chamber, studied the same system and found anomalies in the masses deduced for ^{14}N and ^{11}B when they assumed the recoils seen were alpha particles.[3] More interestingly, they very occasionally noticed very short tracks with a homogeneous range of about one centimeter, which occurred more often when the cloud chamber was surrounded by paraffin, a substance very effective in slowing down fast neutrons to make "slow" neutrons that possessed negligible kinetic energy. Hence, the short tracks could be associated with recoils from neutrons that contributed only mass energy and no energy of motion because they moved so slowly. If the Caltech researchers assumed the short tracks were a composite of ^{11}B and ^{4}He recoils going off in opposite directions from the point of neutron capture, they obtained a heat of reaction, calculated from the ranges of the recoil particles, equal to 2.33 MEV. They could assume that the recoils departed in opposite directions from the point of impact because the slow neutrons were not fast enough to impart a forward direction to the recoil nuclei. To generate more such events, Bonner and Brubaker substituted a thin brass wall for the usual Pyrex cylinder of the cloud chamber, because the boron in the Pyrex was a well-known absorber of slow neutrons.

Meanwhile, in England, Sir James Chadwick, the discoverer of the neutron, and Maurice Goldhaber had been studying the ^{14}N reaction with neutrons using an ionization chamber filled with nitrogen gas. They estimated the total reaction energy produced as only 0.5 MEV, compared with the Caltech value of 2.33 MEV.[4] Goldhaber had the brilliant idea of writing to the Caltech duo to suggest that if they assumed the products were ^{14}C and ^{1}H, rather than ^{11}B and ^{4}He, the reaction energies would agree better. Indeed, the Caltech figure on this basis came out 0.58 ± 0.03 MEV. Hence, it seemed certain that ^{14}C was formed from ^{14}N by slow neutron capture with emission of a proton. This was exciting, not only because slow neutron capture reactions had appreciably greater yields than those involving fast neutrons but also because one could predict that ^{14}C would be unstable, decaying by negative beta particle emission to ^{14}N. The basis for this prediction was that among elements with low atomic numbers (low nuclear charge), it had been found that only one nuclear species was stable for a given mass. Hence, as ^{14}N was stable, ^{14}C, its "isobar" (nucleus with equal mass and different atomic number), would have to be radioactively unstable. The only question was how unstable.

To make an intelligent guess, it was necessary to estimate the difference in mass between ^{14}C and ^{14}N. From this difference in mass, with ^{14}C being the heavier nucleus, energy available in its transformation to ^{14}N could be calculated using the well-known relativistic equation of Einstein, $E = mc^2$ (where E is energy released, m the mass, and c the velocity of light). Bonner and Brubaker reckoned that ^{14}C might be unstable with respect to ^{14}N by as much as 0.2 MEV, from which one could expect a half-life for the ^{14}C of a few days or months. Such an isotope could obviously be of tremendous importance in expanding the application of tracer methodology in biology. So vigorous attempts were made to find it. Bonner and Brubaker found no evidence for such radioactivity in their experiments. Edwin McMillan, who had been looking at an old beryllium-aluminum alloy target, used in the 27-inch cyclotron during deuteron bombardments for several months, found a long-lived radioactivity of low energy which he guessed might be ^{14}C, created by bombardment of carbon accidentally present as a contaminant. He estimated the upper limit of the beta rays emitted as about 0.2 MEV.[5]

This was the situation when I arrived at the Radiation Laboratory at the end of 1936 and found Franz Kurie working on neutron interactions with light nuclei. It was obvious to me that, with the relatively immense neutron intensities uniquely obtainable using the cyclotron, the search for ^{14}C could be pursued more effectively in Berkeley than anywhere else. McMillan, meanwhile, had the same idea and placed a small bottle containing crystalline reagent-grade ammonium nitrate close to the cyclotron to absorb neutrons and, it was hoped, make ^{14}C. Regrettably, after some months it was accidentally knocked off its perch and broken. (As became evident later, there was probably enough ^{14}C to detect in that bottle!)[6]

Franz and I found the short homogeneous range recoils previously reported by the Caltech investigators in such profusion that we could conveniently use them to calibrate the "stopping power" of gases in our cloud chamber and thus calculate energies from ranges with some precision. There was no question but that we were seeing the slow neutron reaction making ^{14}C from ^{14}N. Early in the spring of 1938, soon after Sam Ruben and I had started our collaboration on rare earth neutron activations, we discussed the possibility of finding the ^{14}C in some old graphite targets we had used for our experiments with ^{13}N. Sam had an ideal setup to look for weak activities, the so-called screen wall counter invented by Bill Libby.[7] This device featured a Geiger counter placed along the axis of a surrounding cylinder, the inner surface of which could be coated with the radioactive material to be assayed. In this way, "self-absorption" could be minimized—that is, the radioactive material itself would not be so thick as to prevent escape of the low-energy beta particles. An added feature was that the counter could be placed in a magnetic field produced by a solenoid. The beta rays coming from the cylinder and activating the counter could be bent away so that upper energy limits could be determined as calculated from the magnetic fields required to reduce the count to that of background, determined simply by tilting the device and sliding the cylinder back into the dead space of the counter. Gases could be assayed in it, too. We had a graphite target that had been bombarded for about 1,000 microampere-hours with 8 MEV deuterons. If we assumed ^{14}C had a mean life of days, or even years, we should have found radioactivity in the carbonate we prepared by burning some of the graphite to CO_2, precipitating and drying it as calcium carbonate, and smearing it over a large area of the screen wall cylinder. Instead, we found no radioactivity, thereby pushing the estimate of mean life up several hundreds of years. About a year later, I tried again. This time I filled an ionization chamber, connected to an electrometer circuit, with gaseous CO_2 made by burning another portion of the graphite target. Again the results were negative.

There was still the possibility that ^{14}C was very short-lived, so that all radioactivity had decayed away before measurements could be made. I believe it was Phil Morrison, then newly arrived in Oppie's group, who calculated that a likely half-life would be in the order of days or months, based on the then prevailing notions of beta decay, in which the "spin" difference between the nuclei concerned was related to the half-life in a calculable way.

By "spin" was meant the property of nuclei arising from the fact they could have an intrinsic angular momentum, much as though they had an axis around which they could rotate. In nuclei with even numbers of nucleons, one would expect that the spins of each pair of nucleons would cancel, giving a spin number of zero. In nuclei with odd numbers each of neutrons and protons, the extra neutron and proton could each contribute their basic half-unit of spin to make at least a total spin number of one. By analogy with well-established cases from atomic spectra, transitions from a nucleus with no angular momentum to one with some would be difficult, that is, "forbidden" to an extent depending on the difference in spins, or angular momenta.

I considered an analogous case, the fast decay of ^{6}He to ^{6}Li. ^{14}C was like ^{6}He in that it had two more neutrons than protons. An "even" nucleus with its content of eight neutrons and six protons, it could be expected to have no spin. ^{14}N, an "odd" nucleus with seven neutrons and seven protons, could be expected to have a unit spin. The difference between ^{14}N and ^{14}C could be one spin unit, so in decay the spin had to change by one unit at least. This "forbiddenness," while not great, was sufficient to predict that a *very* short half-life was unlikely.

It was clear to me that I needed to increase the intensity of bombardment by at least an order of magnitude to get visible amounts of ^{14}C if its half-life were very long. Morrison's calculation discouraged the notion that it could be even as long as I seemed to be finding with the bombardments already made. Still, the effort was worth making. A pressing reason for further effort was our continued failure to identify the primary product(s) of CO_2 fixation in photosynthesis using short-lived ^{11}C. Sam and I had a conference late in 1939 in which we gloomily reviewed our three years of effort with the short lived ^{11}C and concluded that we could go no farther unless by some miracle a long-lived radioactive carbon, presumably ^{14}C, became available—if, indeed, it existed. The problem was that the cyclotron was not available for the long runs needed.

It was at this juncture that I received an urgent call to see E. O. L. I ran up the three flights in Le Conte Hall, entered E. O. L.'s office, and found him in a state of great irritation because Harold Urey at Columbia University was disparaging the future importance of radioactive isotopes in biological research. Urey emphasized that no long-lived radioactive isotopes existed for the elements of primary biological importance—hydrogen, carbon, nitrogen, and oxygen. They were available only for elements of somewhat lesser importance, such as sodium, potassium, and phosphorus. In the case of hydrogen, there were no radioactive isotopes at all, the lone candidate—tritium (^{3}H)—being claimed to be stable. Carbon had only one useful isotope (^{11}C), which as we had shown had really only limited applications with its half-life of twenty-one minutes. Nitrogen had just the ten-minute ^{13}N. The case of oxygen was even worse, with its longest-lived isotope, ^{15}O, exhibiting a half-life of only two minutes. In contrast, there were the rare stable isotopes—deuterium for hydrogen, ^{13}C for carbon, ^{15}N for nitrogen, and ^{18}O for oxygen. To make his point even more firmly, Urey cited the work of Schoenheimer and his colleagues using ^{2}H and ^{15}N, which quite dwarfed in significance any tracer work yet done with radioactive isotopes in biochemistry.

Urey could also claim there were means to insure an adequate supply of stable rare isotopes. In his laboratory, studies had been proceeding for some years on exchange equilibria in which isotopes could be separated by minute differences in reactivity. Harry Thode, who had been a graduate student working with Simon Freed at Chicago when I was there, had begun what was to be a distinguished career in isotope chemistry working on equilibria of this type. Other students, such as David Rittenberg and Mildred Cohn, who were to attain distinction as biochemists in years to come, had worked with Urey in the same area of research. (Rittenberg, in fact, was already a collaborator with Schoenheimer in the initial experiments on the dynamic state of living systems.) Urey thus had a solid base for negotiating with industrial concerns (we heard Eastman Kodak Company was one of them) to set up mass production of stable isotopic tracers using exchange reactions developed at Columbia. The case for the superiority of stable isotopes as tracers, on the basis both of their existence in useful elements and their availability, seemed strong.

Such talk was painful to E. O. L. because it hobbled his vigorous efforts to attract funds for expansion of tracer research using cyclotron-produced radioactive isotopes. He asked me what I thought could be done. I described efforts to find the radioactivity expected to be associated with ^{14}C, pointing out that I needed full time on the cyclotron to exploit the probe targets Wilson and I had invented. E. O. L. interrupted to say that I had it, and ordered me to mount a systematic and energetic campaign to find him not only ^{14}C but "any long-lived activity anywhere in this part of the Periodic Table." He said I could have both the 37-inch and the 60-inch cyclotrons and all the time I needed, as well as help from whomever I requested—Segrè, Seaborg, anyone! I came down from Lawrence's office, dazed at my sudden good luck and hurried over to the Rat House to tell Sam. Then I went back to my office in the Crocker Laboratory and began planning an extensive program covering every conceivable way of making long-lived isotopes for carbon, nitrogen, and oxygen. I had available deuterons of up to 16 MEV with intensities of several hundreds of microamperes, 32 MEV alpha particles in microampere amounts, and the high intensity internal ion currents of somewhat lower energies reachable with the probe target technique. In addition, I could obtain 8 MEV protons, as well as the neutrons made in great quantities by both cyclotrons. This formidable armamentarium could be used to exploit a long list of nuclear processes, which I outlined, tabulating for each the target materials and the associated radiochemistry for isolation of expected products. I even included as possibilities the existence of long-lived gamma ray emitting isomers of stable nuclei, although neither experience nor theory encouraged us to think such nuclei could exist. I was particularly interested in reactions that produced isotopes chemically separable from the target material—like ^{14}C from ^{14}N in ammonium nitrate—so that dilution of the radioactive isotope by its stable isotopes would not occur. This was the reason to put two five-gallon carboys of ammonium nitrate in dilute acid solution near the 60-inch cyclotron to catch the large fluxes of neutrons pouring out of the machine during its operation as a deuteron generator.

Having had so much experience making ^{11}C from deuteron bombardment of boron oxide, with its convenient production of the radioactive materials as gaseous carbon monoxide (CO), I planned to begin using the same procedure with alpha particles to make ^{14}C. Also there were the enormous deuteron beams inside the cyclotron with energies more than adequate to saturate the ^{14}C-producing reaction of deuterium with carbon, so I included trials with probe targets covered simply with commercially available colloidal graphite called Aquadag, which I smeared on the copper probe surface. I knew these targets would be poor because of the low heat conductivity of the graphite, but they could serve at least as preliminary indications of feasibility until better designed graphite probes could be fabricated.

I began some trials in the fall of 1939, a few days after the interview with E. O. L. On October 5, I exposed boron oxide to the external 16 MEV deuteron beam of the 60-inch cyclotron for thirty-four microampere hours (μah) and withdrew the radioactive gases formed in the usual way, as in production of ^{11}CO. This sample was given to Emilio Segrè, who put it in his ionization chamber. After a lot of short-lived activity had disappeared, he found only a two-hour activity that decayed with a half-life of 112 minutes, which was assigned to the fluorine isotope, ^{18}F. A few days later, three μah of 32 MEV alphas on boron oxide produced no long-lived radioactivity. The cyclotron then had one of its indispositions, so I had to wait a week before repeating this experiment, again with negative results. I also examined the solid residue left, on the assumption the radioactivity might not have escaped into the gas phase. I did this by adding a little carrier carbon and burning it to CO_2, with which I filled an ionization chamber. Still I observed no radioactivity. It appeared from this result that I would need to bombard the boron oxide with the alpha particles for at least a million μah if the half-life of ^{14}C were very long.

I returned to the fray on October 17 with the first trial using a gas target, methane, and deuterons (20 μah). From this I obtained an enormous yield of ^{13}N (estimated at 100 millicuries) and a little ^{18}F from some oxygen present as contaminant, but no long-lived radioactivity. The 60-inch cyclotron then went back into dry dock. I then resorted to the internal target approach, using the Aquadag probe target installed through the north porthole of the 37-inch cyclotron. I arranged with various crew members to push the probe in far enough to intercept about a hundred microamperes of internal ion current as bombardments went on through January 1940. The total current available (some hundreds of microamperes) could not be used for fear of burning off the Aquadag. Every few days I would take the probe out to see if any Aquadag clung to it, disregarding the radiation exposure I was getting from all the short-lived radioactivities produced in the copper backing of the probe. Often I found the graphite charred and dry, sticking out from the probe surface and on the verge of falling off. I would plaster on more Aquadag, collect whatever had fallen off to add to the final sample and reinsert the probe. In this way, the bombardment went on for over a month. Finally, I stayed up three nights in a row for a final push, accumulating a total of 5,600 μah.

This last phase of the bombardment took place during a great rainstorm. At the same time a class in French drama was meeting to listen to some particularly horrendous Gallic tragedy, complete with screams and guttural groans, emanating from a recording being played on a mezzanine just overlooking the cyclotron control desk. Shortly before dawn on February 19, I withdrew the probe and scraped off what was left of the Aquadag, which looked like nothing so much as bird gravel, shook it into a small weighing bottle along with other collected bits and scraps, and left it in the Rat House on Sam's desk for him to process and examine in his screenwall counter, while I stumbled home to get some sleep. On the way, with rain still pouring down, I was picked up by the police as a likely suspect for a mass murder perpetrated a few hours earlier somewhere in the East Bay. I must have looked the part, with eyes red-rimmed from lack of sleep, unsteady gait from weariness, and a three-day growth of beard. After being taken to the station, where a hysterical survivor of the massacre stared at me but fortunately gave no sign of recognition, I was released and crawled home to our one-room apartment on Oxford Street, where I collapsed and knew no more for nearly twelve hours.[8]

When I awakened, I phoned Sam, who in the meanwhile had found the sample and had been working most of the day to process it. He had burned the graphite to CO_2 over cupric oxide in a combustion tube using a stream of oxygen. The CO_2 was caught in a liquid air-cooled trap and then absorbed in calcium hydroxide to make insoluble calcium carbonate. He had acidified this precipitate with sulfuric acid, distilled over the CO_2 into his screen wall counter, and found activity. I hurried over and watched him repeat the procedure. I had to stay at a distance and handle nothing because I was so contaminated from the contact with radioactivity I had experienced in the previous week's runs. On the second attempt, the alkaline calcium hydroxide trap accidently went acid and some activity was lost. While both of us filled the air with heartfelt maledictions, Sam quickly added more base and then repeated the cycle of acidification and precipitation several times. Each time the specific activity recovered as CO_2 remained constant. By this time, the sample had only some 800 counts a minute, with a background of a few hundred counts a minute, but the effect was significant and adequate to claim that we had a long-lived radioactive carbon, presumably ^{14}C.

At this point, I felt a qualm. The Aquadag I had used was fairly pure colloidal carbon, but there could have been sulfur impurities in it. These could have been activated to give the weakly energetic, long-lived isotope of sulfur (^{35}S).[9] Sam and I discussed this and devised a procedure on the spot to check the possibility. Sodium bisulfite and potassium permanganate were added, the solution acidified with sulfuric acid, and CO_2 collected in calcium hydroxide. Any sulfur present would be oxidized to sulfate ion by this procedure, and so would not be volatile. Again the volatilized CO_2 was radioactive, with no drop in specific activity. The procedure was repeated using potassium triiodide as oxidant in acid solution. Again the CO_2 evolved showed the same radioactivity. At last we were prepared to believe that the active material was ^{14}C, even though by this point we had lost about half of the total radioactivity we had started with, so that our count was now only a few times background. Sam made a preliminary determination of the energy limit with this sample in the screen wall counter, using absorption by thin aluminum, and estimated it to be about 90 kilovolts. The half-life had to be at least 1,000 years. We started daily counts of the sample without removing it from the counter. The search was thus over as of that day, February 27, 1940.

Meanwhile, I had begun a new probe bombardment on February 22, using a better fabricated probe fitted with a graphite button that had a bottom layer of copper. I had been in communication with a Dr. MacPherson at the National Carbon Company in Cleveland, and he had volunteered to make such discs using carbon enriched in ^{13}C, which I hoped would increase the yield of ^{14}C (the normal abundance of ^{13}C was only 1 percent elemental carbon, the main stable isotope being ^{12}C). Harold Urey graciously sent me 630 milligrams of amorphous carbon made from ^{13}C-enriched cyanide by reduction with magnesium metal. Meanwhile, I had compacted some amorphous graphite using a high pressure device in the Engineering Department. I found that this graphite could be bonded readily with a copper base to make a button that could be soldered with phoscopper to the probe surface. However, the carbon powder I had obtained from Urey, with its higher ^{13}C content, was too crystalline to adhere well. So I had a grillwork probe surface made into which I packed the enriched carbon, mixed with a little asphalt-pitch varnish, forming a tacky mixture. This worked better than my initial Aquadag targets and was employed in the next big probe bombardment.

When our initial success occurred, we were immediately tempted to announce it to the world, but we had continuing doubts about the reality of the effects we had observed. After all the radioactivity of our putative ^{14}C was not much more than the background radioactivity in the screen wall counter. For reassurance, we sought out G. N. Lewis and described our experiment to him. He grinned and said, "If you boys think it's ^{14}C, then that's what it is!" We remained doubtful but decided to take the plunge and write a short note for publication. I found that Sam had already prepared a preliminary draft in his first moments of euphoria. I suggested we phone E. O. L. to tell him about our great luck.

E. O. L. was in bed with a cold, trying to round into shape for the award of the Nobel Prize he had just received. He was overjoyed to hear the news and insisted we come and show him our communication. We drove up in Sam's decrepit green jalopy and were ushered into Lawrence's bedroom where he lay resting. On reading our note, he jumped out of bed, heedless of his cold, danced around the room, and gleefully congratulated us. He asked if he could call the newspapers, upon which we suddenly were assailed with doubts again. Mumbling that we saw no reason why he should not, we left. Driving back down the hill, Sam broke the glum silence with ruminations about how he had a wife and kids and hoped we were not wrong, especially with the Nobel Prize celebration scheduled in the next few days. We hurried back to the lab and busied ourselves, Sam with more counting of our one sample and I making more probe targets, while bombarding the second probe in the series.

A few days later, on February 29, there was a momentous ceremony in Wheeler Hall, where the Swedish Consul presented the Nobel Prize to E. O. L. Everyone was there except Sam, who had lost courage to appear.[10] So I sat alone as the chairman of the Physics Department, R. T. Birge, gave the presentation address.[11] He had been working on it in his precise, painstaking manner for some weeks and had been annoyed momentarily to be called and told at the last minute about the earthshaking discovery of ^{14}C. He had planned to feature the recent discovery of a radioactive form of missing element No. 85 (eka-iodine) made by Dale Corson, Emilio Segrè, Joseph Hamilton, and Kenneth McKenzie. After describing this finding, he spoke of the great importance of radioactive isotopes as tracers in biology and possibly as therapeutic agents. Then in a dramatic gesture wholly atypical of him, he stepped back, raised his arm, and portentously announced, "I now, for the second time this evening, have the privilege of making a first announcement of very great importance. This news is less than twenty-four hours old and hence is real news. . . . Now, Dr. S. Ruben, instructor in chemistry, and Dr. M. D. Kamen, research associate in the Radiation Laboratory, have found by means of the cyclotron, a new radioactive form of carbon, probably of mass fourteen and average life of the order of magnitude of several years. On the basis of its potential usefulness, this is certainly much the most important radioactive substance that has yet been created."

As he said this, all eyes turned in my direction, and I experienced a sinking feeling that the whole report might be wrong.

It may be noted that Birge had mentioned Sam's name first. When we returned from our visit to E. O. L. and decided to proceed with publication of our short note, the question of order of authorship arose. To my astonishment, Sam requested that his name be first. Although his performance in working with the low level of radioactivity to establish it as isotopic with carbon had been remarkable, the major contribution of finally making a sample that contained enough ^{14}C to be isolated had nevertheless been mine. Most of the work involved had been needed for production and not for identification. My invention of the probe procedure, and hence the contribution of the Rad Lab in the production of the large bombardment ion currents, had been decisive. There was little that was unique about the chemistry employed, important and difficult as it had been. I reminded Sam we had agreed that in all work on isotope production and characterization I would be senior author while he took seniority in the photosynthesis research. He conceded this was so, but he was worried about obtaining tenure in the Chemistry Department—an aim we both wanted to see achieved—and his appearance as senior author on the note would help greatly. He went on to promise that when I wrote the long detailed paper we expected to publish later, my name would be first.

Of course, as I came to realize later, Sam's being upgraded in this fashion meant that I was correspondingly downgraded. But I did not appreciate this at the time. Instead I decided the name ordering was not all that consequential with only two of us as co-authors. Nor did I take note sufficiently of the downgrading of the Rad Lab contribution. This misjudgment was brought home to me a day or so later when E. O. L. saw a final draft of our note. As he read it, he showed astonishment and then irritation. He turned on me with ill-concealed anger and demanded to know why my name and the institutional credits placed me and the Rad Lab in a position secondary to the Chemistry Department. Taken aback, I could only mumble something about the chemistry being crucial. Whereupon E. O. L. abruptly handed me back the note and strode away. This was to be the beginning of a process whereby my efforts in the discovery of ^{14}C came to be perceived as only supportive. This notion gained credence as time went on, not only in the Department of Chemistry but among others of our colleagues elsewhere[12] and even in Sam's family. Ironically, my undoing was my own doing.

Just why this happened can best be understood in the context of the situation in which Sam found himself. The Department of Chemistry was the country's leading institution of physical chemistry in instruction and research. It included a remarkably illustrious faculty, led by G. N. Lewis, and attracted the best students at home and abroad. It had a unique position not only in prestige but academically as an autonomous college in the university. The corps of young faculty was brilliant and included many who were potentially of Nobel Prize caliber, confirmed later in the cases of Glenn Seaborg and Melvin Calvin. There was intense competition for status among them, and Sam was as ambitious as any of his colleagues in striving for a tenured appointment. The pressures were so intense that confrontations were frequent, sometimes even bordering on outbreaks of violence at parties where tensions surfaced. I am reminded of a remark made by Franz Kurie when we had a discussion with Dr. Sten von Friesen, who was visiting from Sweden. The talk had centered on academic customs in the United States and Scandinavia, and Friesen had described the almost inhuman pressures in obtaining tenured positions at Swedish universities. Franz remarked wryly, "Evidently, the intellectual flower of Sweden is cultivated in a rock garden!" This observation could have been applied equally well to the Chemistry Department at Berkeley.

The atmosphere of aggression was fostered by the attitude of the dean, Wendell Latimer, who was perceived as the real power in the department. Lewis, to be sure, remained the dominant figure, but Latimer in effect ran the department and his support was deemed crucial if one expected to make a successful bid for tenure. His secretary, a Mrs. Kittredge, was a conduit for the application of pressures emanating from the dean, often volunteering her authoritative impressions of how he might be feeling about any given student. Sam was convinced from his contacts with Mrs. Kittredge, and even from a few talks with the dean himself, that his only hope of promotion lay in obtaining active approval by Latimer.

Moreover, there had been an unfortunate episode involving a collaboration between Sam and G. N. Lewis, who had generated an idea for an experiment that might show conclusively that there was a neutron component in cosmic rays. At the time there was much controversy about the nature of the primary radiations in cosmic rays. The experiment was a nightmare to implement. It involved looking for a particular short-lived isotope of iron produced by recoils from neutron capture in solutions of an iron salt. The iron was bonded in such a way that recoil radioactivity made by neutron impact could be chemically separated from the great bulk of iron salt present. Massive chunks of lead had to be erected to shield out radiations other than neutrons and whole bathtubs full of solution set up to be processed by bulk filtrations using industrial-size filter presses. Sam, who was strong and utterly fearless, took on the task, installing tubs and manhandling the backbreaking equipment needed in the outdoor court of the Chemistry building, where freshman lab courses were taught. Only someone with Sam's unique energy and motivation could even have contemplated doing such an experiment, let alone actually have tried it.

Lewis, who had ideas in profusion, seemed not to make a distinction between experiments that were difficult to implement and those that were relatively easy. If the young collaborator he picked found himself with a chore that involved a high chance of success, he was lucky. If success were attained, his stock with Lewis would rise. Conversely, failure, even if unavoidable, could have the opposite effect. An example was the research Lewis undertook to demonstrate that neutrons could be refracted. The experiments were a disaster, effects claimed to show a positive result turning out to be artifacts. Although he had labored mightily and done all that could be done in his eventually hopeless task, Lewis's young associate in these efforts was cast into outer darkness.

Sam's work on a possible neutron component to cosmic rays, being only marginally successful, convinced him, so he told me, that Lewis was not enthusiastic about him, and that his only hope lay with Latimer. The dean did nothing to allay Sam's worry, even exacerbating it by direct statements, as well as through Mrs. Kittredge. There was even a hint that anti-Semitism existed in the Chemistry Department and that Latimer would have that prejudice to contend with if he were to promote Sam's case. It was not surprising, therefore, that Sam acted under a fierce, self-generated pressure, working feverishly and in a highly self-centered manner. Nor was it surprising his relations with his fellow faculty and graduate students were sometimes strained and that he conceived a dislike for Bill Libby, who was equally aggressive and ambitious.

None of this affected my relations with him in our collaboration. We were remarkably complementary to each other temperamentally and I presented no threat, not being in the same department. There was a genuine desire on the part of each of us to improve the other's position so as to assure a long-term collaboration that would exploit the combined facilities of the Rad Lab and the Chemistry Department. It is doubtful whether either of us could have accomplished as much on our own. Nevertheless, my willingness to subordinate my interests was to cost me dearly in years to come.

In the succeeding months, we bent all our efforts to improving ^{14}C production. Research was also done on the recoil chemistry of ^{14}C made in various ways. To get the most spillover of neutrons from cyclotron operation, I positioned the carboys of ammonium nitrate solution next to the deflector, from which a great stream of neutrons poured when the 60-inch cyclotron was operating. Unfortunately, the deflector was often in need of manipulation and the big box holding the two carboys was in the way, constantly having to be hauled in and out of place. The carboys had been soaking up neutrons for several months when one day, I was visited by an angry deputation of cyclotron members, who demanded that I remove the carboys. They had become even more of a nuisance because they had sprung a leak and the acid solution was getting on the hands of the crew when they went in to adjust the deflector.

I had paid little attention to the project because I doubted much ^{14}C could be recovered, even though there might be a good yield from the slow neutron reaction. The recoil ^{14}C nuclei could be calculated to be formed at relatively enormous temperatures, up to many millions of degrees, making possible a large number of reactions with the surrounding atoms and molecules in solution. No one product could be expected to predominate sufficiently to make a simple chemical procedure for isolation effective, which would be required in view of the large volumes of solution to be processed. In addition, the slow neutron yields were not expected to compare favorably with those obtained from direct deuteron bombardment of carbon. The proton emission required to produce ^{14}C from slow neutron capture by ^{14}N had to compete with a much more likely reaction, the emission of gamma rays to produce ^{15}N. The latter reaction should be favored because the emergent gamma rays had no "potential barrier" to surmount, being electrically neutral, while protons were positively charged. There was also the ever present danger of contamination by carbonate in the ammonium nitrate, leading to cancellation of the advantage gained, by minimal dilution with inactive carbon, of ^{14}C produced in the reaction.

I took the box of carboys over to the Rat House, and Sam and I brooded over it awhile before deciding to do the simplest thing possible—merely run a stream of CO_2-free air through the carboys and through a combustion train, then catch the effluent vapors as CO_2, which it was hoped would contain ^{14}C. It seemed too much to expect that most, or at least a significant fraction, of the recoil ^{14}C would end up as CO or CO_2. To our initial dismay, a copious precipitate of calcium carbonate formed, confirming fears that the nitrate, although of reagent grade, had been contaminated with much carbonate. Nevertheless we collected and dried the precipitate. Sam scraped off as much as he could smear over the surface of the screen wall cylinder and placed it in the counter. To our amazement, there were no counts at all. So great was the activity the counter was paralyzed. I had been completely wrong about the yield and the chemical complications to be expected. The discovery that the neutron reaction was so productive ended further attempts to exploit the deuteron reaction as a major source of ^{14}C and with them my preoccupation with the ^{13}C-enriched samples.

About this time, Urey paid a visit to the campus, and I had the opportunity to meet him again. (Our first encounter had been in Columbus two years earlier.) He was still actively engaged in his campaign to get support for the production and use of his rare stable isotope, ^{13}C; he was unaware that ^{14}C had been discovered largely because he had needled E. O. L. and Lawrence had needled me. When I told him about ^{14}C, he showed some distress, realizing that his argument about the absence of a useful radioactive isotope of carbon was no longer relevant. He expressed concern that his negotiations with Eastman Kodak to promote ^{13}C would be compromised, but I assured him that, even with the slow neutron process, there was too little ^{14}C production in prospect to make it a serious competitor for ^{13}C, except for special experiments in which high dilutions, unavailable with ^{13}C, were to be encountered. Again, my crystal ball was clouded. The invention of the nuclear pile was later to produce "buckets" of neutrons and make ^{14}C production so great as to realize its full potential as the most useful isotope available for biological research.[13]

Hard on the heels of the ^{14}C discovery came the demonstration that the mass three isotope of hydrogen, tritium, was not stable as had been supposed, and even claimed by some investigators, who had published estimates of its occurrence in mass spectrographic determinations.[14] Luis Alvarez and Bob Cornog had used an ingenious procedure to raise the magnetic field of the 60-inch cyclotron high enough for short periods of time to detect helium ions of mass three when helium (^{4}He) was used as a source. They found about one part in ten million of ^{3}He in ^{4}He, and no ^{3}H when deuterium was substituted for helium.[15] It was an inescapable conclusion that if ^{3}He was a rare stable isotope, its isobar, ^{3}H, must be unstable, decaying by negative beta ray emission to ^{3}He. In fact, it was relatively easy to show this by looking for radioactivity induced in heavy water (D_2O) bombarded with deuterons. Such radioactivity was found and could be shown to be isotopic with hydrogen by the simple procedure of electrolysis, with the radiohydrogen being evolved only at the cathode. If the original estimates of tritium abundance as an isotope of hydrogen had been correct, ordinary water should have exhibited a radioactivity of several microcuries equivalent per cubic centimeter, assuming a half-life of about thirty years. (Actually, the half-life of tritium was later established as twelve years.)

So, a few months after Urey's pronouncement that no useful radioactive isotopes for the important elements, carbon and hydrogen, existed, ^{14}C and ^{3}H had appeared. Our attempts to fix a reasonably precise value for the ^{14}C half-life were not successful, although in some experiments I performed a year later I was able to calculate that it might be around 2,700 years (not too far off from the accepted value, later determined as 5,700 years). Sam used the yield we observed from our deuteron bombardments to arrive at a value of 2,200 years. There was one deuteron-produced sample from which I actually calculated a value of 4,000 years. In any case, it was obvious ^{14}C had a half-life measured in millennia. This result created much worry for us because Oppenheimer assured us such a long life could not be associated with the decay of ^{14}C to ^{14}N. However, we were consoled to find that algae ate ^{14}C-labeled CO_2. We were inclined to give greater credence to their testimony than to the deductions of theoretical physicists. Finally, there was the remarkable fact that carbon was unique in its chemistry, being the only element in the Periodic Table whose oxide (CO_2) could be volatilized repeatedly from acidified oxidizing solutions.

Meanwhile, we kept our original sample of ^{14}C in the counter, observing that for several months no appreciable decay had occurred. More ammonium nitrate solutions were prepared and placed around the 60-inch cyclotron to increase our stock of ^{14}C so that we could return to the search for the primary CO_2 fixation product in photosynthesis. E. O. L., buoyed up by the prospect of ^{14}C as a source of funding, assigned our chief engineer, Bill Brobeck, to design and fabricate stainless steel vessels equipped with aspirators, to be installed as permanent adjuncts to the cyclotron. There was even talk of forming a company to manufacture cyclotrons solely dedicated to ^{14}C production. A battery of such cyclotrons could produce so many neutrons that, if all were slowed and captured in ammonium nitrate, production of ^{14}C could be in the millicurie range.

Lawrence suddenly remembered that ammonium nitrate was unstable. The classic text on thermodynamics by Lewis and Randall mentioned a catastrophe that had obliterated the town of Oppau in Germany during World War I when a pile of solid ammonium nitrate had exploded. Apocalyptic visions of his 60-inch cyclotron disappearing propelled him down to the laboratory to have the ammonium nitrate cans removed forthwith. I vainly pointed out that ammonium nitrate *in solution* would not explode. Others more authoritative, such as Professor Leonard Loeb in the Physics Department, affirmed this. I also observed that the reaction of nitrogen and oxygen with water to form nitric acid—the reverse of the nitrate decomposition—was not known to proceed at a measurable rate, even though it was thermodynamically expected to do so spontaneously. If it had, then as Lewis and Randall pointed out in the same chapter E. O. L. had read, the oceans should have turned to dilute nitric acid! Later, the famed physicist R. W. Wood happened to be visiting Berkeley and gave a lecture on some recent experiments he had been doing on fulminate explosions. His opinion could be expected to impress E. O. L., and he was asked to add testimony that one need not worry about solutions of ammonium nitrate exploding. He wrote me a strong letter backing up this assertion. It was all in vain—the cans remained banished. Eventually, they were taken to the Chemistry Department, where whatever ^{14}C was in them could be recovered.

Later in the year, Sam proposed to Professor Merle Randall that Randall and his graduate student, James Hyde, collaborate with us in an experiment to test the hypothesis that the oxygen in photosynthesis originated from the water and not from carbon dioxide. Randall and an associate had constructed a tall distillation column to fractionally distill isotopic oxygen in water, concentrating the ^{18}O, or heavy isotope, as heavy water. Hyde operated a mass spectrometer that could be used for assay. Sam and I contributed the expertise in handling the algae, *Chlorella pyrenoidose*. We had available water enriched in ^{18}O to the extent of some 0.84 percent, only a fourfold increase over the ^{18}O content in ordinary water but enough to carry out some definitive experiments, starting with ^{18}O-water or ^{18}O-potassium bicarbonate made from the ^{18}O-water.[16] Our data were consistent only with the conclusion that water oxygen, not that in bicarbonate, was the origin of photosynthetic oxygen. This conclusion was strengthened when we did an experiment on possible oxygen exchange during respiration and found no evidence for such exchange as measured by change in ^{18}O content when *Chlorella* was producing oxygen photosynthetically from normal water and bicarbonate in the presence of ^{18}O-labeled oxygen. Our conclusions were later challenged by Otto Warburg, who wanted to believe that the CO_2 was the source.[17] However, our conclusion has held through the years.

We planned to extend our research to yeast, studying isotopic exchange in respiration, but only one such experiment had been done when pressures of war work intervened. By the summer of 1940, the laboratory was beginning to be drawn increasingly into research related to requests from the army and other government establishments.

In fact, lab personnel had already been, or were soon to be, hired on a contractual basis as federal employees. The immediate effect was to produce a new living standard for me. We were to be treated as civil servants, which meant our being graded by government standards on the basis of experience and training. In my case, the grade for which I was eligible yielded a mind-boggling $5,000 per annum—a pay almost triple what I had been receiving as a staff member of the Rad Lab. Although Don Cooksey cautioned us that such prosperity might be only temporary, I negotiated for the purchase of a fine old Italian viola labeled as made by one Tassini in the early eighteenth century. Included in the deal was a bow with papers from the illustrious house of Hamma, Stuttgart instrument makers, which described it as the work of a great French artisan, D. Peccate.

The changes in living standards extended most dramatically to the operations of the laboratory and attitudes toward supply and material procurement. One morning, Don Cooksey—long frugal guardian of the lab budget—called me into his office and assigned me the job of preparing a shopping list for future chemical operations. He said there would be a considerable expansion in the need for chemicals and chemical equipment and a detailed list was needed immediately. He said that for the moment I was not to concern myself with details or costs. Indeed, it was not clear what would be needed. Somewhat frustrated at being asked to proceed in the dark, I decided to draw up as varied and exotic a requisition as I could imagine, regardless of the shock it might cause Don. Consulting the general catalogue of the Central Scientific Company, I wrote up a long list, including at least one of everything, available or not. I recall in particular a "Podbielniak fractional distillation apparatus" with gold-plated seals and ground joints, which cost in the neighborhood of a thousand dollars and which I included mostly out of curiosity to see what such an apparatus looked like.

When I presented the bulky requisition totaling many thousands of dollars to Don, I expected a reaction of shock and dismay. Instead, he pored over the list calmly and intently until he had gone through it completely—a procedure lasting almost half an hour, while I shifted from one foot to another in wordless apprehension. Then he looked up at me quizzically and remarked, "Martin, I don't think you have understood the situation. For instance, here is an order for rubber tubing. [I had specified several miles each of all sizes.] You know our supplies of rubber are cut off in the Pacific. Don't you think we should triple this order?" Only a year or two earlier, Don had led an intense discussion on whether one of two possible types of heavy duty storage batteries should be purchased, a decision involving a saving of a dollar or so.

Sam's stock in the Chemistry Department kept rising, much to our satisfaction. He was giving seminar talks on various phases of biochemistry as he avidly absorbed the literature, and his enlarging group of graduate students was busily engaged in extending research Sam and I had started on the exchange reactions of metal ions with biologically interesting chelating reagents, as well as probing mechanisms of oxidation of organic acids.[18]

Proceeding from these earlier studies, Sam and his students examined in detail mechanisms operative in the synthesis and oxidation of several three-carbon acids. I found time from my commitments at the Rad Lab to participate in some of these inquiries. We studied the synthesis and oxidation of oxalate, propionate, and lactate labeled in carboxyl groups, and found that during oxidation by alkaline permanganate of propionate to one mole each of carbonate and oxalate, only about 25 percent of the carbonate formed came from the carboxyl carbon. It appeared that in highly alkaline solution, the bond between the two carbons other than the carboxyl carbon was broken preferentially, and we formulated a mechanism for this surprising result.[19] In another series of experiments, we found that oxidation of the propionate using acid dichromate yielded the expected rupture only of the bond between carboxyl carbon and its adjoining carbon atom.

Another subject of inquiry for us was the role of chlorophyll in photosynthesis. The numerous theories proposed fell into two categories. The first type held that chlorophyll actually participated chemically, being oxidized and reduced cyclically. In this process, hydrogen atoms could be expected to leave and return to chlorophyll during active photosynthesis. The second type proposed that it acted solely as a sensitizer, like dye in photographic emulsions. We thought that by studying the exchange of hydrogen chlorophyll with tritium-labeled hydrogen of the medium in which algae were photosynthesizing, it would be possible to decide between these two alternatives. Sam's group did some experiments attempting to detect labeling of chlorophyll during algal photosynthesis in tritium-water (HTO).[20] They found none, thereby indicating that the first alternative—chemical participation of chlorophyll as a redox component in photosynthesis—was unlikely. There were two caveats, however. It could be argued that during extraction of the chlorophyll in our experiments, tritium label contained in it could have been lost by exchange with hydrogen in the solvent used (80 percent ethanol). However, we demonstrated that no such exchange occurred when purified chlorophyll was incubated with tritium-labeled ethanol. Another objection, which James Franck brought up, especially in some correspondence with Sam, was that tritium, being so much heavier (three times more massive) than the normal hydrogen of mass one, might be discriminated against as a tracer for hydrogen and not adequately reflect movement of hydrogen atoms. We did not think such isotopic effects would be sufficient to explain our completely negative findings, but concluded that more experiments with 100 percent D_2O (heavy water containing only deuterium) could settle this issue.

A special problem—synthesis and oxidation of fumaric acid using ^{11}C-labeled CO_2 and HTO as starting materials—was assigned to Mary Belle Allen, who had been accepted by Sam as a graduate student. She attacked her problem with energy and soon acquired the skills needed to carry out rapid syntheses of the labeled fumarate. She produced a very respectable piece of work in which she showed that during the oxidation of fumarate by acid permanganate, the carbon-hydrogen bond of the formate remained intact, as did those of all intermediates formed.[21] To effect this research, she had worked out a rapid and efficient synthetic procedure, making ^{11}C-labeled cyanide from $^{11}CO_2$, which she reacted with ethylene dichloride to form succinonitrile from which the free succinic acid was produced by hydrolysis. She then used the enzyme fumarase to dehydrogenate the succinate to fumarate.

The interest in exchange experiments extended to isotopic tracers other than carbon and hydrogen. In work done by Albert Frenkel with us, the centrally bound magnesium atom of chlorophyll did not interchange with magnesium ions in 80 percent acetone.[22] This prompted us to initiate a systematic program on metallo-organic reactions that could be studied using exchange with labeled metal ions.[23]

Paul Nahinsky used some of our miniscule stock of ^{14}C to do the first experiment with this long-lived isotope, a very preliminary effort in which he synthesized ^{14}C-carboxyl labeled propionic acid by a Grignard reaction of $^{14}CO_2$ with ethyl magnesium bromide. Recovering the propionic acid by a so-called Duclaux distillation, he subjected it to alkaline permanganate oxidation and confirmed previous results obtained with the short-lived ^{11}C isotope.[24]

By fall 1940, we had decided to summarize all the results we had obtained on the isolation and characterization of ^{14}C in a full-length paper, which I outlined and wrote. In its final version, the credits were listed as "M. D. Kamen and S. Ruben, Radiation Laboratory and the Department of Chemistry, University of California, Berkeley, California." Sam took it to Mrs. Kittredge for typing. According to Sam, she invoked the dread specter of Latimer and left him with no option but to ask that the order of the names be changed. I reacted angrily, citing his previous promise. The order remained reversed, although the Rad Lab somehow still received top billing, retaining its original position first in the credits. This at least put the proper emphasis on the contribution of the laboratory and indicated the original order of authors. The paper was sent to the *Physical Review* in early December and published in February of the following year.[25]

There was an attempt to right the matter in a short abstract that appeared in the accounts of a meeting of the Physical Society, where I finally achieved top billing.[26] In this abstract, we included my pessimistic and erroneous prediction that the use of ^{14}C would be confined to research in which high dilution of tracer material was unavoidable. We also noted that the sample had not decayed in six months. The half-life remained a mystery for decades, being explained in various ways.[27] An adequate theoretical treatment was not possible until late in the seventies, when Henry Primakoff and his associates presented their resolution of the problem.[28]

The ineradicable impression of the ^{14}C authorship fiasco on E. O. L. was borne in on me a few months later when both Sam and I received invitations to a conference on applications of isotope tracer methodology, organized by MIT and Harvard. We had a great mass of material to present, certainly enough to warrant both of us attending. We prepared lengthy papers for presentation, sending in abstracts to be printed in the proceedings of the meeting and published in the *Journal of Applied Physics*. Senior authorship was divided equally.[29] When I approached E. O. L. with a request for a travel subsidy, he cut me off abruptly, saying he saw no reason why both of us had to go. Sam's participation would be sufficient. This action, which disappointed me greatly and excited comment later at the meeting, clearly demonstrated that E. O. L. had fastened firmly on the erroneous notion that my contribution to the research had been secondary.

Meanwhile, the ominous happenings in Europe and Asia had begun to claim our attention, rendering personal frustrations trivial. The laboratory had already undergone great changes since the freewheeling days of only a few years before. The plans for a gigantic new machine, the 184-inch cyclotron, were well advanced, and ample construction funds were being raised. Whereas a child's wagon had been sufficient in the good old days to carry lab supplies for a week, ten-ton trucks now rumbled over the dirt road up Strawberry Canyon to the hills overlooking the campus, where a massive complex of buildings was rising. Personnel increased by orders of magnitude. New faces replaced old and soon far outnumbered the few remaining old retainers. E. O. L., burdened by the great new administrative problems he had acquired, was hardly ever seen in the old laboratory. There were two major new buildings, the Crocker and Donner Laboratories, already in existence and operating to meet the many new commitments arising from cyclotron construction and the medical program.

I found myself almost wholly engaged in meeting the demand for ever more radioisotopes in ever greater quantities, and could no longer spend much time in the Rat House. It became apparent that our collaboration would remain dormant until the war emergency was ended. Although we had in ^{14}C the tool needed to make a breakthrough in our quest for the elusive first product of photosynthesis, we were denied the opportunity to use it.

Frustrations of a much greater order were to plague E. O. L. as the deepening war crisis began to make more and more demands on Rad Lab facilities, eventually forcing the completed 184-inch magnet and massive supporting electrical installations to be dedicated to use as a mass spectrograph for production of enriched uranium-235. Vannevar Bush was implementing his remarkable plan to contract research for defense purposes to universities and industrial laboratories,[30] beginning with the establishment of the National Defense Research Committee (NDRC) and its rapid growth into a full-fledged operation, with eventual evolution into the Office of Scientific Research and Development (OSRD).

Contracts given to the Rad Lab under agreements with NDRC required me to mount several large-scale efforts for production in unprecedented amounts of short-lived radioisotopes, such as ^{11}C and ^{38}Cl, as well as longer-lived isotopes, such as ^{24}Na, ^{32}P, ^{55}Fe, ^{59}Fe, and ^{64}Cu. I was given authority to recruit talent as needed from the pool of students on the campus. One of my helpers was Gerhart Friedlander, later to enjoy a successful career as a nuclear chemist. During his tenure as my assistant, I managed to extract one piece of constructive research from the drudgery of isotope preparation. With Friedlander's help, I set up large batteries of carboys containing carbon tetrachloride (CCl_4) to produce sulfur 35 by slow neutron capture.[31] Except for a half-life of approximately eighty days, little else was known about it. I measured yields for all the important nuclear reactions for its preparation, as well as the upper energy limit for its emitted beta rays. I found a more accurate value for the half-life of eighty-eight plus or minus three days. A procedure for the efficient extraction and production of bulk quantities of sulfur activity from neutron-irradiated CCl_4 was worked out. In addition, I observed, with Friedlander's help, that a resonance level existed for slow neutron capture by the target chlorine nucleus.

It could be concluded from these data that the proper assignment was to mass thirty-five.[32] Later I demonstrated this directly, using silver chloride samples, enriched in chlorine isotope of mass thirty-five, to show that the production of radioactive sulfur paralleled the ^{35}Cl content and not that of ^{37}Cl.[33] I believe this was the first instance in which separated isotopes were used to establish isotopic assignments.

These demands, generated by the approaching war emergency, left little time either at home or for social activity in general. Esther went to work across the Bay in the Office of War Information. Chamber music sessions became infrequent. Their major organizer, Waldo Cohn, left after the tragic premature death of his wife, Elma, thereby creating a great void in our musical circle. Later he pioneered the crucially important application of ion exchange chromatography to the isolation and preparation of pure nucleotides while enjoying a highly successful career as a biochemist at the Oak Ridge National Laboratory, where he also organized and directed the Oak Ridge Symphony Orchestra.

8

Pearl Harbor and the Manhattan Project

(1941–1943)

On SUNDAY mornings human flotsam washed up from the storms of the previous night's debauches could often be found in the hushed ambience of the Faculty Club reading room. Sunday, December 7, 1941, was no exception. Into this atmosphere of quiet convalescence burst Glenn Seaborg—not a man known to be easily excited—with the news that the Japanese had bombed Pearl Harbor. Al Marshak and I had just drifted into the club. We paid little attention, assuming that he was having his little joke. After all, the Sunday morning headlines announced the continuing peace negotiations between high-ranking Japanese envoys and the White House. But Glenn was insistent, adding that radio Pundit H. V. Kaltenborn had analyzed the news. Some of us went up to his room to hear for ourselves. There was nothing on the radio but the usual Sunday morning pap for many minutes. Just as Glenn was about to concede that he might have been hallucinating, the morning news broadcast was interrupted by an almost hysterical announcer confirming reports that the attack had occurred. At that very moment, half the Pacific fleet was at the bottom of Pearl Harbor, and there were scenes of frightful carnage at Hickam Field. It was expected that an emergency session of Congress would be called to hear the president, with every indication that a state of war with Japan would be declared.

In mute dismay, we stared at one another, aware that bombs might be raining down on us at any minute. San Francisco, across the Bay, was an ideal target, undefended as far as we knew. Its largely wooden construction and hills, perfect terrain for updrafts, would guarantee wholesale destruction in a fire raid. (The major damage in the great 1906 earthquake had been caused by fire.) Terrified, Al and I rushed downstairs and to our homes, where we remained glued to our radios for the rest of the day. Bulletins kept pouring in, together with declaration of a "brown out" that evening. Ironically, a group of us were scheduled to give a performance of Bach's cantata "Jesu, Bleib Bei Uns" ("Jesus, Do Not Leave Us") that evening. We had been rehearsing for a month for this event under the direction of David Boyden of our Music Department. I doubled as concertmaster and viola soloist. The rendition that night could not have been more sincere.

Next morning, we heard from E. O. L. that the laboratory was essentially on a war footing. Whatever work we were still doing that was not directly related to the war effort would have to end immediately. Actually, many of us had been spending most of our time on such efforts for several months as we received directives from the NDRC. In the Chemistry Department, much the same situation existed, with chemical warfare service contracts pouring in. Sam and his little group of graduate students were maintaining some of the momentum built up in our research over the previous two years, but most of the work had been wound up and results were either published or in press. C. B. van Niel had prepared a draft of a position paper on the general nature of CO_2 as a metabolite, of which Sam, Stan Carson, Jackson Foster, and I were co-authors. This paper, published as a kind of swan song, spelled out for the first time the evidence, obtained not only in our laboratories but also elsewhere, that decisively repudiated the old notion that CO_2 was merely a waste product of metabolism in animals and plants.[1]

The progress of the war offered no basis for the optimistic notion that it would be over soon or even that our side would win. There was, however, the anesthesia of working hard. As before, my job remained largely an effort to produce radioisotopes in bulk for use elsewhere on the campus and in the country. As an example, there was an assignment to produce a sample of the lethal gas phosgene, with an enormous radioactivity, equivalent to many grams of radium. This could be done using either ^{11}C labeled carbon monoxide or ^{38}Cl-labeled chlorine. The classic procedure was to expose carbon monoxide and chlorine gases to light. The two gases reacted to produce phosgene. Samples of the infernal mixture I made this way were so radioactive they emitted a blue glow. The phosgene was needed, among other things, for research by Don Yost and Richard Dodson at Caltech, and the assignment was the beginning of a friendship with Yost and Dodson that became one of my positive gains from the NDRC (later OSRD) work.

The possibility that a super bomb could be fabricated was not yet being taken seriously. Most of us thought that the uranium chain reaction could be a basis for power production, but not for a bomb. There were cogent arguments against the latter. Among the products of fission, for example, were isotopes of rare earth elements. Often effective absorbers of slow neutrons, these might be expected to "poison" the propagation of the chain reaction. It seemed unlikely that none of the rare earths produced in fission would show such properties. Another hitch was that for an explosion to occur, the reaction mass would have to be held together a relatively long time for the enormous energy released to be confined in a small volume. These pessimistic predictions fed the hope that such a bomb would never be possible or be produced. However, there were rumors that Fermi had been so "optimistic" as to suggest that the probability of bomb production might be as high as 10 percent.[2]

Apropos of the fears about chain reaction "poisons," I was asked by E. O. L. to investigate the "capture cross section" (probability of absorption) of slow neutrons by the ^{13}C isotope of carbon. At Chicago there was a proposal to take up the approach, pioneered by Leo Szilard and Enrico Fermi in the early experiments on the neutron pile, using graphite as a moderator—that is, using elementary carbon rather than hydrogen to slow down the neutrons flooding out from the initial fission process. The ^{12}C isotope was known to have a normal cross section, but measurement of the same parameter for ^{13}C was rather difficult. It could be done most simply by measuring the ^{14}C radioactivity produced when slow neutrons (created, for example, when fast neutrons from fission were passed through paraffin) were captured by ^{13}C.

I was the logical candidate to undertake this experiment, which I did early in 1942. I placed samples of elements with known cross sections, such as gold and antimony, in a paraffin box, together with different materials containing carbon and nitrogen. I hoped to obtain a better value for the ^{14}C half-life than we had estimated previously. To complete the setup, all samples were provided with duplicates surrounded by cadmium metal to cut out slow neutrons and thus establish the proportion of radioactivities ascribable only to slow neutron activation.

My notes show that on April 16th the box was bombarded with neutrons from 14 MEV deuterons hitting beryllium in the 60-inch cyclotron. The exposure lasted for eleven hours, amounting to a total of some 1,300 μah. From the radioactivities induced in the gold and antimony samples, I estimated a ^{14}C half-life of 4,000 years. I expected to estimate the slow neutron capture probability for ^{13}C from the ^{14}C activity found in the ^{13}C samples (graphite enriched in ^{13}C). I think I found it not to be anomalous and submitted a report, but my notes do not record the result. Obviously from the fact that the graphite-moderated pile worked well, ^{13}C could not have functioned as a "poison" by soaking up the neutrons needed to maintain a chain reaction in the uranium pile.

When the pilot plant at Oak Ridge went into production, it became clear that the uranium produced was not sufficiently enriched in ^{235}U to be useful for bomb fabrication. A need to devise a process to recycle the ^{235}U seemed urgent. ("Alpha" was the system that initially produced the low-quality material in the "race tracks"—massive rings of gargantuan mass separation machines, which in the aggregate could easily be considered the eighth wonder of the world.) A group was organized under the general supervision of F. ("Pan") Jenkins, an authority on spectroscopy in the Physics Department, with James Carter and myself as group leaders. We were charged with perfecting a procedure (to be called Beta) for rapid recycling with minimal loss of the ^{235}U produced by the initial mass separations at Oak Ridge.

It is interesting to note that chemical projects at both Berkeley and Chicago were led by physicists. Only at Columbia, where Harold Urey was in charge, were chemists entrusted with overall responsibility. At Chicago, Arthur Compton's project, dealing with plutonium production from uranium pile operation, had a particularly crucial chemical component. Nevertheless, the direction of the chemical isolation work remained with physicists. Nowhere were project pressures more extreme and resultant personality conflicts so violent as among the chemists at Chicago. The assignment was enormously difficult, some might have said impossible—the chemistry of plutonium had to be learned and isolation procedures worked out when no plutonium was available except in microscopic quantities. But the rewards for success—ultimately achieved by the group led by Glenn Seaborg—were as great as the difficulties.

The traumas of the directors in keeping the peace among fiercely contending factions in the ranks of the chemists created some turnover, beginning with Sam Allison and ending with James Franck. Even Franck, a veteran of combat in World War I, complained that the mere presence of one of his group leaders caused him gastric upset. Psychosomatic disturbances plagued some of the personnel. Propelled into a new area of research for him—the radiochemistry of fission products—Charley Coryell felt the pressures so keenly that he had to be hospitalized at various times, but went on to lead a group that made notable contributions to the successful prosecution of the pile project.[3]

The biggest problem in making the project move was the security syndrome created by the attempts of Army intelligence ("G-2") to apply a narrow concept of secrecy. General Leslie Groves, the officer in overall charge of the Manhattan Project, was proud of his strict "compartmentation." Few of the scientists who were expected to make the project work could adapt easily to a system in which free inquiry and consultation with colleagues was forbidden, however. In testimony I later described the situation as follows:

> [There was] a peculiar and very strict rigid compartmentalization. [One was not] supposed to know, for instance, that an old friend of yours was working on the project. . . . It was never really understood during the war what was meant by security. The relevant questions were: (1) are we working on an atom bomb? and (2) how far have we gotten with it? Groves' insistence on compartmentalization required breaking a long-standing habit of scientists. . . . But in any case we did our best to conform to what we thought the Army's idea of security was.[4]

The question of what constituted security was never resolved. The consequences were to be tragic, not only for me but for many others, the most famous being Oppie. A few years later, unevaluated raw records of unsophisticated intelligence agents were peddled by former G-2 employees for use by unprincipled politicians bent on attaining power, under a smoke screen of patriotic devotion to the "best interests of the United States." The fact was that all security matters required exercise of highly informed judgment based on technical knowledge. In my case, as with Oppie, there was a strong predisposition to judge me negatively because of my associations. Conversations in which I had been allegedly overheard discussing "forbidden" topics were construed as evidence of my disregard for "security."

It was difficult to take the strictures of Army intelligence in these matters seriously. It was forbidden to mention terms such as "plutonium," "uranium," and "neutron chain reaction." Special code words had to be memorized for the elements on which we were working and the locations and names of the various projects, as though such naive procedures would confuse any professional espionage agent. It was especially difficult to act as though the whole subject of the nuclear chain reaction was unknown to anyone except project personnel. *Chemical Abstracts* carried descriptions of sophisticated research on the process, originally published in French, German, and Russian journals.[5] (The fact that open descriptions of presumably top-secret work came from Russia was hardly ever mentioned by the superpatriots bent on conducting witch hunts during the hysteria of the Cold War, or by the Army intelligence agents who found employment feeding them doctored transcripts of field reports.)

It seemed to me that project security could best be achieved by minimizing the publicity attached to nuclear research and thereby lessening the curiosity that was bound to arise when whole towns were constructed to support factories rumored to be engaged in such efforts, as at Hanford and Oak Ridge. I recall the reaction of the Berkeley Ration Board when it was petitioned to approve a request for a tremendous amount of coffee to be supplied to the steam table at the lab. The lab representative pleaded eloquently for permission to obtain the coffee, but was turned down. As he picked up his briefcase to leave, he was said to have remarked that the "machine" would be badly affected. At this, the board members perked up their ears and cried, "If the machine needs it, that's another matter!" The coffee requisition was granted. Berkeley knew that there were mysterious goings-on "up on the hill" and that they had something to do with nuclear research. Without the sophistication needed to deduce what progress might be possible toward a specific objective (such as an atomic bomb or power plant), such knowledge in no way threatened security.

The problem of instructing intelligence personnel as to what might constitute a breach of security was apparently insoluble, however, and the only safe procedure seemed to be a blanket prohibition on discussion of fission. This required that agents in the field report anything they thought suspicious, thereby setting their judgment over that of the scientists they were monitoring. The friction between scientists and Army monitors was aggravated by the wholly different lifestyles of the two groups. Many of the former led private lives that would have raised eyebrows in church circles. Some G-2 agents were moreover restive at being assigned to relatively innocuous surveillance, rather than being where the "real action" was.

My troubles with security were somewhat special, because I had such a wide circle of acquaintances and such varied social activities. It took much more than the routine procedures to cover adequately whatever I might be doing. Early in the project, it became obvious that my phone was tapped and that G-2 was watching the cottage on Buena Vista Way. Often the agents sat a little way up the street, where their presence was noted by housewives made anxious by the absence of husbands off to war. The strangers were made even more suspicious by their unaccountable habit of keeping the engines of their cars running, a practice sure to create attention in a time of gas rationing. The police refused to intervene when called, giving no explanation for their apparent lack of interest.

At least once the agents had to "blow their cover" by showing their identity cards. On another occasion I was out of town on project business and had given a woman friend permission to use the cottage. She entertained a male friend one evening, during the course of which they went out to dinner. When they returned, the place was in a shambles. Drawers and bureaus had been ransacked, but nothing had been taken. The gentleman involved was in a sweat because he had left behind a valise with secret documents pertaining to war business, but he found it intact. It was not easy for me to explain this away without exciting suspicion that the lab was involved in supersecret activity.

There was a whole chain of confessions from neighbors seeking my advice and reassurance. I was amazed to discover what an exotic crowd they were. A woman living across the street entered the cottage in a state of great agitation one afternoon and told me that her estranged husband, whom she was suing for divorce, had set agents on her trail. (She was having an affair with a bartender in Richmond, whose baby-blue Buick coupe could be seen parked outside her house several nights a week.) The pressures of this situation had driven her to drink. I had to convince her without pointing suspicion at my lab work that she was not being watched. Up the street the local sector and block wardens were living in a homosexual arrangement, which they feared the agents were spying upon to discredit them as proper representatives of civil defense. For them, too, I had to provide reassurance.

One of the big "secrets" G-2 was anxious to keep concerned the connection between work at Oak Ridge and at the Rad Lab in Berkeley. Frequent transfers of personnel occurred, and they inevitably excited comment. On one occasion I was standing in a crowded bus traveling up Euclid Avenue when I heard a woman ask a companion about the whereabouts of a friend who had left town. "Oh, X —— has gone to Tennessee," she was told. "There is a big factory there doing war work connected with what's going on at the Rad Lab!"

Another time I was required to do some troubleshooting in Oak Ridge in connection with a process being set up there, based on our work in Berkeley. I needed a strong sample of a radioactive material to check one step in a proposed procedure. I knew, but was not supposed to know, that a neutron pile was being operated just down the road at the "Clinton" works in Oak Ridge by the Chicago Group ("X-10") and that it might solve my problem. I asked Charley Coryell, whose connection with X-10 I was also supposed not to know about, if there was any possibility of my getting a sample of radioactive sodium (^{24}Na) by neutron bombardment of salt. At that time, the biggest secret imaginable was how many neutrons were being produced by the pile at Oak Ridge. Indeed it was supposed to be a secret that there was a pile at all. Charley said the radiosodium could be made, but that in view of the security situation I would have to appear wholly unaware that a pile irradiation could be done. Nor was I to indicate I had been in communication with him. Somehow, it was arranged that I be interviewed by the top brass running X-10. One afternoon I was ushered into a room where the director and his aides sat in solemn conclave. I was told that they would expose some five grams of sodium chloride in their "machine" for several hours. This would make enough ^{24}Na to meet my request (which was only for a few microgram equivalents of radium activity).

A day later I was escorted to a car with closed blinds to prevent my seeing where we went. After a bumpy ride of nearly half an hour, we arrived in front of a big black building. A worker escorted by Army personnel presently emerged, laboriously trundling a heavy lead cylinder on a hand truck. The container was maneuvered off the truck and into the trunk of the car, which sank down on its axle owing to the weight of the lead. I was amazed that the weak sample I expected should require so much lead shielding, but asked no questions. I was driven back, again with the blinds down, and deposited with the lead container in front of the laboratory where I was to make my experiments. A crew of workers hauled the sample inside, where I was left alone with it.

First, I had to estimate how much radioactivity there was. To do this, I lifted off the top so that I could insert a radioactivity probe. I was shocked to see a purple glow. Immediately, I dropped the top back down, realizing there must be many gram equivalents of radium radioactivity in the sample they had prepared down the road. In other words, the flow of neutrons from the pile had to be millions of times greater than what I was accustomed to seeing in the cyclotron. As I had been told the amount of salt and the bombardment time and knew the yield of the neutron capture reaction, it was a simple matter to estimate just how big the neutron production was. G-2 had divulged the Big Secret without realizing it.

E. O. L. and his Army guard came by while I was still in shock. I blurted out that the cyclotron had become passé as a neutron source and told him why I had reached that conclusion. Lawrence strode on, dissembling any interest in the news, but shortly afterward I heard that an investigation had been instituted to find out the source of the leak to me.

Often enough there was confusion at top levels. Back in Berkeley a little later, I was called in to a high-level conference in Lawrence's office, where E. O. L. and General Groves sat with the top officials of our engineering and industrial contractor, the Tennessee Eastman Company. It appeared that someone had to be sent to Rochester, N.Y., to confer on a crucially important matter with company engineers. E. O. L. turned to me and said that I should go there forthwith. To do what, I asked. I would find out when I got there, he said, glancing at his watch, and remarked that I could catch a plane in two hours. Alarmed, I protested that I could not possibly leave on such short notice. E. O. L. merely reiterated his statement; I had to get going immediately.

Rushing home, I made a few quick phone calls to friends to wind up some chores at the cottage, stuffed two shirts and some socks into a briefcase, and rushed by cab to the airport. Some hours later, I landed in Rochester, to be met at the plane by a worried-looking group of Tennessee Eastman chemists. After a few minutes of sparring, their spokesman warily asked me why I had come. I replied in amazement that I had been told that *they* would tell *me* why I was there. Baffled, we adjourned to the nearest bar and had a supper clouded by apprehension as to what might be in prospect. I stayed in Rochester for several days. Then, having run through my shirts and socks, and receiving no further instructions, I decided to return. I had been back in Berkeley a few days when I met E. O. L., who seemed surprised to hear that I had been out of town. I never did learn why I had been sent to Rochester.

Project business involved frequent trips East. Lab personnel had top priority, so that we could even "bump" congressmen off flights. In New York and Chicago I often stayed with colleagues and friends. One of these was H. H. ("Hy") Goldsmith, who was employed on the Manhattan Project at Columbia. He had wangled a vast penthouse in a posh residential building on West 72nd Street in New York at a ridiculously low rental. Hy was an avid admirer of Isaac Stern's and Isaac was often to be found there, since he used Goldsmith's apartment as headquarters when in New York. Chamber music sessions frequently took place there, with eminent musicians taking part. One of these, Robert Mann, was then beginning to organize the Juilliard Quartet. I also met many of the leading lights of the musical world—Virgil Thomson, David Diamond, Lehman Engel, and others—when I stayed with the composer Arthur Berger on East 28th Street.

One amusing incident that featured musical connection occurred on a trip when I was returning west, sharing a sleeping compartment with a project lawyer. We boarded the streamliner in Chicago, making our way through the dismally depressed tourist sections to our elegant first-class accommodations. We had been seated only a few minutes in the dining car when none other than Henri Temianka appeared at the door. His face lighted up when he saw me. Behind him was the conductor, looking a bit confounded, as Henri turned and told him something. The conductor left. Henri came to our table, sat down, and told us he had been unable to find accommodation in the tourist section, having given up his seat to his parents, with whom he was traveling. He faced the prospect of standing all the way back to San Francisco. In desperation, he had decided to investigate the possibility that some opening might have occurred in the first-class section, but had been intercepted by the conductor, who told him tourists were not allowed there. Henri brazened it out by claiming he was visiting a friend. One can imagine his delight and surprise when it turned out there actually was a friend in first class! For the rest of the trip, we shared our quarters, Henri using them during the day.

Making enough ^{235}U by separation of the uranium isotopes necessitated scaling up the cyclotron by many orders of magnitude and then designing and building prototypes of such monsters that could be operated routinely and successfully by essentially illiterate hillbillies. Mass spectrometers were still lab instruments of extreme fragility and erratic performance, which could be operated only by highly-trained technicians. To assert that they could be built on the scale needed and operate more or less on a push-button basis was to invite invitations to don a straitjacket. Nevertheless, the task was eventually accomplished. The Rad Lab physicists and chemists and their engineering counterparts achieved miracles. The whole endeavor appeared endangered for a while, however, by the fact the machines were not producing uranium samples sufficiently enriched in ^{235}U. As mentioned, our group at Berkeley was assigned the task of investigating procedures for recycling the material to obtain adequate enrichment.

Our problem was horrendous. We had to find a means of washing out the separation apparatus to recover all the uranium material scattered around in the vacuum chamber after a run (only a few percent actually reached the collection chamber), reprocess it to make highly reactive uranium tetrachloride (UCl_4) of high purity, bring it to the right consistency as a crystalline material, and take no more than a day for the whole process. Each batch from the Alpha machines might have to be recycled as many as forty times.

Numerous conferences at the highest levels did nothing to convince us that there was an adequately sophisticated comprehension of the difficulties confronting us. Take General Groves, for example. The whole idea of the Beta process seemed to elude him, at least at first. When he had heard a long discussion of process factors in which we concluded that the maximum loss per cycle could be no more than a few hundredths of 1 percent, he broke in to chide us for being so concerned about loss. Uranium cost only a few dollars a pound, he pointed out, so the ^{235}U, present at 1 part per 140, should cost only a few hundred dollars a pound. E. O. L. tactfully broke into the embarrassed silence to explain that if a bit of material had to be recycled some fifty times and as much as a few percent were lost each time, there might be a total loss of as much as 100 percent. If one did not make essentially perfect recovery, mountains would labor and produce molehills.

An important member of our group was our chemical engineer,

Robert Q. Boyer. Unconventional and highly intelligent, he conceived and brought into operation an electrochemical procedure using rotating nickel electrodes to change the oxidized uranium salts to a reduced state as the tetrachloride (UCl_4). It was essential that the uranium be reduced for the continuous process ideally suited for rapid recycling. We also experimented with tantalum "bayonet" heaters to develop flash evaporators that could withstand corrosion and quickly concentrate the big volumes of wash solutions for handling by the electrolytic reductors.

Another idea for synthesis of usable uranium compounds was the production of the pentachloride (UCl_5). The great inorganic chemist Charles A. Kraus and his group at Brown University had been working on details of its production, using the reaction of carbon tetrachloride with uranium oxide. The reaction required a high temperature under pressure in a so-called bomb tube. The product was further purified by vacuum distillation, being collected on a "cold finger" cooled by liquid air in the center of the distillation tube. Sam Weissman and Dave Lipkin in the Chemistry Department did these distillations. The head of our analytical chemistry section, Clarence Larson, determined the composition of the final product. When he analyzed the putative pure uranium pentachloride, he kept getting ratios of chloride to uranium that approached 6.0, rather than 5.0. At first we did not believe him, pointing out that the chemists at Brown would have noticed such a discrepancy. Nevertheless, we soon found that we had in fact discovered the hitherto unknown uranium compound containing six chloride atoms (UCl_6). (The only compound of hexavalent uranium we knew to be reported was the hexafluoride (UF_6) which was to play so important a role in the successful diffusion process developed at Columbia University for making the explosive ^{235}U samples.)

A few weeks after we had written a report describing this discovery, I happened to be in E. O. L.'s office reviewing some project material I had been given permission to read. The evils of compartmentation became more evident when I found that a group of British chemists had reported the synthesis of UCl_6 a year earlier. UCl_6 might have been a better starting material than UCl_4 because it volatilized around 120° C, whereas UCl_4 required temperatures over 400° C. The advantage in designing a machine to use UCl_6 rather than UCl_4 appeared obvious, inasmuch as it would involve much lower power dissipation and fewer corrosion problems. Moreover, on heating UCl_6 as a source material, there would be less likelihood of depositing sludge in the form of UCl_4 on the slits defining the ion beams, a problem when using UCl_4. However by the time we had unearthed UCl_6, the design of the Alpha machines had been frozen using UCl_4. It may be debatable whether the use of UCl_6 would have eventually simplified the design problem and separation operations, but it did seem to us chemists that use of a readily volatile material could only be an improvement.

Our efforts with UCl_5 led to a memorable adventure with Kraus, a legendary figure in American chemistry, along with his famous contemporaries, Albert A. Noyes and Gilbert N. Lewis. (It will be recalled that at my Ph.D. oral in Chicago I had been quizzed on the general nature of the work Kraus and his students had done at Brown University.)

Kraus came to Berkeley to hear our story about UCl_6. The first evening there, he invited Jim Carter and me to his room at the Durant Hotel, where we observed at first hand his remarkable ability to ingest great quantities of alcohol without visible effect. The aftermath left us unable to be of much help in a session scheduled later that evening with Bob Thornton to plan a detailed Beta procedure. An attempt to impress Kraus with the hospitality and alcoholic resources of Berkeley, first at a cocktail party presided over by Pan Jenkins and later at Trader Vic's famed rum-dispensing emporium in Emeryville was largely unsuccessful. He characterized the supposedly lethal concoctions at Trader Vic's as so much "fermented fruit juice." Standing on the steps of the car in the train on which he departed for the East the same night, he invited Jim and me to experience "real party atmosphere" at Brown University.

This we did on a trip to troubleshoot planning difficulties with our engineer associates at the Stone and Webster Company offices in New York. We continued to Providence where we spent an uproarious afternoon and evening as the guests of Kraus's lieutenant Paul Cross, a younger and more robust version of Kraus. I barely caught the night train to New York at 2 A.M., and stumbled into Grand Central Station later that morning much the worse for wear and under the impression that the trip had taken only a few minutes, rather than five hours.

Our major troubles came in attempting to achieve some rapport with our engineering associates at the Tennessee Eastman plant in Oak Ridge. E. O. L. thought that inasmuch as the company chemists were successful industrial engineers, they must know more than we did about the processes to be developed. There never had been anything like the Beta process in chemical engineering experience, however. For example, the basic requirement that all material had to be cycled continuously without loss meant that there could be no hold up. Nevertheless, when Jim Carter and I went on our troubleshooting expedition East, we found that the Tennessee Eastman engineers had totally ignored this feature of the process and had designed a conventional storage tank to hold Beta product for later processing. This would have required the Beta machine to labor for several months before a second cycle could be run. The catastrophic consequences of waiting for forty or fifty cycles were obvious. Production of suitably enriched material would have been delayed for years.

It is not clear whether the ^{235}U needed would ever have been supplied if it had been necessary for the Beta process to succeed. Fortunately, later development of a thermal diffusion process by Phil Abelson rendered the Beta process as originally conceived superfluous. The positive good I derived from my own contacts with the engineers was a better appreciation of what industrial chemistry was all about. I learned to respect the remarkable fortitude and courageous stoicism my engineering colleagues displayed when confronted with the orneriness of chemical systems.

9

Tragedy and Transition Again

(*1941–1945*)

"DON'T LOOK BACK. Something may be gaining on you." So said Leroy "Satchel" Paige, baseball player and folk philosopher. I should have heeded his advice in the seventh year of my Berkeley experience. But I was too busy looking forward. Intimations of the new biochemistry were reaching even our backward campus—especially the Rat House, where Sam and I were avidly reading the papers of Fritz Lipmann and Herman Kalckar.[1] We were beginning, however dimly, to sense something of the nature of the basic ideas of biochemistry, as opposed to the physical chemical systems with which we were familiar. We even considered ways of bringing Lipmann to Berkeley, but my approaches to the laboratory brass proved ineffectual. "We can't hire every Tom, Dick and Harry!" responded one of our highly placed administrators.

Reading the articles of Lipmann and Kalckar, we saw that the CO_2 fixation process might not work on the basis of a simple equilibrium, as we had been supposing, but probably involved a much more complex system in which many equilibria participated. Perhaps the "phosphorylation" processes Lipmann and Kalckar described in their reviews explained how carboxylation could occur against a free-energy deficit of many kilocalories.[2]

Ruben had also been brooding over the classic papers by Otto Warburg and W. Christian on the oxidation mechanisms in the transformation of phosphoglyceraldehyde to phosphoglyceric acid,[3] and of Lipmann on the production of acetate and CO_2 from pyruvate.[4] By analogy with the mechanisms presented by Warburg and Christian, he suggested that adenosine triphosphate yielded a phosphate to an aldehyde, which then reacted with CO_2 in a reductive process to form a keto acid and free phosphate. The further reduction of this keto acid to carbohydrate was pictured as requiring another phosphorylation of the keto acid, and its subsequent hydrogenation to the free hydroxy-carboxylic acid with liberation of inorganic phosphate. Sam published his scheme later that year.[5]

I had reservations about these proposals because I could not see how light energy could be used to produce energy-rich phosphate—that is, how light energy could be converted to chemical energy as phosphate bonds. Nor could I easily reconcile the negative results we had observed attempting to find radioactivity in hexose phosphates with these suggestions. If phosphorylated intermediates were involved we would have expected that they would be in equilibrium with their product hexose phosphates, so that the latter would have been radioactive. I also had doubts that an aldehyde was the acceptor of the phosphate for the initiation of CO_2 fixation, even though Sam had found some indirect evidence supporting this notion from experiments on cyanide inhibition of the dark uptake of CO_2.[6] The eventual significance of these speculations for me was that they started me to thinking about the general relation between phosphate metabolism and biochemical energy storage—the central theme of my subsequent research career for several decades.

Meanwhile, the pleasures of life on Buena Vista Hill in the Maybeck complex discouraged dark broodings or forebodings. Ben Maybeck, the great old man himself, and his peppery little wife, Annie, lived in a little cottage up the road behind the "studio" in which we had taken up our residence. Often he could be seen striding down the street like some white-bearded Old Testament patriarch, swathed in a toga and wearing a red beret at a rakish angle, followed by Annie, whose role was to protect him from such petty annoyances as might arise in the affairs of the considerable Maybeck estate.

During the winter months, it was a delicious feeling to burrow snugly in the cottage and listen to the wild crashing of the torrential rains outside. Even catastrophes could be fun, as when the lashings of the wind would blow down power lines, creating electrical shorts and melting connectors. One night we even heard the hissing of molten copper dropping on the roof (which was wholly fireproof and leak-proof thanks to the inspired invention of a pink "bubble stone" Maybeck and an associate had contrived out of gunny sacks dipped in cement, made porous by blowing air through the mix).

Early spring saw the onset of a balmy ambience, redolent with the odor of eucalyptus and smoke from wood fires. Until summer came, we were all treated to a uniformly mild, steady fluctuation of temperature in the range optimal for human life. Summers were occasionally hot and humid for a few days, but nothing like the hellish weather I had endured for weeks on end back in the Midwest. Fall brought a new sparkle, with a repetition of the spring atmosphere. It was ecstasy to be alive and living in the Berkeley hills, where the only challenges to survival were the parties that broke out frequently as urges to neighborly hospitality became overwhelming. Nor was there any noticeable diminution in activity on the musical front in Berkeley or across the Bay.

I was wholly unprepared when Esther suddenly confronted me one night with the announcement that she was leaving to take up residence with a former classmate from Bakersfield, a woman who had come to San Francisco to work in the office where Esther had found employment. She was adamant that our marriage had come unraveled and could not be revived. I had noticed, it is true, that she had of late seemed withdrawn and distraught. She did not say it directly, but she implied strongly that I was more interested in my work and outside activities than what went on at home. I sensed that she had a case, but was too hurt and angry to react coherently.

After Esther left, I became depressed and sought companionship wherever I could find it. Soon I fell in with a new and exciting group of leftist intellectuals and bons vivants, clustered around a couple called Ralph and Bonnie Gundlach. In search of a more liberal milieu, less cramping to his political beliefs, Ralph had left the University of Washington in Seattle, where he had been an associate professor of psychology. With Bonnie, a teacher and practitioner of modern dance, he had soon attracted a group among whom was J. R. ("Ric") Skahen, who was in the Quartermaster Corps, enrolled in the Army premedical program. Ric, whom I was to know for many years as a devoted friend, moved in to the cottage to provide companionship and share living expenses.

As the war progressed, sympathies formerly monopolized by the

Spanish Loyalist cause were extended to Russia. I went to many meetings through my associations with the Gundlachs. But I shied away from political affiliations, just as I had done in Chicago. Because of my skill as a musician, I was asked to appear at fund-raising benefits sponsored by organizations such as the Anti-Fascist Refugee League and the Soviet-America Friendship Association (if that is the correct title). At one of these, a cocktail party at the home of the Sterns to celebrate Isaac's return from an USO tour, I was introduced to the Russian consul and vice-consul. Isaac mentioned I was on the staff at the Rad Lab. On hearing this, the vice-consul, one Gregory Kheifetz, inquired if I knew Dr. John Lawrence. I replied that I saw him practically every day. Kheifetz then told me that he had been instructed to communicate with John about the possibility of using radioactive phosphorus (^{32}P) to treat a leukemic official at the Russian consulate in Seattle, but had been unable to get through the university switchboard. There was some urgency as the request had come from a Dr. Burdenko, who held a post in Russia equivalent to that of the surgeon general. Burdenko had seen an abstract in the medical literature describing the work of John Lawrence and his associates in using radioactive materials, particularly ^{32}P, to treat leukemia and other blood disorders.

I volunteered to tell John about the Russian request and did so the next day. John seemed happy to hear of it, but indicated there was some question about supplying the ^{32}P until he had a complete case history. Later, Kheifetz phoned to say that he had heard from John Lawrence and thanked me for my intercession. Then, as he was scheduled to leave shortly for Russia and wanted to show his gratitude in some concrete fashion, he suggested I have dinner with him and his successor, a Mr. Kasperov, in the next few days, whenever convenient for me. I accepted and met the two for a relaxed and convivial two-hour dinner in San Francisco at Bernstein's Fish Grotto, then a well-known restaurant.

I remained quite unaware of the fantastic dossier my ever-watchful followers from G-2 were compiling on me as a "security risk," replete with innuendos of moral turpitude and leftist associations. Very likely, I caused the G-2 agents as much concern as Oppie did. (Indeed, my association with Oppie was doing me no good either, as will become clear.)

Some time late in the summer of that fateful year of 1943, I got a bright idea. It occurred to me that some of the tremendous quantities of radioactive carbon monoxide (^{11}CO) I was making for the phosgene projects could be used to support a proposal to work on the metabolism of phosgene, about which little or nothing was known. Some ^{14}C was also available from the old stainless steel cans that E. O. L. had ordered removed from the 60-inch cyclotron. I suggested to Sam Ruben that we resume our work together on such a project and even divert some of the radioactive carbon produced to take up the search for the primary product in photosynthesis again.

Sam took fire at the idea and eagerly began planning ways to conclude his war project as soon as possible. This was a study on dissipation of phosgene on the beaches of Marin County, commissioned by the Army with a view to immobilizing beach defenses in a possible invasion in the European theater. Every week Sam and his associates drove to Marin County with tanks of phosgene, which they exploded on the beach. They would then measure dispersal of the toxic gas and obtain data on how long it might take a gas barrage to dissipate. It was backbreaking, tedious, and wholly uninspiring work, from which Sam wholeheartedly wished to be liberated. He told me that Professor Latimer had assigned the project to him because it had not been moving fast enough with the previous group to which it had been given.

One Friday noon, sometime in October of 1943, I met Sam at the Faculty Club. He looked wan, gray, and tired. His right arm was in a sling. Alarmed, I asked what had happened. He explained that he and his group had been working around the clock to finish the beach project. He had not slept for several days. Driving back from one of the sessions in Marin, he had fallen asleep at the wheel, run off the road, and broken his right hand. The student working with him, sitting in the seat next to him, had lost some teeth, but otherwise there had been no great harm done. I urged him to slow down, go home, take the weekend off, and get some rest. He agreed this would be a sensible thing to do.

I was never to see him again. The following Monday, I heard he was in the hospital. He had returned to the lab that morning, still weak from lack of sufficient rest, and had attempted to transfer some phosgene to his vacuum line. This operation required that a glass vial containing liquid phosgene under pressure, obtained from the Bureau of Standards, be immersed in liquid air to freeze out the gas. Then the tip could be filed and broken off, and the opened vial connected with a rubber tube to the vacuum system. A cold trap was present to freeze out the phosgene gas diffusing into the line as the vial was warmed to room temperature. Although forced to work awkwardly with his left hand, Sam characteristically insisted on performing this operation himself, rather than letting one of his understudies do it. While his associate stood watching, he plunged the vial into the liquid air. There may have been a slight crack in the glass. (I heard that the Bureau of Standards was still using soft glass for the vials of phosgene it was sending out.) In any case, the glass cracked on being cooled suddenly and some of the warm, lethal liquid splashed into the liquid air. It immediately sprayed into the room in fine droplets. Sam, standing directly over the liquid air flask, must have ingested some of them.

Acutely aware that he might have received a lethal dosage, he did not panic but walked slowly to a chair, sat down, and called the hospital. He knew that he had to be immobilized completely if there were to be recovery. Perhaps only a sublethal dose had been ingested. A hospital crew hurried over and carried him to an observation ward. Tragically, nothing could be done. The poison took its usual course. A massive edema developed overnight, and Sam was dead by the next day.

So perished a most gifted young scientist—a victim of his own compulsive drive, amplified by the pressures of the struggle for academic tenure. The accident with its fatal outcome was totally numbing and overwhelming. The funeral was attended by a large crowd, including Latimer among other faculty. Some wept openly. It was one of the few times I saw Sam's family. There was no adequate insurance coverage for his widow and three small children—one still a babe in arms. It was found that Sam had neglected to fill out the necessary forms given project personnel, so no federal insurance compensation could be obtained. I understand that after considerable effort on the part of Lawrence and others, the university was prevailed upon to make up some of the difference between what his family would have received as compensation and what they got. It was hardly adequate for Helena Ruben, who was left with the heavy task of bringing up her fatherless family.

After a few days, I recovered sufficiently to go to our old laboratory and gather up the research protocols and data Sam and I had labored so hard to produce, and which we had expected to use when the time came to resume our photosynthesis research. I felt strongly that the best way to honor his memory was to assure that the research we had both planned would continue.

As the year ended, my divorce became final. The work on the Beta process proceeded, but I continued to plan research into photosynthesis, and carried on a correspondence with James Franck about some points of dispute he had raised with Sam concerning the mechanism of cyanide inhibition. Most of my time was still taken up with implementing the details of the Beta process, however. Time passed more or less uneventfully, until suddenly one day early in July 1944, I was summoned into Don Cooksey's office and confronted with an order to leave the project immediately. Don, looking gray and devastated, wordlessly handed me the typed notice. As I read it unbelievingly, he kept repeating how he had cautioned me not to "talk so much in the club." I firmly believed I had not transgressed as far as conversations at the Faculty Club were concerned and could make no sense out of such drastic action. In utter dismay, my world a shambles, I stumbled out of Don's office and the Rad Lab. I was never to return. It would be years before I would learn some of the history behind the Army's action. At the time, I had no clues whatsoever and certainly got none from Don.

Somehow, I reached home, where Ric Skahen found me a few hours later, sitting in mute despair by the fireplace. I was unable to tell him what had happened, but he seemed to sense it. Eventually, when I had recovered sufficiently to mumble something about having been fired, he asked me if I had heard from E. O. L. I replied that the whole sordid business had apparently been dumped on Don Cooksey. Ric then went into a tirade against Lawrence, whom he charged with disloyalty for allowing the Army to have its way after all my years of service and accomplishments. He asserted that E. O. L. could have intervened, but obviously had chosen not to do so. I had to admit that it was strange that I had heard nothing from E. O. L. As Ric talked on, my despair first changed to fury then to a dull apathy. I was beginning to realize that my career prospects, so promising and so bright only a few hours before, had vanished. Without explanation or apparent cause, my future was a black nothingness.

News of my sudden departure from the project spread rapidly and aroused much angry comment among the colleagues who had known me for so long. I heard that protests surfaced not only from a few of my friends, who even braved official disapproval in coming to express their sentiments, but also from so prestigious a person as Professor Mark L. Oliphant, who had been a member of Rutherford's original group at Cambridge and was on assignment to the Rad Lab as a liaison officer with the British effort. The whole incident was viewed as but another example of the irreconcilable antagonism between G-2 and the scientists. Nevertheless, with the war still on, work had to go forward. My departure was not a critical loss. In time, as the project administration knew well to expect, the whole uproar blew itself out.

I realized that I could not stay in limbo. With no prospects and no source of income, I still had commitments, not only to feed myself and attempt some kind of comeback, but also to continue helping my father, who depended on the meager sums I had started to send him. I felt a deep sense of shame about what had happened and a complete inability to explain the misfortune that had befallen me to him or to my sister.

I bestirred myself to start looking for a job. Many industrial laboratories in the Bay Area had approached me at various times with offers of employment as a research chemist skilled in tracer methodology. The first application I made was to Dr. Otto Beeck, the director of a research organization at the Shell Development Company in Emeryville. Beeck was most enthusiastic, but only a few days after he had assured me that there was a place in his laboratory for me, he sent a letter informing me that he had been mistaken. The same pattern of initial acceptance and rejection occurred each time I tried to find a job. I learned later that G-2 had put me on a master list of unemployables for the information of possible employers.[7] The same futility dogged efforts of my friends in the Life Sciences faculty at the university to obtain some sort of appointment there, where my presence and skills were earnestly desired. The research I had helped implement with such colleagues as Mike Doudoroff, Zev Hassid, and particularly H. A. ("Nook") Barker had lain in abeyance while I worked on the project, and could be reactivated. But, Dean Lipmann informed me, the university could not hire me while the war effort continued. Strangely, the Draft Board made no move to enroll me in the military.

With my resources depleted, I was in desperate straits. Joe Miner, a neighbor working in the Kaiser shipyards in Richmond, suggested I try there. He thought I might find employment as a test inspector, and that was the position I finally obtained. As both Joe and I worked the day shift, I could ride to and from work with him. We would leave before dawn and return in the late afternoon.

A whole book could be written about life in the shipyards during the war. My job as a test inspector required checking the specifications, installment, and operation of ship gear. This work, which involved clambering over and climbing under boilers, generators, winches, refrigerators, and so on, was dirty and tedious, sometimes even hazardous. Accidents at the yards were numerous and the clinic was constantly jammed. As a member of the yard elite, I was entitled to wear a soft hat, but I soon learned this was an invitation to sudden annihilation, as workers took delight in dropping wrenches and other heavy objects from great heights onto the craniums of those luckless types foolish enough to flaunt their status in this way.

The underlying motive for maintenance of a separate inspector staff by the yards was the need for protection against the arbitrary, and sometimes capricious, actions of the inspectors brought in by the Navy, the Maritime Commission and the Coast Guard. The only group whose word was final was that of the inspectors assigned by the Maritime Commission. Most of those were devoid of real knowledge about ship machinery, but highly sensitive to fancied insults or reflections on their competence. Yard representatives, such as myself, had to exercise great tact in disputing their findings. The pronouncement of one such worthy could hold up approval of a million dollars' worth of ship. On one memorable occasion, this happened over a particularly trivial matter. The winch motor installed for a davit assembly was provided by the Westinghouse Company rather than General Electric, as specified. Although it was identical in every respect to the G.E. model, the inspector refused to clear the vessel for sea trials. Agonized complaints by our chief inspector finally reached his opposite number on the Maritime Commission. An order was issued from his office to his subordinate to pass the ship. The underling refused, hotly declaring: "I ain't frontin' fer no monkey in no front office!"

Indeed, although the Kaiser Company achieved miracles in shipbuilding, fashioning complicated structures using essentially illiterate help, there were plenty of instances in which the Maritime Commission inspectors were justified in disapproving the finished product. Sometimes the mistakes were hilarious. I recall one instance where a high-pressure steam line was connected by error to a privy. The thought of what would happen when the first hapless character to sit there pulled the chain kept everyone in stitches. Sometimes, mistakes in construction surfaced in more serious ways. One ship went to sea on its first trial and was found to have an unbalanced propeller. I found out about it when I was thrown from a bunk in which I was catching up on my sleep before starting on my rounds. The vibration set up in the ship appeared certain to tear it apart, and we returned to the yard at quarter speed and tied up to a dock, where concrete blocks and steel chain were trucked in all night to load down the forward hull so that the stern could be tilted up above the water line and the propeller shaft assembly removed. This tedious and dangerous procedure was decided upon, I heard, because the only dry dock in the harbor belonged to the Moore Shipbuilding Company, a rival concern that would have charged an exorbitant fee.

The propeller was reworked, remounted, and the shaft assembly replaced. Another night passed as the cement blocks and steel chain were unloaded from the holds. The ship, relieved of the dangerous strain on its midline, stopped its ominous creaking and resumed its normal tilt. Again we cleared the Golden Gate at dawn and sailed out toward the Farallon Islands. I lay in the bunk waiting for the test to start. Again the engines revved up to full speed, a terrible vibration set in, and I found myself on the floor. The whole procedure of loading and dismantling at the dock had to be repeated. On the third trial, there was still some vibration but everyone was too weary to worry further and the ship was judged adequate for duty. Sea trials were a welcome diversion from the tedium of work in the yards. Everyone looked forward eagerly to these tests and hoped to be picked for the trial crews.

I was aware that I was still being watched while I worked at the yards. For example, I went on a holiday to visit friends in Santa Cruz and noticed a character in a gray tweed suit, whom I had seen before, trying to look unconcerned as he followed me there and back. The phone still emitted squeaks and whistles. At times, friends and acquaintances would inform me that they had been questioned. But I had ceased caring. With the resilience of youth, I was beginning to think I might still get back into research. I wrote letters to colleagues I thought might help, such as Don Yost, Linus Pauling, and James Franck. I learned later that some of them had queried E. O. L., who had urged them to do what they could. They were unsuccessful in finding me employment, however.

At a very low point during the depression that held me in thrall, I visited the FBI in San Francisco, hoping to get some clues as to why I had fallen into such disgrace. I went to the FBI office and was hospitably received by the agent there. He listened to my story, apparently sympathetically, and then volunteered the information that all security in the "project over there" was the monopoly of Army intelligence. The FBI was not involved, he asserted. He could not give me any help but did invite me to have lunch with him. In retrospect I cannot imagine what I expected from this interview. Obviously, the FBI were not going to tell me anything, but would record whatever I told them, for transmittal back to G-2 (although one might have some reservations about this in the light of disclosures in later years about the friction between the various agencies of law enforcement).

A few days after my fruitless visit to the FBI, I was with a party at Spenger's Restaurant at the foot of University Avenue in Berkeley when Bob Boyer, the engineer in the group Jim Carter and I had headed at the Rad Lab, rushed over to greet me and inquire how I was doing. I was glad to see him, if somewhat apprehensive that he might be doing himself no good by consorting with me so openly. But Bob was not one to worry about surveillance. He asked me if I had any notions as to why I had been fired. When I said I had none, he volunteered that there were rumors that my troubles had started because of a "dinner I had with some Russians." I was startled to hear this, because my meeting with the Russians had seemed quite innocuous.

His remarks started me thinking, however. I remembered just before the fateful interview with Cooksey attending the cocktail party at the Stern's. I suddenly recalled, too, that I had seen my faithful followers from G-2 on the occasion of the dinner at the Fish Grotto, but I recalled nothing that had been said that could be imagined to have broken project security. Still, there was the coincidence that my dismissal had followed almost immediately.

Later in the year of my disgrace, Nook Barker suggested that even though I could not obtain an appointment at the university, there might be no objection to my working in his laboratory in my spare time. He was anxious to resume some of the research on CO_2 fixation and utilization we had started before the war. I grasped eagerly at this opportunity, although it was not clear when I would get time to sleep. I would be working with Barker every night after an exhausting full shift at the yards. We arranged that I come to Barker's laboratory in the Life Sciences building every night and measure samples he had prepared in various experiments during the day. Protocols and problems were discussed when we met on my arrival late in the afternoon. He could obtain some ^{14}C-carbonate from Tom Norris, one of Sam's former graduate students still in the Chemistry Department. In this way, we began the first experiments in biology using ^{14}C.

A major constraint was the small quantity of ^{14}C available, no more than a few microcuries in toto.[8] We therefore decided to select problems that required minimal amounts, preferably for experiments in which the tracer material could be recovered sufficiently undiluted to reutilize. An ideal problem presented itself in a strange fermentation in which a certain strictly anaerobic microorganism metabolized glucose. This bacterium, *Clostridium thermoaceticum*, produced acetic acid from sugars and thrived at high temperatures, as implied by its name. It had been isolated by some workers at the University of Wisconsin, who found it in the population of thermophilic bacteria that inhabit composts and manure piles. In such stagnant environments, with no circulation of air or dissipation of the heat produced by the fermentations of the bacteria, internal temperatures could sometimes rise sufficiently to cause spontaneous combustion. The bacteria had to adapt to perform under such conditions. The fermentation of *C. thermoaceticum* was not only atypical in that it required relatively high temperatures (around 60°C), but also in that it produced no CO_2, only acetic acid, when it broke down glucose. In the usual fermentations such as encountered in yeasts making alcohol and CO_2 from sugars, equal molar amounts of two-carbon compounds, such as ethanol or acetate, and one-carbon compounds, such as CO_2 or formate, were produced. In *C. thermoaceticum*, the one-carbon fragments were missing.

Based on Barker's previous work on such fermenters, a possible explanation was that one-carbon fragments were initially produced, as in a normal fermentation, but then recombined to make additional two-carbon compounds. For the particular case of *C. thermoaceticum*, one could write a two-step process in which one molecule of glucose (a six-carbon compound) was fermented to two molecules of acetic acid (the two-carbon compound) and two molecules of CO_2 (the one-carbon compound) which with the addition of hydrogen joined to make another molecule of acetic acid.[9]

When the experiments were performed, our predictions were found to be correct, in that both the methyl and carboxyl carbons of the acetic acid produced were approximately equally labeled. Moreover, our calculated results agreed with the prediction based on the simplest assumption, namely that two molecules of glucose-produced CO_2 were used to make an extra molecule of acetic acid. This experiment—the first done in a biological system using ^{14}C as tracer—was the forerunner of a whole field of subsequent research by Harland Wood and his associates, not only confirming that CO_2 could supply both carbons in acetate production but also uncovering a new class of biocatalysts—the cobalt-binding corrinoids, related to vitamin B12.[10]

In the next few months, with the collaboration of two of his graduate students, B. T. Bornstein and V. Haas, Barker and I extended our research to an examination of the mechanisms for production of long-chain fatty acids from acetate, results of which we published.[11] A problem of another sort also engaged our attention. This was the anomalous enrichment of atmospheric oxygen in ^{18}O, compared with oxygen in water—the so-called "Dole effect." From the geochemical evidence available and the experiments Sam and I had done using ^{18}O-enriched water and CO_2, it seemed clear that all atmospheric oxygen should have originated from the photosynthetic splitting of water rather than from CO_2. Yet the oxygen of the atmosphere was known to have an isotopic content of ^{18}O more like that of carbonates than of ocean water. There was a possibility that in the experiments we had done, a rapid exchange reaction between CO_2 and water could have taken place inside the chloroplasts, and thus rendered unsure our assumption that the slow exchange we saw outside the cells was the same as that inside them. We examined all the data available on exchange reactions and attempted to rationalize the measured ^{18}O content of natural water and oxygen. After making allowances for all the possible sources of isotopic enrichment, we were still left with a significantly larger ^{18}O content in atmospheric oxygen than in water, which we could not explain.[12]

In the meanwhile, Malcolm Dole and Glenn Jenks had communicated to us the results they had obtained (then in press) using a procedure that seemed unambiguously to support the conclusion that oxygen originated from water.[13] They allowed *Chlorella* suspensions to photosynthesize in ordinary water equilibrated enzymically with carbonate. Using the minute differences in natural ^{18}O content between water and CO_2 as their assays, they again found that the ^{18}O of the water more nearly approximated that of the photosynthetically produced oxygen than did that of the carbonate. They got the same result with two land plants (sunflower and coleus). The Dole effect thus remained anomalous. Barker wrote to Harold Urey to solicit his views in the matter and received a reply stating that he, too, had no explanation for the Dole effect. From this letter and conversations I later had with Urey, it appeared likely that exchange reactions induced by ultraviolet radiations in the upper atmosphere could account for the isotopic anomaly, as originally suggested by Dole.[14] The photosynthetic origin of oxygen could be considered well established.

Other investigators confirmed our original experiment.[15] An especially searching experiment using isolated chloroplasts producing oxygen in the artificial system based on the "Hill reaction," in which the chloroplasts evolved oxygen from water in the presence of a "Hill oxidant," yielded the same result as had been seen with whole plants. In this experiment, internal isotopic exchange was ruled out as a mechanism for invalidating the results previously reported.[16]

The major significance for me of this effort by H. A. Barker and myself to rationalize the Dole effect was to bring my name to the attention of that remarkable scholar G. Evelyn Hutchinson, who was to provide me with a bridge to my new career as it came to develop a few years later at St. Louis. I should mention that in the article on the Dole effect with Barker, I appeared as senior author. Working with him I experienced none of the aggression that I had felt at the other end of the campus. The contrast in attitude between Barker, W. Z. Hassid, M. Doudoroff, and R. Y. Stanier on the one hand and many of the younger faculty in Chemistry on the other was painfully apparent. I doubt that I would have been able to renew my scientific career so expeditiously during the years of my disgrace if I had not been rescued by Barker and given such material and spiritual support by him and his colleagues in Life Sciences.

During this time, I made every effort to keep away from old haunts on the east side of the campus as well as from old friends there, to avoid embarrassing them as well as myself, not to mention inviting suspicion by G-2. One day, however, I happened to need something at the Rat House and made one of my infrequent trips there. To my extreme discomfiture, I ran into E. O. L. He seemed embarrassed, too, but inquired what I might be doing. I curtly mentioned that I was carrying on some research with ^{14}C on some problems initiated with friends in the Life Sciences. He expressed pro forma interest, but quickly got around to urging that I make every effort to find employment away from the Bay Area. I said that I would be happy to have his help and any suggestions he might make, but that I had not been able to get a job anywhere except in the shipyards. Our brief encounter ended with him turning away with the remark that he would continue to see what could be done.

In the months following my dismissal from the project, the initial shock began to wear off. I busied myself in the work at the shipyards and at Barker's lab, and continued to enjoy the warm friendship of loyal supporters from the biological faculties, musical circles all over the East Bay, and good-hearted people in general. I even resumed attempts to improve my understanding of women. In fact, life was not all that unpleasant. The extent to which my friends had rallied to my defense had given me the needed assurance to begin looking to the future and the prospect of better days again. The feeling of panic and hopelessness I had endured during the first few months of my exile from the Rad Lab was very much in abeyance by the middle of 1945.

It was then that I received a surprise visit from Robert Thornton, a mainstay of E. O. L's group. It was a surprise not only because I had the impression that no one from the project was allowed near me, let alone a person in Thornton's position, but because he came with an attractive offer of academic employment. He said that the Medical School at Washington University in St. Louis was anxious to acquire my services to take over supervision of a cyclotron and its use in medical and biological research. The cyclotron had been built by Bob and Alex Langsdorf in the period just prior to 1942. Bob had returned to Berkeley at E.O. L.'s request to work on the project and appeared unlikely to return to St. Louis. In any case, E. O. L. had supported the proposal warmly. I learned later that Arthur Compton, slated to take over as chancellor at Washington University, had phoned E. O. L. to ask his advice on my proposed appointment and had been told by Lawrence that there not only was no question that I had the necessary qualifications, but that he was confident that any investigation that might develop would disclose no grounds to doubt my loyalty as a U.S. citizen.[17]

The war in Europe was beginning to wind down. Although prospects for academic employment were likely to improve, I could see little reason to expect any offer more attractive than the one Thornton carried. I was to be given an appointment as associate professor on a tenure track and to monitor tracer research in one of the nation's outstanding biological and medical establishments. The salary was low (originally around half what I had been making at the peak of my career on the project) but adequate, and there was the prospect of steady increases.

I accordingly decided to accept the offer. By now it was late spring of 1945. The news that I was leaving Berkeley saddened my many friends, but everyone was happy that I seemed to be coming out of academic exile, and my departure brought with it an occasion I shall always remember and cherish. I had been invited to a party at the home of Tanya Uri, a well-known pianist. When I arrived, I found to my astonishment that a great farewell party for me was in progress. Dave Boyden and Tanya had arranged, unknown to me, that practically everyone who knew me be present. The party was planned as a complete surprise and much thought and effort had gone into making sure secrecy was maintained. Stands had been set up for a performance of Haydn's *Farewell* Symphony and a complete complement of players was present to play it. I had been told to bring my viola, which I had done, expecting only to perform some piano quartets or quintets. Now, I found myself playing with a great ensemble of friends. As the symphony progressed, with its gradual departure of players, stand by stand, I savored this truly unique farewell.

A few days later, I boarded the train for Los Angeles, where I was to pause for another farewell to California before heading on to my new career in St. Louis. A small group of friends saw me off. A particularly wrenching note was the news that President Roosevelt had just died. Not only was the timing symbolic of the end, for me, of the good old times, it was the close of an era for everyone.

In Los Angeles the Temiankas had arranged for me to be invited to a celebration of the liberation of Paris at the home of the Polish-born composer Aleksander Tansman, long a Parisian. We were met at the door by Tansman's wife and ushered into a room full of Hollywood greats. Among those present were Charles Laughton, Elsa Lanchester, Dudley Nichols, and a grande dame of the French stage, Mme. Pitoëff, who held court in the living room. The musicians included the conductor Georg Szell, the cellist Alex Compinsky, and Henri Temianka. During the evening, we played a newly written string trio by the host—a bit of a challenge, inasmuch as I had to read at sight a difficult and exposed viola part with the composer standing behind me and breathing down my neck. The performance was adequate, apparently, although I remember nothing about it now, except Szell's occasionally pounding out a note on the piano to correct someone's mistake he had spotted from the score he had in front of him. Later, Laughton gave a reading from Shakespeare's *Measure for Measure,* with copious asides by way of footnotes.

A few days after this, Isaac Stern and Alexander Zakin, his accompanist, came into town to play a concert in Whittier. They included a memorable tribute to the memory of the dead president, in the shape of an unprogrammed slow movement from a Brahms sonata. After the concert we all drove back escorting the motherly old lady who had managed the concert. Much to our disquiet, she drew from Isaac the promise that he would go to bed early and get proper rest. As soon as she was safely out of sight, Isaac grinned and became all business as he set out to organize an all-night party, as we had hoped. One of the group was sales manager on the West Coast for a big liquor company, and also knew how to get meat cuts and other delicatessen, although we had no ration stamps. The whole procedure was accomplished at a place called, inappropriately, the Ethical Pharmacy. Plentifully supplied with the necessities, we adjourned to Isaac's hotel room where a great party raged, including an unscheduled performance by Isaac of the unaccompanied Bach Chaconne.

With this great send-off, I left the next day for St. Louis.

10

Postscripts on Lawrence and Oppenheimer

My impressions of E. O. L. and Oppie remain vivid even after almost half a century. Products of wholly dissimilar backgrounds and cultures, a more unlikely pair of scientific colleagues can hardly be imagined. Nevertheless, as Oppie remarked, "Between us there was always the distance of different temperaments—but even so we were very close."[1] It was only under the pressures of the Cold War and the estrangements created by the controversies over the future course of nuclear controls that the basic differences between Lawrence and Oppenheimer surfaced.

Oppie—highly cerebral and introspective, by turns arrogant and charming—was continually plagued by a sense of insecurity. He possessed extraordinary analytic powers, but little manual ability. E. O. L.—less cerebral and highly intuitive—showed practically no self-doubt and remarkable mechanical skills. They shared a common drive to be center stage. With the theoretical acumen of the one complementing the experimental skills of the other, there was a basis for an intimacy that minimized the gulf separating them intellectually and culturally. Their strengths and weaknesses were amplified and exaggerated because they played key roles in the birth and development of the Nuclear Age.

When I first met him in 1936, E. O. L. was at his best as a charismatic leader and organizer of research. He created a level of morale among us in the Rad Lab hard to imagine as ever having existed elsewhere in a scientific research establishment, except possibly in the early days of Rutherford at the Cavendish. He accomplished this largely by personal demonstrations of experimental skills and sincere concern for his subordinates. He could play the cyclotron like an organ. The heady euphoria he created among us was wholly indescribable; it had to be experienced to be appreciated. He was our leader by virtue of a remarkable intuition about experiment, unquenchable conviction as to the course he was pursuing, and the ability to take charge at critical times. Although there were numerous provocations, he rarely lost control.

His self-assurance had disadvantages when extended to his perceptions of people. E. O. L. entertained few doubts about the rightness of the life pattern that he had been brought up to in semi-rural South Dakota. His forebears were hard-working, respectable Scandinavian settlers who preached the virtues of frugality and self-discipline. He was ill at ease among Bohemian types,[2] whom he was able to tolerate, but never accepted wholeheartedly as associates. He could make allowances for unconventional behavior in others as long as there were redeeming qualities, such as intellectual brilliance and research productivity. Perhaps his early intimacy with Oppie can be explained in this way.

Eventually, E. O. L. began to assume obligations and commitments that drew him away from the bench and from his home. There were needs to find financial support for his increasingly insatiate urges for expansion. He was almost always on the road by 1939. The early magic of the Rad Lab had largely vanished, and a new, less personal ambience, unpleasant in its demands for aggrandizement, was paving the way for the emergence of "Big Science."[3] It would be no exaggeration to credit E. O. L. with its invention.

The changes these developments wrought are most evident in the comments of early retainers who felt they had somehow lost contact with E. O. L. Many left; a few of those who remained drew closer to him. Don Cooksey, wholly devoted from the beginning, became the essential buffer between E. O. L. and the daily affairs of the laboratory, fielding complaints and expediting whatever course E. O. L. might project. Luis Alvarez and Ed McMillan, later E. O. L.'s brother-in-law, moved into positions of leadership. As time went on, E. O. L.'s distrust of those whose lifestyles he disapproved of became more pronounced. As a practical matter, he had to be concerned that some of the Rad Lab personnel whom he had accepted as members of his staff in the carefree early days might prove liabilities in the future, particularly as conservatively minded administrators in government and industry would come to assume importance in managing projects.

From the beginning of his career, E. O. L. had always respected success and admired those who succeeded. He had a suspicion that those who failed to meet this criterion were basically flawed. This notion became dominant as the scale of operations at the laboratory expanded. It was not long before he found himself unable to empathize at all with "losers."

At home, he set standards for the conduct of his children that created problems for his remarkably patient and understanding wife, Molly. It seems clear that E. O. L.'s increasing later inflexibility in personal relations helped generate stresses that may have been a factor in his untimely death at the age of fifty-seven of ulcerative colitis. While the etiology of this malady is debatable, a strong psychosomatic component is plausible. Lawrence's family history shows a consistent pattern of longevity. His brother John, born on January 7, 1904, is still living and in excellent health at the time of writing. Both parents were long-lived. The enormous pressures on E. O. L. were largely self-generated. He was never to experience defeat, as did Oppie, and died at the peak of his career.

There was much camaraderie in those happy times in the thirties. At lab picnics Lawrence liked to join in rough beach sports, such as touch football games, which often took on a slight touch of mayhem. There was no restraint in the treatment he received when he carried the ball, and he accepted this with a total lack of resentment.

Once he conceived a passion for skiing, which he thought all of us in the lab should take up. It would be good for us to get out in the open air and enjoy the liberating effects of exercise, he declared. He was not alone in this belief. Berkeley had its share of ski enthusiasts. Every winter the buses and trains took the maddened faithful to the mountains and soon enough brought many of them back on litters, broken and crippled in varying degrees.

So it was with E. O. L. After one weekend in the Sierra Nevada, he appeared on crutches, having sprained a knee. His enthusiasm had not noticeably abated, however. The next weekend he was back on the slopes, but proceeded to pull a ligament in the other knee. Completely crippled, he made his way painfully on crutches through the Rad Lab the following day. He hurt everywhere, and he showed it. His face was lined, his chest sunken, and his whole manner uncharacteristically one of utter dejection.

Taking his stand behind Lorenzo Emo, who was operating the cyclotron, he watched the proceedings moodily. It happened that we were in the throes of "outgassing" the cyclotron, which had just been reassembled and put back in working order. The first phase in its rehabilitation required intermittent application of power until no gas was driven out of the metal surfaces to destroy the operating vacuum. The operator was required to push the power button gingerly, watch the vacuum meter, and quickly turn it off as the pressure built up. Lorenzo was not particularly conscientious and, to E. O. L., was maddeningly dilatory in his reactions. Soon Lawrence could not contain his impatience. He lunged over Lorenzo's shoulder to turn off the power, arousing all the dormant devils of pain in his joints. It was one of the few times I was to see E. O. L. blow up. He read the riot act to the shaken Lorenzo and ordered him from the control desk. Then he angrily hobbled out of the room. The next day there appeared a small notice on the bulletin board advertising a pair of skis "for sale cheap."

I do not know how sorely he was tried by the demand that I be banished from the project in 1944. He was out of town, and Don Cooksey, who had a genuine affection for me, had the onus of carrying out the Army's orders. When E. O. L. returned, he made no effort to explain to me what had happened or to see me.

In later years, I learned that Army intelligence had been insistent that I be removed from any contact with the project. Some officers apparently even entertained the suggestion that I be done away with "by accident if connections were not severed" and perhaps Lawrence was worried that I might come to harm. There is little point in such speculations, however. In the final analysis, he saw me as a liability in furthering his future projects. He could not know that in accepting Army demands that I be dismissed from the project, he helped sentence me to a decade of harassment and lost opportunities for career advancement.

At the beginning of the project, E. O. L. made considerable trouble for security officials, holding that first priority should go to getting the job done,[4] but with time his attitude changed. The crushing responsibilities of his later years hardened his attitudes about people and increased his suspicions of their motivations, a process debilitating for him. His embittered reactions in the Oppenheimer matter are an example.

The tortured and complicated history of Oppie has not escaped the notice of myth makers as a model for a modern Passion Play. The ingredients are all present—the urge to make a new world, inspire fervent disciples, struggle with inner scruples and doubts, experience debasement and humiliation after heady triumphs, undergo Golgotha and crucifixion, and be reborn as the victim of reactionary fanatics. It is clear there is no need for more improvisations on the record, but rather for clarification of events that have inspired the literature on Oppie.

I had much in common with Oppenheimer. We came from very similar cultural backgrounds, and we were both firstborn sons of Jewish families, with all the pressures for achievement that implied. We both had received an intensive education in classics and humanist literature. I was thus drawn to Oppie from the moment I met him, and I could appreciate his strengths, as well as his failings. I had them, too, to some degree, though I did not presume to think of myself as his intellectual equal. There were few who could.

Oppie was not omniscient, however, strive though he might to give that impression. His omnivorous curiosity and drive to master whatever subject drifted into his ken caused him to spread himself too thinly and left him with occasional embarrassing gaps in understanding. I discovered this in two areas where I had some expertise—music and chemistry. In the former, I recall one incident in which Oppie was heard to dismiss Franz Schubert as "second rate." It was clear that he could not have heard Schubert's great works, such as the later piano sonatas, the cello quartet (Opus 163), or the G-major quartet (Opus 161). In fact, he admitted he had not when I pressed him on this point. As for chemistry, I was amazed to learn some years later that Oppie had majored in chemistry at Harvard. He certainly showed little insight into chemistry when the matter of the identification of ^{14}C arose. Dogmatically asserting his own preconceptions about the stability of ^{14}C, he expressed strong doubts that Ruben and I had discovered a radioactive isotope of carbon. He saw no basis for the long half-life we were observing. In the then primitive state of nuclear theory, there was no accounting for the extraordinary slow rate of radioactive decay exhibited by ^{14}C, and Oppie was stubbornly reluctant to admit that the theory might be wrong. The chemical facts impressed him not at all.

These were minor failings, however, when one considered Oppie's deductive abilities in the area in which he had chosen to specialize—quantum mechanics and electrodynamics. His performance in these abstruse and difficult realms of physical inquiry could be overwhelming and frequently earned him the admiration of his colleagues, who excused his occasional fits of arrogance and dogmatic assertiveness.

The confusing twists and turns that characterized Oppie's dealings in his private life were much less easy to rationalize. They were eventually exploited by political opportunists to discredit and crucify him, along with numerous others—victims of his bad luck, atrociously poor sense of timing, and impulsive reactions. A single case study in which I became entangled through Oppie's initiative is illustrative.

Late in 1938, I was sent to represent the Rad Lab at a symposium honoring Harold Urey. The meeting was held at Ohio State University in Columbus. Among those attending, there was a participant with brochures and other material supportive of attempts to create an organization in the United States modeled after the British Association of Scientific Workers, the seminal figure in which was J. D. Bernal, the eminent x-ray crystallographer. Bernal's writings were the Bible of the BASW, whose guiding assertion was that society dictated the course of scientific research. It sought to improve salaries and work conditions for scientists by trade union methods, based on a Marxist analysis of the relation between science and society. A number of American scientists, some of considerable stature, such as Arthur H. Compton, subscribed to its broad humanist aspects. My own reactions, the product of growing cynicism following my experiences in Chicago, were negative. I felt that the situation in America was just the opposite of that in Britain, in that there was no strong tendency of American scientists to affiliate with Marxist idealists. Moreover, U.S. scientists—particularly chemists—were largely employed in industry. The majority of them were uninfluenced by whatever liberal opinions might be held by their colleagues in academia. Regardless of the merits claimed for such an organization, I saw no practical basis for efforts to form it. Nevertheless, I agreed to carry back the organizational material sent by the nascent ASW to Oppie and Edward Tolman, who were the most active among leading scientists on the Berkeley campus in encouraging liberal ideas.

On returning to Berkeley, I found Oppie in a depressed mood. A California version of the House Un-American Activities Committee had sprung up, chaired as I recall by State Assemblyman Sam Yorty, and was exciting concern in the university community. It had come down especially hard on Oppie and Tolman, who had been active in helping teachers organize a union. Oppie felt that his participation in any attempt to help the ASW might be its "kiss of death." The situation was not quite as desperate as he seemed to think. It did not appear to me that he was wholly justified in taking the stand that the case for the ASW should not even be heard. A few weeks later, an eminent British scientist visiting the campus happened to express the same sentiments at a lunch Oppie gave in his honor, and was astonished to hear Oppie vigorously rationalize his stand against efforts to create a campus branch of the ASW.

Just two years later Oppie found himself in sympathy with a trade union attempt to organize scientific personnel for purposes of collective bargaining. Reversing his previous stand, he had become enthusiastic about the case for an affiliate of the CIO—the Federation of Architects, Chemists and Technicians (FAECT)—in its campaign to organize a branch at the Emeryville factory of the Shell Development Company. A group of dissident employees with doctoral degrees were put off by the personnel policies of the company. They alleged preference in promotions and advancement was given to British and Dutch nationals, and were receptive to attempts to organize the employees at all levels of technical expertise, from dishwashers to professional chemists, as a FAECT union. This attempt was opposed by another group of professionals, who objected to being lumped together with unskilled technical help and preferred a company-sponsored union. The two sides were locked in a struggle to win a National Labor Relations Board election for recognition as sole bargaining agents. The FAECT group had come to Oppie, who was attracted to their idea that a university chapter of the ASW be formed to support their case with the NLRB. A representative of the opposing group had also visited the campus and presented their arguments.

I was surprised to hear Oppie extol the FAECT stand and urge Al Marshak and myself to join in a meeting at his home to hear the case for the FAECT. I reminded him that this represented a complete turnaround from the opinion he had voiced when I had showed him the ASW material a few years previously. However, he argued the situation had changed since 1938 and was now more favorable for such organizational activity. This made no sense at all. We were already engaged in war work at the Radiation Laboratory and subject to strict security regulations regarding our employment there. Any attempt to penetrate the lab by an outside organization, such as the CIO, would certainly be resisted by the authorities.

His enthusiasm was not to be denied, however. Not too willingly, Al and I attended the party at Oppie's Eagle Hill house. A small group of physicists from the laboratory and the department were present, cozily distributed in an informal circle on the floor, and the two representatives of the FAECT were presented by Oppie with an introductory benediction. We heard an eloquent description of the situation at the Shell Development Company and arguments for the desirability of university support. Oppie asked us all to air our views in turn about the proposal that there be a campus branch of the ASW to bolster the FAECT effort at the Shell Development Company. When my turn came, I embarrassedly asked if permission had been obtained from E. O. L. for people such as Al and myself to be approached on this matter. Oppie said he had not done so, but would immediately. The effect, however, was to cast a damper over the previously warm group feelings, and the party broke up in some disarray.

The next day, Oppie came to see me at the lab, visibly upset and worried. He had been shocked to find that Lawrence was most vociferous in his disapproval of what he had done. E. O. L. demanded to know which of his staff had attended the meeting and why he had not been informed. Oppie denied that anything subversive was involved and demurred about identifying anyone. Instead, he said, he would inform those involved and leave it to them to see E. O. L. if they wished. This irritated E. O. L. even more, and it seemed to imply that he was not in control.

Oppie had placed me (and Al) in an impossible position. If I went to E. O. L., he would have doubts about me, no matter how I protested my innocence. If I did not, there was the chance that he would learn that I had been there anyway, and I would incur an additional onus. Much worse was the fact that I henceforth became an object of suspicion to Army intelligence, which, I learned later, had monitored the meeting and knew who had been present. I had no alternative but to go to E. O. L.'s office and tell him my side of the story. It was ironic that I, who had opposed the whole idea of the ASW from the beginning, now appeared a proponent. In the course of our discussion, I mentioned that people such as Arthur Compton belonged to the ASW, whereupon E. O. L. said that he did not doubt that the organization was well-intentioned, but that communists, because of their single-mindedness and ruthless organizational ability, eventually took over such idealistic enterprises. He urged me to "get out," whereupon I somewhat hotly asserted that I had never been "in." He repeated that I should get out, and added that Oppie had given him much trouble in the past with his fuzzy-minded efforts to do good. In fact, he had experienced much difficulty in getting Oppie cleared for the project, notwithstanding that his talents were badly needed. I left with the definite impression that E. O. L. still had reservations about my reliability.

In later testimony, Oppie showed his unfortunate, but by no means unique, tendency to remember selectively when he remarked that, "some time after we moved to Eagle Hill, possibly in the autumn of 1941, a group of people came to my house one afternoon to discuss whether or not it would be a good idea to set up a branch of the Association of Scientific Workers. We concluded negatively, and I know my own views were negative."[5] This testimony is astonishingly at variance with my recollection. Again, in a letter to E. O. L. dated November 12, 1941, he stated: "I had hoped to see you further before you left but will write this to assure you that there will be no further difficulties at any time with the A.S.W. . . . I doubt very much whether anyone will want to start at this time an organization which could in any way embarrass, divide or interfere with the work we have in hand. I have not yet spoken to everyone involved but all those to whom I have spoken agree with us, so you can forget it."[6] This gives a totally erroneous impression. Taken together with the testimony cited, it would lead one to think that he was an outspoken opponent of the ASW. In fact, he oscillated, being against it at the wrong time and for it later, at a most inopportune time. Certainly, there were some of us who were not allowed to "forget it."

This incident well illustrates how Oppie unwittingly caused so much trouble for others. In all such debacles, Oppie had nothing but the best of intentions and the poorest of judgments. This combination resulted in behavior that was later interpreted by those who desired to do so, or else who could not see it as rational, as evidence of a subtle and malicious nature. This kind of inference is well exemplified in the testimony of Wendell Latimer, who held that Oppie had exercised enormous influence over all those with whom he came in contact on the project, an influence Latimer regarded as insidious and mysterious.[7]

In the years after the dropping of the first atomic bomb, Oppie felt that he had earned a place in the Establishment, but I doubt he was ever accepted in a real sense. There were always those who mistrusted his origins, and intellectuals in general, and who were ready to pull him down. The higher he rose, the more they resented him. In the end, they succeeded, writing one of the darkest pages in American history.

I count it my good luck to have known both E. O. L. and Oppie. On balance, the gains outweighed the losses in my case. I owe the start of my career to E. O. L.'s generosity and some of my most happy and unforgettable cultural and social experiences to Oppie's friendship.

11

A Career Builds Under a Cloud

(1945–1948)

THERE COULD not have been a better place or time for me to begin a new career than Washington University and St. Louis in the spring of 1945. The liberal traditions of the city, nurtured by refugees from the harsh repressions of political reaction in Europe in the mid-nineteenth century, were carried on by the Pulitzers in the columns of the *Post-Dispatch,* and by the perhaps even more progressive *Star-Times.* The morning *Globe-Democrat* could hardly be termed reactionary either, even though its stance was to the right of the other papers. I was thus provided with a haven against the tide of cold war hysteria rising in the country, and particularly in Washington, D.C., as political expediency evoked latent anti-intellectualism.

I was quickly made aware, however, that the cloud under which I had arrived in St. Louis had not dissipated. The FBI, which had inherited the security files of the Manhattan Project, was busily shadowing me and my new associates. I was to remain an object of suspicion to its agents, subject to invasions of privacy in the form of telephone taps and inspection of mail, as well as grillings and other harassment.

Despite this, there was every prospect of a brilliant career for me in isotopic tracer methodology, a tool crucial in developments that would eventually usher in the era of molecular biology, bringing with it understanding of the molecular basis of genetics and elucidation of metabolic pathways in living systems. At Washington University there was a medical school faculty of scientific stature unsurpassed anywhere in the country. The roster included Carl and Gerty Cori in Biochemistry (soon to be Nobel laureates), Oliver Lowry in Pharmacology, the Nobelist Joseph Erlanger and George Bishop in Physiology, Carl Moore in Hematology, Edmund Cowdry in Anatomy, and Willard Allen in Obstetrics and Gynecology, to mention a few. With some of these I would eventually enjoy collaborations. The junior staff boasted such as Alfred Hershey (later a Nobel laureate) and Sol Spiegelman in Bacteriology, a department presided over by the forward-looking and paternal Jacques Bronfenbrenner, and Mildred Cohn, just arriving in Biochemistry, with her gifted physicist husband Henry Primakoff. On the general campus, there were islands of strength in Physics, Zoology, Botany and—soon to come—in Chemistry. Led by Joe Kennedy, whom I had known at Berkeley, a remarkable collection of brilliant young investigators, including Dave Lipkin, Lindsay Helmholtz, Art Wahl, and—especially happily for me—Sam Weissman, my old companion from Chicago, was coming from Los Alamos.

When I arrived I was taken in hand by Arthur Hughes, chairman of the Physics Department, a warm and genial Welshman whom I knew as an international authority on x-rays. The details of my recruitment had been handled by Hughes, who had the major responsibility for continuation of cyclotron activity as it converted to peacetime operation from its crucially important role as the major producer of plutonium in the Manhattan Project. The cyclotron program was under the aegis of a committee that included the avuncular Sherwood Moore, head of Radiology and its associated Edward Mallinckrodt Institute, from which the funds to construct the St. Louis cyclotron had originally come.

It seemed I was to be a Moses who would guide the great faculty of the university into a Promised Land.[1] My presence and expertise were represented as the keys to continuing the institute's program of research into cancer and allied diseases. In addition I was expected to give lectures on techniques in tracer chemistry.

Reality crowded close on the heels of expectation. I found several clinical research teams eagerly awaiting me. One of these, headed by a junior faculty member of the Department of Medicine, Dr. Edward Reinhard, and ably assisted by young clinician in Radiology, Dr. Louis Hempelmann, was embarking on an extensive experimental program to evaluate the efficacy of radioisotopes, particularly radiophosphorus (^{32}P), in the treatment of leukemias and polycythemia vera. This research had been pioneered by John Lawrence and his associates at Berkeley, where I had been the chemist for the program and had prepared all the needed tracer materials. I was thus familiar with the assay procedures and supply problems. A major difficulty was the standardization of assay procedures so that dosages of radioactivity could be accurately controlled and comparisons made with results obtained at other institutions doing similar clinical research, such as Berkeley and MIT. We had used an electroscope method at Berkeley, which was originally calibrated against Bureau of Standards samples of uranium salts. This failed to agree by a factor of two with determinations made at MIT by Dr. W. Peacock, who was using a procedure based on measurements with a Geiger counter. The importance of resolving this discrepancy dictated giving this relatively uninspiring, routine work the highest priority among all the projects crowding in on me for consideration. There resulted a year of intense effort cross-checking samples. At the same time, I maintained a correspondence with Peacock and with Dr. Cornelius Tobias who had taken charge of the biophysical aspects of the Berkeley program. Eventually, the problem was resolved, and I published a short note on the assay of ^{32}P as applied to clinical research. It was an addendum to a massive article by Ed Reinhard and his associates definitively describing the advantages and limitations of this form of radiotherapy.[2]

Coincidentally, I also collaborated with a group of doctors at Professor Cowdry's Barnard Skin and Cancer Hospital, led by Dr. C. J. Costello, in which we studied ^{32}P uptake in mouse skin treated with methylcholanthrene to induce carcinogenesis.[3] I found such collaborations fascinating, as I had an opportunity to watch surgical procedures and thus get a much better notion of the problems that arose in applying tracer methods clinically. I also had the good fortune to fall in with Dr. Albert Lansing, who was studying calcium metabolism in those fascinating little animals, the rotifers. I learned much about cellular differentiation and later became a co-author of a paper describing our collaborative research.[4]

A project for the study of cancer metabolism utilizing ^{14}C was submitted early in 1946, and with much fanfare was allotted the first shipment of ^{14}C to be provided by the isotope division at Oak Ridge, an event duly celebrated in the local press. Everyone assumed that my presence on the research team as co-discoverer of ^{14}C was a major factor in the allotment.

Among the first assignments I received on arrival was a request to organize a course on tracer technology and radiochemistry for the senior faculty, including an introduction to nuclear physics. I thus found myself—a very junior staff member—lecturing to the illustrious department heads of the medical school. While this was going on, I also was busy writing an account of isotope effects relevant to the new science of "biogeochemistry," which I had promised G. Evelyn Hutchinson before leaving Berkeley. This assignment involved long hours of work in the library, fortunately air-conditioned. In the wake of Harold Urey's discovery of the heavy hydrogen isotope, deuterium (^2H), there had been an explosive development of research on distribution of isotopes in the biosphere. A noted Russian scientist, W. I. Vernadsky, had suggested that because there was a preferential selection of elements for use by living organisms, there might well be a similar phenomenon involving the isotopes of a single element.[5] Thus, for example, there might be isotopic differentiation for elements such as hydrogen, nitrogen, oxygen, carbon, and sulfur that would be greater than one might expect solely from purely physicochemical processes such as evaporation, diffusion, and absorption. I had to examine the existing data and determine whether such variations in isotope content could be correlated unambiguously with biological activity.

Early in the fall I sent Hutchinson a draft of my manuscript, in which I presented a survey of the current methodologies and data. I concluded that no certain grounds for Vernadsky's generalization yet existed, as most of the effects observed might be adequately explained without invoking biological intervention. I was delighted to receive a highly complimentary reply. The final version was published as the introductory chapter in a monograph on biogeochemistry.[6] In later years, as assay instrumentation and sophistication in methodology increased, the use of isotopes in biogeochemistry became a well-developed science. My essay, the first such effort in the field, was soon relegated to the status of a "classic," much of its content having become obsolete.

Another, and very basic, research collaboration arose from the interests of Dr. Carl Moore in blood diseases. Dr. Moise Grinstein, a research associate from Argentina and expert in porphyrin chemistry working with Dr. Moore, came to see me about the possibility of labeling the "heme" group in hemoglobin so that the metabolism of blood could be followed in normal and diseased patients suffering from various anemias. In particular, it was hoped to compare the average lifetimes of erythrocytes as influenced by disease conditions.

David Shemin and David Rittenberg at Columbia had published what would become a classic study on the lifetime of erythrocytes in a normal human (Shemin himself), using the amino acid glycine labeled in its amino group with ^{15}N. It was a particularly heroic effort on the part of Shemin, who had to eat enormous quantities of evil-tasting amino acid supplements. Administered as the amino acid in the diet, ^{15}N-labeled glycine soon gave rise to labeled heme in hemoglobin and acted as a reliable indicator of the synthesis of new blood cells, into which the heme was incorporated. The red cells, so labeled, remained in circulation for some 120 days, then lost activity as they were broken down and removed, the labeled heme being metabolized and its products excreted.[7]

Shemin and Rittenberg, using model reaction schemes known from organic chemistry reasoned that glycine labeled with ^{14}C either in its alpha carbon or in its carboxyl carbon would be an adequate tracer molecule for heme synthesis. As carboxyl-labeled glycine was commercially available, Grinstein and I decided to use it rather than go to the trouble of synthesizing the alpha carbon–labeled material. To our amazement, however, no activity appeared in hemoglobin when the carboxyl-labeled glycine was used.[8] The fact that the alpha carbon, not the whole molecule of glycine, was used was crucial for the eventual brilliant elucidation of heme synthesis by Shemin and his colleagues in later years. I began to see that nature often accomplished its syntheses in ways other than those that would occur to an organic chemist. The requirements of the living organism to accommodate syntheses with the complex integrative processes of cellular metabolism prohibited simple solutions. I remembered the warning Ibsen gave Peer Gynt, "Take the long way around."

Before my arrival at Washington University, Arthur Hughes had been asked to look into the possibility of constructing a mass spectrometer for assay of stable isotopes, so that the tracer program could be broadened to include them in addition to radioactive isotopes. There was available a magnet ideally suited for focusing ions through a 180° angle, the basic feature of the spectrometer tubes used by Dempster and his associates at the University of Chicago. To my delight, Harry Thode, an old friend from graduate days there who had spent some time with Harold Urey at Columbia on ^{15}N isotope enhancement and exchange reaction, had gone to McMaster University in Hamilton, Ontario, where he was also using a Dempster-type instrument. Another scheme exploiting a different principle, in which the tube was bent at a more acute angle, popularized by A. O. Nier at Minnesota, could also be considered. To investigate its potential, Hughes and I journeyed to Iowa State University at Ames, where Harland Wood and Lester Krampitz had built a Nier-type spectrometer that they had been using to trace reactions in bacteria, using ^{13}C. I thus began a lifelong friendship with these two investigators, who were later to make many seminal contributions to the understanding of intermediary metabolism.

In the course of troubleshooting to get our mass spectrometer to work, I wrote to Nier and also to Harry Thode, who had previously used a Dempster-type spectrometer at Urey's lab in Columbia University. I learned in the process that Thode had actually performed some experiments on the source of photosynthetic oxygen and had confirmed the earlier experiments with ^{18}O that we had done at Berkeley.

Al Schulke and Harry Huth, two knowledgeable cyclotroneers, handled the chore of cyclotron management and operation that supported the extensive program of isotopic tracer research at Washington University and elsewhere.[9] Paul Aebersold's isotope division at Oak Ridge, based on the neutron pile, soon took over the production of isotopic material, so that the Washington University cyclotron was gradually phased out as a major source of supply, although it maintained this position for another few years. Eventually, it was turned over completely to the Physics Department and was dedicated to physical research.

It was becoming clear to me, however, that with all the demands made on my time for collaborative research, there was no opportunity to continue along the lines Sam Ruben and I had begun in Berkeley. Even if there had been time, I lacked space, students, and support personnel, such as were available to competing groups springing up elsewhere, attracted by the rewards in prospect for successful isolation and characterization of the first product of CO_2 assimilation in photosynthesis. Such success seemed all the more certain because the long-lived carbon isotope, ^{14}C, was available in adequate supply from the isotope division at Oak Ridge. In particular, E. O. L. and Latimer back in Berkeley were determined that such research should continue in the Rad Lab and Department of Chemistry there. They had selected

Melvin Calvin to head a reconstructed group, which included Andy Benson, who could supply continuity with the work that had gone before as well as expertise, making up for the lack of the original protocols, which I had taken to St. Louis. At Chicago, in Franck's laboratory, another group led by the eminent microbiologist Hans Gaffron and by E. W. Fager, a fine organic chemist, was vigorously pursuing the same research.

In the years to come, these two groups engaged in a fierce competition, finally won by Berkeley, bringing with it eventually a Nobel award to Calvin. The secret of success for Calvin and Benson in their demonstration that ribulose diphosphate was the long-sought-for acceptor of CO_2 and phosphoglycerate the first product in photosynthesis was their adoption of paper chromatography as an identification and isolation technique when it was still not acceptable among more conventionally minded biochemists. Erwin Brand, in those years a leading authority on the amino acid composition of proteins, derided paper chromatography as "spots on paper." The pressures on Calvin and his group in achieving recognition of the correctness of their findings based on paper chromatography were extreme. Before they were finally successful, Calvin suffered two cardiac episodes. The bitterness engendered by the controversies during those years claimed Fager as a victim, too—he left chemistry to start a new career in ecology at the Scripps Institute of Oceanography.[10]

The highly competitive nature of such research provided me with yet another reason to strike out in new directions. I started thinking about aspects of metabolism associated with the new concepts of "group transfer" and "high-energy phosphate" put forward primarily by Fritz Lipmann, with whom I entered into a vigorous and fruitful correspondence. In this exchange, we discussed the original suggestions by Sam Ruben (see chapter 9 nn. 2, 5), which were closely similar in most respects to those made by Lipmann in a 1945 paper with Constance Tuttle. By the end of that year, I still had too little knowledge of phosphorus chemistry and metabolism to be productive in developing specific research in photosynthesis. I could see, however, that energy conversion would become the basic problem even if that of CO_2 assimilation were solved. I also realized that I needed to acquire a thorough understanding of organophosphate chemistry and, particularly, nucleotide biochemistry and metabolism.

As luck would have it, there was another colleague with the same need—Sol Spiegelman, who was about to take up a teaching appointment in bacteriology at the medical school, having just finished his doctorate. We both had backgrounds in mathematics and the physical sciences, and had similar ideas about the future course of biological research, as influenced by the need to develop the quantitative aspects of biology. Coming from the more rigorous pursuits of physical science, we were struck by the generally deplorable intellectual condition of graduate students in the biological sciences. It seemed to us that the academic disciplines, as organized in the universities and bolstered by the educational practices in secondary schools, selected against the recruitment of students in biology with a bent for logic and rigorous analytical thought. We were convinced it was time for a change. Joined by Burr Steinbach from the Zoology Department, we formulated a plan to improve the training of biologists. It was evident that there would be need for a new breed of biologists able to exploit the new physico-chemical techniques coming out of nuclear science, as well as the increasingly sophisticated analytical methods derived from molecular spectroscopy and chromatographic absorption analysis. In short, we wanted to train what would come to be called molecular biologists.

These ideas, at least a decade ahead of their time, were not greeted with enthusiasm by senior faculty in the medical school, who looked askance at proposals that would involve radical inroads on resources for preclinical instruction. On the campus, however, there was strong support, even to the extent of volunteering space and instructional facilities. A curriculum was proposed to implement the plan. There were to be three main sequences. One would provide a series of courses in chemistry and physics that would give students the solid background basic to an understanding of isotopic tracer techniques and the newer analytical procedures of various chromatographies, as well as a knowledge of theory and practice in all the molecular spectroscopies, ranging over the full electromagnetic spectrum from high-energy gamma rays through x-rays to the ultraviolet, visible, infrared, and microwave regions. The second sequence would deal with theoretical biology and statistical methods, so as to enable students to evaluate intelligently the statistical data often encountered in biological systems. Finally, there would be a thorough indoctrination in the methodology of genetics and genetic analyses of recombinant systems.

It was clear that such a curriculum would make great demands on both faculty and students and would be interdisciplinary, cutting across the established academic departments. The enthusiasm for the plan was great enough on the campus to support recruitment of special talent. The outstanding example was our success in bringing Franco Rasetti, one of the original Fermi group, to Washington University as a visiting professor. His interests were breathtakingly broad, embracing all aspects of science from nuclear physics to archeology and entomology. (I recall Emilio Segrè telling me many years later that he had learned a good deal of physics chasing butterflies with Rasetti in the Apennines.) Rasetti had refused to join the Manhattan Project and had sat out World War II doing neutron capture experiments. After the war, he was amazed to receive a letter from a Jesuit institution, Laval University in Quebec, offering him an appointment as professor of physics—the first such professorship there. As he remarked later, "I had never heard of them, but then they had never heard of me!" It appeared the university had, to its mystification, acquired a spectroscope, and had been informed by its only resident scientist, a chemist, that a physicist was needed to operate it. Rasetti accepted the offer and arrived to find that he could indulge another of his many interests—paleontology. Quebec, so I am told, sits on the greatest outcropping of Precambrian fossils on the North American continent. In a very short time, Rasetti had chopped enough rock to become one of the world's top experts on trilobites. It turned out later that the Ozarks south of St. Louis contained Precambrian strata second only to those in Canada, and Rasetti reasoned that whatever might happen to our program, he could still pursue his paleontological studies.

He informed us that he had made a major discovery in academia—the league of Jesuit universities. By planning his tenures as visiting professor, he could manage to be at a Northern Hemisphere institution such as Laval in the winter and at a Southern Hemisphere university, such as Tucuman in the Argentine, in the Northern Hemisphere summers. He thus had the opportunity to ski the year around.

While Rasetti was with us at Washington University, Sol Spiegelman and I were excited to hear that he had read the little book on integral equations by the famous Italian mathematician Vito Volterra, which had much significance to me because it provided a treatment of a problem analogous to one that arose in the use of tracers. Volterra, so Rasetti told us, had become interested in the fact, brought to his attention by his son-in-law, a marine biologist stationed in Naples, that Italian fishermen had experienced lean years immediately after the cessation of World War I. They had anticipated big catches because there had been so little fishing during the war years, but found instead that there was a fish shortage. A possible explanation emerged when Volterra examined the proposition that a certain rate of fishing was needed to get the biggest population of prey in the presence of predators. If there were no prey, the predator population fell to zero. If the numbers of prey increased, the predator population would follow suit. Eventually, the predators would eat so many prey that the numbers of prey would fall, followed by a diminution in the population of predators. A certain rate of fishing thus had to be maintained to get maximal catches. Volterra worked out a general treatment of the situation, which led to an equation quantitatively describing a precursor-product relationship identical to that between two compounds one of which was formed from the other in labeling experiments.

As might have been anticipated, no department in the medical school, where much of the faculty talent needed was to be found, would yield its autonomy or—even more important—access to its budget to support our proposals for training biologists. So the idea came to naught. Washington University thus lost its chance to steal a march on other universities in anticipating the rise of molecular biology.

Close cooperation between the campus and medical school was difficult to achieve in any case. The difficulties that could arise were illustrated in my own case soon after I arrived. I required permission to accept and supervise doctoral research of graduate students, because I was not a member of a department authorized to conduct graduate instruction. In the medical school, no department could reasonably accept me on the basis of my expertise and training. It took nearly two years before I received a compromise appointment as an associate professor of chemistry, whereby I spent half my time at the Mallinckrodt Institute, where Sherwood Moore was still willing to pay my whole salary ($5,000 per annum) and provide me with laboratory space, and the other half teaching on the campus in the Department of Chemistry.

This appointment did not provide the academic support I needed, because there were few graduate students in Chemistry interested in biological tracer research. Moreover, the distance between my laboratory and the campus (several miles across Forest Park) made attendance of students at courses on the campus that would be required of those I might attract difficult. I still had to rely on collaborative work with faculty at the medical school who could support graduate students and postdoctoral fellows.

My need to learn phosphate biochemistry while mounting a significant research program was nevertheless satisfied admirably by joining Sol Spiegelman who had an active group of graduate students and postdoctoral fellows. We decided to follow the ^{32}P content in various fractions of cellular phosphate in the presence of administered ^{32}P-labeled inorganic phosphate, hoping to detect synthesis of protein and associated enzymes during growth and differentiation, and correlations between flow of phosphate in and out of specific nucleotides or other examples of organic phosphate. We hoped thereby to obtain clues as to the nature of the molecular mechanisms and associated enzymic activities controlling these processes.[11]

Our program attracted the attention of the Committee on Growth of the National Research Council, and we received a grant of $20,000 in support of our research, including salaries and equipment purchases—a first for Sol and myself. This was funded by the American Cancer Society, which had decided to support fundamental research in the biological sciences, inspiring similar programs in the governmental research institutes that would eventually unite to become the Department of Health, Education and Welfare. Joined by the National Science Foundation, these research institutes were to become the major factor in the rise of American biochemistry and molecular biology to the commanding position it was to enjoy in the coming decades.

By the beginning of 1947, we had to conclude that we had no convincing correlations between any particular fraction of cellular nucleotide or organic polyphosphate and initiation of protein synthesis. However, when we examined phosphate turnover in the absence of cell growth and protein synthesis, we found that no flow of phosphate occurred out of the "nucleoprotein" fraction—that is, the ^{32}P-labeled phosphate remaining in the cell residue after successive extractions with water, cold trichloracetic acid, alcohol, and hot alcohol–ether mixtures—whereas this fraction lost radioactivity in significant amounts when protein synthesis began.

Starting with this correlation, and after much thinking about published data on cell growth and adaptation, as well as our own findings, we proposed a theory about mechanisms of heredity, which was published as a leading article in *Science*.[12] We described a process by which genes in the nucleus continually produced partial replicas of themselves as nucleoproteins that migrated into the cytoplasm. There they continued to replicate at varying rates, becoming the ultimate mediators of gene expression. We suggested that they competed with one another for sources of cell material, the final cellular products depending on the rate of replication and destruction of these replicates. If the nuclear genes produced them faster than they disappeared, classical Mendelian selection would result; otherwise there could be anomalies ("non-Mendelian" distribution of heritable traits) such as were being described under the general category of "cytoplasmic heredity." Such ideas were not novel, having been prefigured by such eminent geneticists as Sewall Wright at Chicago and C. D. Darlington in England. Wright had even coined the term "plasmagenes" to label such cytoplasmic replicates, which later came to be known as messenger RNA.[13]

Our article, with its emphasis on cytoplasmic factors in inheritance and the mechanisms of competitive interaction excited much comment. Sol had been invited to a symposium on quantitative biology at Cold Spring Harbor in the summer of 1946, where he gave a preliminary account of our research, and there was a more detailed presentation from us both at the symposium the following year.[14] *Time* picked up the suggestion about competition between plasmagenes in the *Science* article and earned us some notoriety by printing a dramatic account under the provocative title "Tempest in the Cells."[15] More publicity came our way in a series of articles in the *Post-Dispatch* by Evarts A. Graham, Jr., the son of the distinguished head of surgery at Washington University Medical School.

In all these exercises, I had come to admire Sol as a scientist and person. Our collaboration was like the one I had enjoyed with Sam Ruben in Berkeley. Again, hard circumstance decreed that it end before its potential was fully realized. A misunderstanding of major dimensions between Sol and some of the leading figures in the senior medical school faculty resulted in his departure.

The disposition of the grant we had received caused us some concern. I accordingly wrote to the executive secretary of the Committee on Growth, acquainting him with the unfortunate turn of affairs, and shortly received a generous commitment to allow Sol the transfer of equipment paid for by the grant to wherever he might go. Eventually, the eminent University of Minnesota physiologist H. O. Halvorsen took Sol under his wing as a special research associate, enabling him to continue a career that later earned him international recognition as a leader in molecular biology and cancer research.

The research undertaken with Sol's group had given me occasion to look into the chemistry of all the carbohydrates fermentable by yeasts, among which were the 4-carbon tetroses and their oxidation products, the C_4–saccharinic acids. These happened to be objects of interest to my old mentor, Professor J. Glattfeld, who it will be remembered, had striven so heroically to help me steer clear of science just fifteen years before, when I started my college career at Chicago. Now he received a letter from me asking for information on his favorite compounds and, if possible, authentic samples of them. It must have been quite a shock to him to find that his erstwhile charge had turned into an organic chemist, but he replied most cordially and sent some helpful reprints.

However, my transformation was more drastic than Glattfeld might have imagined. I was becoming a biochemist. The need to prepare enzymes and authentic samples of intermediates involved in glycolysis known to mediate fermentation of carbohydrates propelled me into the biochemical literature, where I clawed out the procedures advertised as effective in making such preparations. I began to pay my dues as a novice entering the arcane practice of enzymology. It should be remembered that no commercial sources could be relied upon for supply of such materials in those days.

First, I had to suppress some squeamishness about engaging in the butchery required in some procedures. K. Lohmann and P. Shuster, who had discovered ATP in the early thirties, presented a method that began with clubbing a rabbit senseless, then proceeding with haste to decapitate, skin, and eviscerate the carcass, followed by dissection of the muscle and chemical isolation using precipitation with heavy metal salts. After several such episodes, I became convinced I would rather be a microbiologist.

Another hurdle was the tendency of authors in the biochemical literature to be somewhat vague about, or even to omit, essential details. Possibly they thought such information was common knowledge. I soon learned that much intuition, as well as ordinary organic chemical expertise, was required. I recall one instance in which I attempted to make a particular phosphate ester. The procedure called for several kilos of yeast to be suspended in buffer and undergo breakdown by addition of toluene and incubation at 37°C. After two days of hard work, I ended up with a barely perceptible blue haze on the side of the beaker. I finally managed a reasonable yield after several more tries, although I cannot remember now what the secret of success was. Gradually, I acquired sufficient intuition and manual expertise to make me into a plausible facsimile of an enzymologist.

The opportunity to return to photosynthesis research with my newly achieved insights into enzymology and phosphate chemistry before the yeast work wound down came in a most unexpected way—I acquired my first graduate student in the person of Howard Gest, whom I had met during a visit to Oak Ridge to touch base with Waldo Cohn and other cronies. Waldo had organized a small chamber orchestra for my entertainment, and Howard had participated, playing the double bass most proficiently. Perhaps it was this contact that had started him thinking about graduate work under my aegis at Washington University. Originally a major in bacteriology at UCLA, Howard had spent two summers working with the formidable Max Delbrück at Cold Spring Harbor, where he had gone to follow up a newly developed interest in bacterial virology. He had followed Delbrück to Vanderbilt University to start graduate work there, but the war had intervened and he had spent the next three years on purifications and characterizations of uranium fission products. In 1945, at the relatively late age of twenty-four, he wrote to me indicating a desire to return to work in microbial chemistry. With his background in isotopy and bacteriology, he seemed a fine candidate for the work I had in mind. I received enthusiastic letters of recommendation from Harrison Davies and Charley Coryell and also a somewhat lukewarm endorsement from Delbrück. I learned that Howard and Max had had a falling out only a few months after Howard's arrival in Nashville; thereafter, Max noted, there was a "considerable deterioration" in the quality of Howard's work, which previously had been quite satisfactory. Howard had become quite rebellious, and Max ascribed this behavior to "emotional instability," but expressed the opinion that "under more secure circumstances and in new surroundings there would be a good chance he would do all right."[16] I interpreted these statements in Howard's favor. He looked very much like one of my kind and a young man I could get along with. This judgment was to prove correct through all the decades that followed.

When Howard arrived, I obtained a dispensation from the ever generous Jacques Bronfenbrenner to accept him in the Bacteriology Department. We were also able to obtain a fellowship for him from the

Committee on Growth. With the hurdles of financial support and academic placement cleared, I suggested that Howard first read Eugene Rabinowitch's monograph for a general background in photosynthesis and go to C. B. van Niel's laboratory at Pacific Grove the following summer to obtain a thorough indoctrination in handling microorganisms. Meanwhile, he began acquiring experience in analysis of phosphate composition of algae, working with the methodologies Sol and I had been developing in the yeast research.

Quite unexpectedly, a recruit to our program appeared in the person of Mary Belle Allen, who had left Berkeley after Sam Ruben's death and finished her doctoral thesis at Columbia University. Having obtained a National Research Council fellowship, she elected to spend a year with us. We decided on a modest project to see if we could use various inhibitors, such as cyanide and azide, to differentiate the dark fixation of CO_2 in photosynthesis from the unrelated reversible equilibria involving CO_2 in respiration. This work was a continuation of research Ruben and I had published some years before in an attempt clearly to distinguish fixation products of the photosynthetic system from those that might appear due to respiratory activity and complicate interpretations of labeled products found in algal CO_2 metabolism. As the year progressed, it became clear that the research was not likely to produce definitive results. Mary Belle left to take a position at Mt. Sinai Hospital in New York. Later, she distinguished herself as a member of the team Daniel Arnon created at Berkeley, which discovered algal and green plant photophosphorylation.

Meanwhile, our experiments on the algae *Chlorella* and *Scenedesmus* had indicated that phosphorus uptake in the light was much greater than in the dark. Mary Belle and I had found, using cyanide at low concentrations to inhibit photosynthesis, but not respiration, that this excess uptake was not related to respiration. Howard and I could not demonstrate any strict proportionality between phosphate uptake and overall metabolic acitivity, but using organisms, internally labeled by prior growth in ^{32}P-labeled phosphate media, we could show that light stimulated a flow of phosphate between acid soluble and acid insoluble fractions. A major nuisance in all this work was the gluttonous behavior of the algae in response to the customarily high concentrations of inorganic phosphate used in growth media. The cells absorbed whatever phosphate they were given, piling up excesses in intracellular inorganic phosphate and labile organophosphate esters.

These produced large enough fluxes of phosphate in and out of the cells to obscure the differentiations we were trying to establish in the various cellular fractions of organic phosphate.

Howard's attendance at van Niel's course in 1947 had generated in him an enthusiasm for bacterial photosynthesis and a conviction that the photosynthetic non-sulfur bacteria would be superior to algae as test objects for our research. He quickly convinced me, and we extended our observations to *Rhodospirillum rubrum,* the classic organism in this group of bacteria. The results were much the same as with algae. We found that for all microorganisms used, the gross phosphate distribution was altered by experimental conditions prior to analysis. In particular, the method of culture and washing procedures were critical in determining what eventually appeared in the various cell fractions. Organisms exposed to media with the high inorganic phosphate content recommended for optimal growth piled up excess labile phosphate easily removed by washing, whereas in lowered phosphate media no such washable phosphate was noted. I suggested, therefore, that Howard return to Pacific Grove in the summer of 1948, there to investigate conditions for growth in minimal phosphate.

In the summer of 1948, I had had yet another windfall, which considerably expedited recognition for me internationally as an authority on isotopy applied to biological research. The Hitler madness in Germany had produced a flood of great scientists to the United States along with many other intellectuals. Among them was Kurt Jacoby, who had guided the Akademische Verlag to eminence as a leading scientific press. Exiled to America, he had resolved to carry on in an English format as the Academic Press. Joining with Walter Johnson's enterprises in New York, Jacoby soon achieved a respectable output of quality books on various scientific subjects. He was quick to sense that the time had come for a definitive text on radioactive tracers in biology reflecting the explosive modern developments that had occurred since the earlier texts of Georg von Hevesy, Otto Hahn, and Fritz Paneth. His inquiries led to me, and I was receptive. I had a large accumulation of original research material from my experience in Berkeley, and it seemed that I would have time during the summer to write the book.

A symposium on heredity in microorganisms was scheduled for Cold Spring Harbor that summer, and Sol and I had been invited to attend. Picking up my material and bracing myself for an intense session of writing in the uncluttered atmosphere of the Carnegie Institution at Cold Spring Harbor, I set forth late in June. For the next two months, I closeted myself in a dank, dark room and wrote. I must have had the material at my fingertips, because I finished the whole text that summer. I worked with such intensity that I was rarely seen outside the building, and some alarm was expressed about my sanity, not to mention my health. One day, I was seen to emerge in a swim suit and stand blinking in the sun on the pier. The Delbrücks were rowing past at that moment and urged me to join them. We had gone about half a mile, with me lounging languidly in the boat when Max abruptly turned and ordered me into the water. Everyone deplored my avoidance of exercise and outdoor activity, for which the summer at the institute was so well suited, he announced, and he had determined to do something about it. Needless to say, I did not stir from my cell again without first carefully scouting the terrain.

On returning to St. Louis, I sent the manuscript to Jacoby, who submitted it to Coryell and Lionel Goldring at MIT for review and editing. In November, corrected galleys were sent to press and the book appeared the following May under the title *Radioactive Tracers in Biology*. In the interim, I had received a request from Louis Fieser to have it kick off a new series of monographs Fieser and his wife, Mary, were editing for Academic Press. I was flattered to have the request, as I was in some awe of the Fiesers, who occupied a most prestigious position in organic chemistry. The correspondence led to a new friendship and many sessions enjoying their hospitality in Cambridge and at their laboratories at Harvard University.

The appearance of the book created something of a sensation, quite to my surprise and to Jacoby's joy. It elicited a number of rave reviews and sold so many copies that the staff at the Academic Press referred to it as their *Gone with The Wind.* I was especially pleased to have letters from Georg von Hevesy and the Dutch microbiologist A. J. Kluyver, who praised it inordinately. Fritz Paneth wrote a particularly enthusiastic review.[17] Only a few years before I had as a beginner read Paneth's own classic treatise on radio elements as indicators.

I was already drifting away from the field of tracer research, however. I was not temperamentally disposed to keep abreast of new developments. I did make some effort to revise the text as need arose. Each new edition was essentially a new book, even to a change in title.[18] I found the task onerous and unrewarding in the next decade, finally ceasing the effort with the third edition late in the fifties, despite Jacoby's urgings that I continue.

Being a major figure in the newly fashionable field of tracer methodology demanded stamina. Seminars and symposia proliferated, and the number of experts to support them remained small. For some years, at least up to the early fifties, the same group of practitioners made their weary way from one meeting to another, surviving much concentrated hospitality.

At one such meeting, we formed a riotous group at a roadhouse many miles from Madison, which we did not leave until early morning, when we regrouped on the pier jutting into Lake Mendota and donated large quantities of nonprotein nitrogen to the limpid waters. They were probably already sufficiently polluted. Folke Skoog, plant physiologist at the University of Wisconsin, had in his lab literally hundreds of little bottles containing isolates of blue-green algal species, exhibiting all colors of the rainbow and representing the biggest collection of new species ever assembled. Lake Mendota experienced occasional overgrowths, or "blooms," of smelly, viscid blue-green algae, which were objectionable from a tourist view point. One of the regents of the university remembering his course in introductory biology, thought that these could be minimized or eliminated by lowering the phosphate content of the lake. He suggested that large quantities of iron salts be dumped into the lake to precipitate the phosphate. The lake had a number of areas of little or no circulation, into which iron was poured and the results awaited. Blooms erupted in such profusion as had never been seen before! The algae loved all that iron. Skoog, sampling the blooms, came up with numerous new strains.

The most memorable of these tracer symposia for me was one chaired by Harold Urey at the New York meeting of the American Chemical Society in September 1947. While at a symposium at Cold Spring Harbor on microbial inheritance, I decided to come into New York and visit my friends at the Columbia Medical Center—no small undertaking in the heat of August on the Long Island Rail Road. I arrived after a long and tiring journey that took most of the morning to find the halls and laboratories deserted. I stood in the hall brooding over my stupidity in not remembering that few of my confreres were likely to be in town in August, when I happened to notice an elderly gentleman passing.

The sudden thought he might be Erwin Brand, the famed Columbia biochemist, inspired me to chase after him and introduce myself. His reaction was choleric. "So you're Kamen—you're a bad guy!" he cried. While I stood dismayed, he inquired heatedly why I never answered my mail. I discovered that he and John Edsall had sent me an urgent message a month earlier inviting me to join a symposium on Preparation and Measurement of Isotopes, under Urey's aegis and under the auspices of the combined Inorganic and Physical Sections of the American Chemical Society. The letter required a quick answer, because I had only been invited at the last minute. It had included specific instructions on how to reach Edsall at his vacation cottage in New England. To Brand's intense annoyance, there had been no reply.

I explained that I had been at Cold Spring Harbor all summer and that no mail had been forwarded to me. Whereupon Brand remarked that now that he had me, I could sit down forthwith and write an abstract of my talk. Astonished at this turn of events, I inquired what the topic might be. "Identification of Intermediates: Criteria of Purity," he said and told me I would get no more than twenty minutes, including discussion, to present it. I asked how it happened that I had been assigned this topic. Brand replied with a grin—for the first time—that Urey had assured him and Edsall that I could talk on anything and that they should wait until they had all the speakers lined up before inviting me to speak on whatever was left. As I sat clawing out the abstract with Brand hovering over me, I ruminated sadly on the fix I had gotten into merely following an innocent impulse to see some friends. Brand relented a little and even became cordial, treating me to a lobster dinner later that summer.[19]

Symposia I sought to avoid were those dealing with developments in photosynthesis, as there was always the prospect of being caught in the middle between the warring groups led by Calvin and Gaffron. As might be expected, whenever one group unearthed results at variance with those of the other I was asked my opinion. The earliest such confrontation was caused by a claim from the Calvin-Benson group that succinate was the major CO_2 fixation product in early photosynthesis. This was in complete disagreement not only with what Sam Ruben and I had reported and Gaffron's group had found, but also with later findings in California. I wrote to Calvin suggesting that the appearance of label in succinate was an artifact caused by their conditioning of the algae, which they had subjected to many hours of dark anaerobiosis before exposure in light and air to the labeled CO_2. Such long periods under fermentative conditions caused appreciable lags in the onset of photosynthesis in a following period of illumination. In fact, the label followed the carbon pathway of fermentation, so that most of the tracer carbon could be expected to end up in carboxyl groups of succinate, and did. The fixation observed had nothing to do with photosynthetic dark fixation.

This was but one example of many at the time illustrating the pitfalls of tracer research in its infancy. The introduction of a new and revolutionary methodology is invariably accompanied by error stemming from the excitement of the novel until experience brings with it a more critical attitude. Until solid results take over, the more powerful the method, the more catastrophic the errors.

The California group did the job in the end. Their contribution to the solution of CO_2 fixation in photosynthesis was a milestone in tracer research. I was happy to cite it in my book, along with the biosynthesis of cholesterol by Konrad Bloch and the biosynthesis of heme by David Shemin, as a brilliant example of the power of tracer methodology.[20]

The frenetic and successful action in which I was swept up on the scientific front was equaled in music. A few months after my arrival, I was recruited as leader of the viola section in the St. Louis Philharmonic Orchestra, an organization of semiprofessionals and amateurs with an impressive tradition. Its concerts were attended by a vociferous and enthusiastic following, and as the new first viola, I found myself the featured subject of a story in the *Star-Times*.

Among the exciting personalities I met in St. Louis was Bateman Edwards, the head of the French Department at Washington University and one of the most accomplished musicians I have had the good fortune to know. Musicians of note who happened to be in town were usually to be found at his home. I recall one evening with Leonard Bernstein, then just at the beginning of his great career. Bernstein and Edwards raced through a four-handed piano rendition at sight of a new score that Bate had just found in the university music library, bawling out the vocal lines and pounding out the orchestral accompaniment for several hours. It was a long-neglected opera by Rossini, *La Cenerentola* (Cinderella), which thus received its first performance in St. Louis, and also probably its impromptu premiere in this country, before its ultimate revival at the New York City Opera some years later.

There was a never-ending social whirl centering around the newspaper crowd from the *Post-Dispatch* and the *Star-Times,* whereby I met and came to know not only the editorial staffs, including such luminaries as Ernest Kirschten, Tom Sherman, and Robert Blakely, but also that remarkable cartoonist and graphic commentator, Daniel

Fitzpatrick. His opposite number on the *Star-Times* was another gifted cartoonist, Dan Bishop. Then there were all the parties thrown by the faculty of the medical school and on campus. It was at a party given by Burr Steinbach, who was professor of zoology, that I saw my future wife Beka Doherty, a handsome and willowy redhead. Beka was then married to Arthur Hepner, a reporter on the *Post-Dispatch,* and was a journalist of repute in her own right. She was lounging on a couch, observing with wry amusement the antics of the party-goers milling around me (the guest of honor). With her, equally bemused, was Ethel "Ronnie" Ronzoni, a faculty biochemist and the wife of George Bishop, the eminent neurophysiologist. Soon I became a regular guest at both the Hepner and Bishop establishments, where I made many new friends among artists and humanists, notably H. W. and D. J. Janson, husband and wife, who were each to make their names in art history. I was delighted to find we all had a common bond—an allegiance to Isaac Stern, whose appearances in St. Louis were occasions for happy meetings of what we called the "Stern Marching Society."

From Berkeley days, there was Alex Langsdorf, who had returned to St. Louis and married a rising young artist, Martyl, the daughter of Martin and Aimee Schweig, whose home was the scene of many chamber music sessions. Martin and his son, Martin Schweig, Jr., presided over the operation of the most prestigious photography studio in St. Louis. Alex's father was the dean of the engineering school at the university and his mother a vital cog in the formidable force of women who had organized the League of Women Voters and, when occasion demanded, visited the mayor and his henchmen to keep them industrious and virtuous. Among these remarkable ladies were Edna Gellhorn and Tess Loeb.

Scientifically and socially, I had no cause for complaint. But there was an ominous counterpoint to these happy times in unceasing pressure from the FBI, still bent on unmasking me as what they perceived I might be—a foreign agent. I had neither privacy nor any assurance that something like the Berkeley fiasco might not happen again. Of course, the situation was rather different in St. Louis. Soon after his arrival Chancellor Compton had gone to the trouble of telling me that he had made a thorough investigation of my past and was convinced I was a loyal American. On all sides, I had new and powerful friends, who would be supportive in any showdown with the Feds. Still, I was apprehensive.

I was called downtown to undergo grillings by the FBI several times. A pair of agents would question me for several hours, checking my answers by reference to a thick book, which I took to be my dossier. No two agents were the same from one session to the next. It was evident that the FBI remained convinced that by persistent probing they could eventually uncover something incriminating. The dossier grew, and so did my apprehension. I had become a latter-day Damocles—a guest at a sumptuous banquet over whose head hung a sword held by a thin thread.

12

Times of Discovery and Dismay

(1948–1951)

As THE summer of 1948 drew near, the attachment between Beka Doherty and myself deepened. She had left St. Louis shortly after our first meeting at the Steinbach party, but I had often seen her on trips to New York. After her divorce from Arthur Hepner, she had taken on the demanding post of medical researcher at *Time*. Despite a complete lack of scientific training, she had quickly displayed a knack for analysis of research objectives and evaluation of newsworthy experimentation, coupled with amazing intuition about motivation in the subjects of her stories. She soon attained prominence as a science and medical journalist. Beka's reputation was such that she had been chosen by Bennett Cerf of Random House and the eminent pathologist C. P. Rhoads to collaborate with Rhoads on a book for the layman on cancer. Helping her occasionally with some of its technical aspects gave me added opportunities to see her.

On the scientific front, I was becoming increasingly fascinated by the phenomenon of bacterial photosynthesis. The fact that bacteria could grow using light energy in a fashion analogous to that of green plants and algae was a relatively recent discovery. The celebrated physiologist T. W. Engelmann (like myself a musician-turned-scientist) had demonstrated in the 1880s that certain bacteria showed a tendency under the microscope to migrate toward illuminated portions of the visual field (a phenomenon called "phototaxis"). By an ingenious arrangement of prism and light source (gas flame or sunlight), Engelmann found he could superimpose a "micro-spectrum" on the droplets in which the living bacteria swam, and he was able to demonstrate that they collected in those portions of the spectrum that coincided with the wavelength preferentially absorbed by their photosensitive pigments. This was the first demonstration of "action spectra," and a key observation in establishing the existence of bacterial photosynthesis.

I became convinced that research on bacterial photosynthesis might be productive of new insights into the general mechanism of conversion of light energy and its storage as cellular material. According to the "unitary hypothesis," largely developed by C. B. van Niel, all photosynthesis, plant or bacterial, began with the same general reaction, called "water splitting."[1] The absorption of photons by the photosensitizers—chlorophyll in green plants and algae, and its close analogues, the various "bacterial chlorophylls"—resulted in fission of the water molecule, HOH, to produce short-lived products that could be represented as H and OH. Electrons had to be pulled off the oxygen moiety of water along with an equivalent number of positively charged hydrogen ions, or protons, to maintain electrical neutrality. In other words, an electron-rich hydrogen atom (H) was produced that was a powerful reductant (donor of electrons), leaving the electron-deficient OH as a powerful oxidant (acceptor of electrons). The light energy captured in this fashion would be retained only if the recombination of these two products was prevented. They had to be separated and stabilized in less than a billionth of a second, or they would react with each other, producing only heat.

The important difference between bacterial and plant photosynthesis was that with bacteria no oxygen was evolved. Either the system for making molecular oxygen was missing, or the system for handling OH, called *y*, was different. Some means was needed to take care of *y*OH. This could be done by providing special hydrogen atom donors (reductants) from outside the bacterium to react by donating electrons (and protons) to the *y*OH. If the system for oxygen evolution were missing, it might be assumed that the bacteria harbored the precursors of oxygen in a sufficiently long-lived form to last and react with the relatively sluggish hydrogen (electron) donors that supported photosynthetic growth in bacteria. There was a great variety of reductants, ranging from simple inorganic compounds, such as H_2S, S, and other reduced sulfur compounds, to complex organic hydrogen donors, such as organic acids, alcohols, and so on.

It seemed to me that there was a good chance of revealing the nature of *y*OH in view of all these "handles" provided by the various systems for handling electrons and protons, many of which were already well worked out in nonphotosynthetic systems. Moreover, one might make extracts of the bacterial photoactive organelles, called chromatophores,[2] which could carry out a reaction analogous to that of plant chloroplasts discovered by Robin Hill.[3]

The prospect of wrapping up the two kinds of photosynthesis in one elegant package suggested experiments that might establish the fate of *y*OH. Moreover, the experiments Howard Gest and I had done on ^{32}P turnover in *R. rubrum* and the algae had shown that light absorption energy could be stored somehow in organic phosphorus compounds. Hence, ATP probably had a role to play, at least in the photoassimilation of CO_2. A first objective, then, was to find out what enzyme systems were present in the chromatophores that would support phosphorylation, working with light-mediated reactions.

Howard and I knew it was essential to establish defined nutrition conditions for photosynthetic growth in *R. rubrum* that would not complicate understanding the effect of light because of formation of extraneous phosphorus compounds not essential to growth. We had to devise growth media with completely defined chemical composition. They would be required to provide the carbon, nitrogen, and sulfur sources, as well as only essential amounts of inorganic phosphate. Also, there would need to be whatever trace nutrients were required for essential vitamins, and minerals such as iron and magnesium. This research could be expected to be tedious. Neither of us suspected that such apparently routine research would lead to a seminal discovery.

I wrote to van Niel asking that Howard be permitted to spend another summer at Pacific Grove in the Hopkins Marine Biology Laboratory to work on this project, with the possibility that I would join him there for a part of the time. (However dull the experimentation might prove, at least we both would escape the dread St. Louis summer.) Van Niel graciously consented to our occupying some of his precious space, even though it was greatly limited during the summer, when students flocked from everywhere to take his famous course.

At this point, I was lucky enough to recruit Jack M. Siegel, who, like Howard, had left the Manhattan Project laboratories at Oak Ridge to return to academia as a doctoral candidate. He came as a bona fide graduate student in chemistry, so there was no academic difficulty about his electing to work with me on a biochemical problem—reactions associated with metabolism of alcohols in bacterial photosynthesis. I had just the right project for him, or so I thought. A provocative demonstration of a simple removal of hydrogen from an organic compound coupled to CO_2 assimilation had been reported by Jackson Foster, one of our old group of collaborators back in the Berkeley days. Foster had noted that certain strains of bacteria related to *R. rubrum,* members of the group called non-sulfur purple bacteria (Athiorhodaceae) and classified in the genus *Rhodopseudomonas,* could convert the three-carbon alcohol, isopropanol, directly to acetone, with simultaneous assimilation of carbon dioxide.[4] In this process, two hydrogen atoms were removed from the central carbon of isopropanol. If substantiated, it would provide strong evidence for van Niel's general scheme for photosynthesis, even though it might only be a special case. (It was unlikely that the simple role of the organic compounds as mere hydrogen donors, and not also sources of carbon, could be generalized from this single case.) I suggested that Jack reinvestigate Foster's claim, using isotopic labeling techniques to see whether any carbon, or only hydrogen, came from the isopropanol. Jack obtained some of the original Foster strains from van Niel while taking the course at Pacific Grove and also applied enrichment culture techniques he had learned there to isolate other strains capable of photosynthetic growth in isopropanol.

I was still busy with my various collaborations at the medical school and had also taken on an interesting project suggested to me by L. S. Tsai, a visiting scientist at the Department of Chemistry on leave from Yenching University in Peking, China. Tsai was interested in a claim by Paul Nahinsky, of our original group in Berkeley, that the -CN group of nitriles (RCN) existed in an exchange equilibrium with free cyanide ions (CN^-). Tsai and I worked on this problem for several months and were unable to substantiate these observations. This was regrettable as, had they been true, a simple synthetic method would have been available for labeling the carboxyl group of organic acids by incubating their nitriles with labeled cyanide, separating the reactants, and hydrolyzing the nitrile to the acid.[5] Tsai left for China in late May 1948. Many years later, when China was reopened to Americans, I tried to find him, but without success.

As the summer approached, I was thus busy with meaningful and

• Relatives and friends on Chicago's South Side, sometime in 1921. My mother is standing, center. Seated immediately to her right is my father, who has arranged a remote control to set off a magnesium powder flash. My sister Lillian (aged two) is unconcernedly reaching for something on the table. I am the nine-year-old squinting in dread of the flash.

Don Cooksey took this snapshot of Franz N. D. Kurie outside Le Conte Hall, on the Berkeley campus of the University of California, early in 1937. Kurie is preparing to haul away a week's supply of cyclotron needs, easily contained in a child's wagon. Big Science had not yet come to the Rad Lab.

• The author posed to show the effects of a week in the wilds, photographed by Don Cooksey in the summer of 1938. This picture was subsequently passed around at a party at Di Biasi's restaurant in Albany to commemorate the discovery of ^{14}C and other notable events of 1940–41, and was signed on the back by many of those present.

De Biasi Party July 31, 1940

Paul C. Aebersold

Betty Helmholz

Chuck Cooksey

Ernest O. Lawrence

Blitz – Frieg

Donald Cooksey

"Gate jumper" Cornog

Harry Walker

Elizabeth Sloan

David Sloan

Louys Ems.

Milicent Sperry

Marjorie Shultt

John H. Lawrence

Whiskey Al

Freck

Kenneth MacKenzie

Molly Lawrence

John A. Harrie

Helen Harrie

Elizabeth Cooksey

Carl Helmholz

Cornelius Tobias

Margaret Lewis

Bob Wilson

Jane Scherger

Milton Tucker

Bob & Joan Livingston

John + Helen Barkus

Jules & Phyllis Halpern

Señor + Senora Luis Alvarez

Ernie and Tebby Lyman

Wm Foley 3rd

Elna & Winfield Salisbury

Edwin M. McMillan

Flower

Bill Brobeck

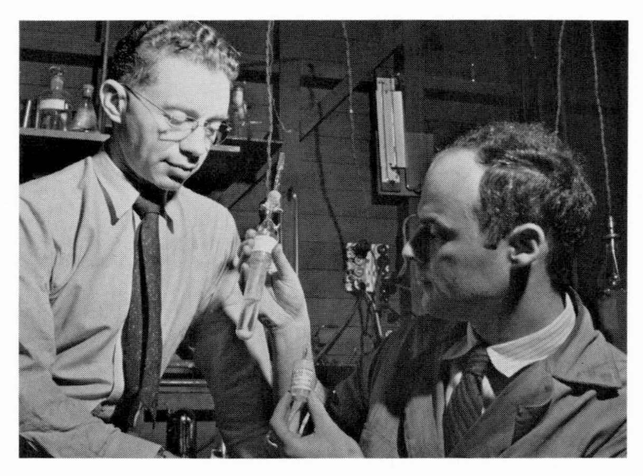

• Sam Ruben, seated, and Zev Hassid, standing, in a pose meant to illustrate an operation performed in one of our experiments on CO_2 uptake in photosynthesis. Photograph by Hansel Mieth, *LIFE Magazine*, © Time, Inc.

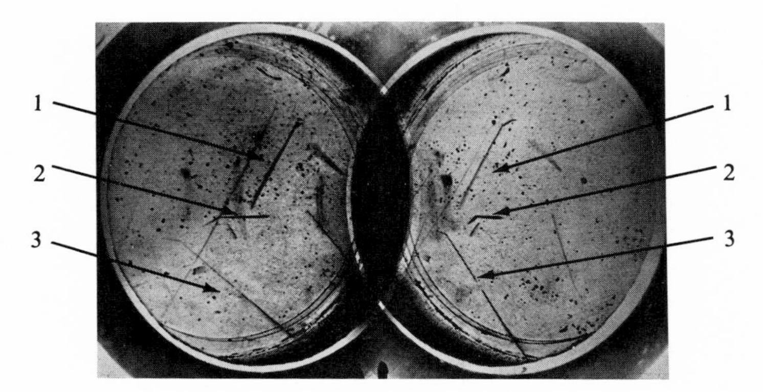

• The only extant photographs showing production of ^{14}C as visualized in a cloud chamber. Two views, taken at different angles to give a stereoscopic projection, show typical events in a cloud chamber exposed to the neutrons from the 37-inch cyclotron. Matched pairs of tracks are indicated by numbers. Note particularly the pair marked 3, which show the long track made by the light proton (^{1}H), and the short mate, made by the heavier ^{14}C. The process involved is the expulsion of a proton from the ^{14}N nucleus, forming a ^{14}C nucleus. Occasionally (not seen here) very short tracks with a stub on the end, corresponding to the formation of ^{14}C by slow neutrons, also appeared. In these instances only the proton had enough energy to make a visible track. These photographs were made by F. N. D. Kurie and the author during the research reported in *Physical Reviews* 53 (1938): 212.

• The author at the probe port and target chamber of the 60-inch cyclotron at the Donner Laboratory, sporting a not unusual 3-day beard, the result of many long hours without sleep. A photograph taken by Don Cooksey early in 1941.

• With Sol Spiegelman (right) at Cold Spring Harbor in 1947, attending a conference on cytoplasmic inheritance. Photographer unknown. Much interest was excited by our work on phosphate turnover in yeast. This provided some evidence for the existence of genetic copies of nuclear genes, which we called plasmagenes.

- With Charles D. Coryell (left), famous for his early work with Linus Pauling on the magnetic properties of hemoglobin, among other things. We were attending a conference on isotopes and their applications held at the University of Wisconsin, Madison, in the late 1940s. Photographer unknown.

- The author with Harland Wood (center), the discoverer of heterotrophic CO_2 fixation, and Hungarian Nobel laureate George Hevesy (right), originator of the use of isotopes as tracer elements in chemistry, at an isotope symposium at Cold Spring Harbor in the summer of 1948. Photographer unknown.

Smear First—Ask Later!

• Cartoon by Dan Bishop in the *St. Louis Star-Times,* September 4, 1948, at the height of the furor sparked by the investigations and charges of the House Un-American Activities Committee.

- A session at Robert Emerson's lab at the University of Illinois, Urbana, in December 1952. Left to right, the English biochemist Robin Hill, Stanley Holt, Emerson, and the author. We are discussing procedures devised by Emerson to measure the yields of photosynthetic processes, as well as the reactions Leo Vernon and I had discovered at Washington University in which cytochrome *c* was photo-oxidized by plant chloroplasts. Photographer unknown.

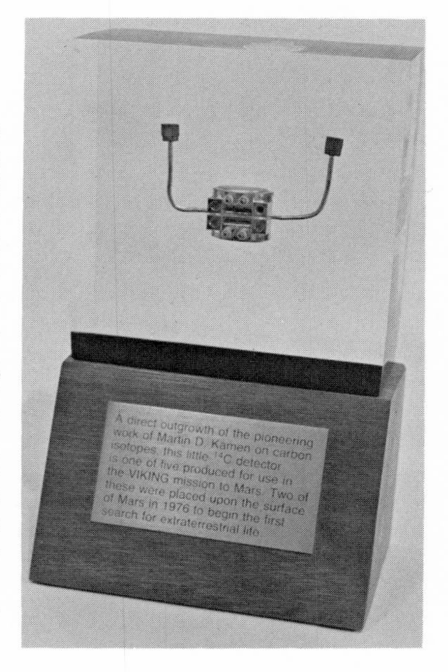

- A memento presented to the author on the occasion of his sixty-fifth birthday by Harold P. Klein, director of life sciences at the Ames Research Center. The inscription reads: "A direct outgrowth of the pioneering work of Martin D. Kamen on carbon isotopes, this little ^{14}C detector is one of five produced for use in the VIKING mission to Mars. Two of these were placed upon the surface of Mars in 1976 to begin the first search for extraterrestrial life." Photograph by Maurice Lecovre. (Black-and-white reproduction from color original.)

exciting research with Howard and Jack, while planning a fruitful few months in Pacific Grove. There was also the prospect of a reunion with my old friends in Berkeley, whom I had not seen for several years. Invitations to lecture were pouring in, including many abroad—at an isotope symposium in Paris, the newly formed Australian National University in Canberra, and the proposed Weizmann Institute in Israel, for example. I had been busy filling lecture dates at seminars all around the country during these early years at Washington University and had been approached a few times about the possibility of moving to new positions. The approach from the Australians included the proposal that I might consider joining the faculty at their new university, and had been initiated by my old friend Mark Oliphant, who had gone to Canberra as the first director of the Australian National University's Division of Physical Sciences.

The invitation to participate in the Paris meetings had come from Moise Haissinsky and Mme. Irène Joliot-Curie. Both were considered "undesirable leftists" by the State Department's visa section. Mme. Joliot-Curie—a Nobel laureate like her mother Marie—had been grudgingly given a limited visa to visit the United States. Thanks to this, I had the prized opportunity to spend an afternoon with her at the Drake Hotel in Chicago (my diary notes the date as Friday, March 23, 1948). We exchanged views on who might be contaminating whom, and I discussed with her the content of a paper I proposed to present in Paris with Joseph Kennedy and his graduate student Orlo E. Myers.[6]

However, I never got to attend this meeting or go abroad at all for the next seven years because, as I shall relate, my passport was withdrawn by the State Department. Oliphant, too, himself was having trouble with our State Department, which refused him a visa to visit the United States in response to lecture invitations he had received.

Later I received a card from the participants at the meeting in Paris expressing their regret that I had been prevented from attending and thanking me for the paper we had sent. I noted that I had missed a banquet at a three-star restaurant. The card was signed by a most remarkable collection of science greats and assorted talents, including Frédéric and Irène Joliot-Curie, Georg von Hevesy, Linus Pauling, C. A. Coulson, C. H. Ingold, J. Govaerts, A. Pullman, P. Daudel, Charles Coryell, M. Haissinsky, and many others. Two decades later, I finally got to that restaurant and celebrated the occasion with the Joliot-Curies' son, Pierre, and daughter-in-law, Anne, the third generation of

Joliot-Curies to maintain the great family tradition of outstanding husband-and-wife teams in science, in this case in research on the early reactions in green plant photosynthesis.

The affair of the Weizmann Institute began auspiciously in early 1947, when I accepted an invitation to lecture there. At a meeting in New York, there was even a suggestion that I might take over direction of the isotope division, with the young English investigator I. Dostrovsky as my assistant. (Dostrovsky later became head of the Israeli Atomic Energy Commission.) After receiving the official written invitation, I made my first attempt to obtain a passport. It arrived promptly and I prepared to make the trip. Suddenly, however, I received a summons to call at the downtown office of the FBI. There I was politely but firmly questioned for nearly six hours, mostly about the Fish Grotto episode. The two agents conducting the questioning referred continually to a large volume, evidently my dossier, which they consulted as the hearing proceeded. I became aware for the first time that the record of the conversation with the Russians, as quoted from their transcript, contained numerous incoherences and much nonsense. For instance, we spent some time trying to make sense out of the word "hitted," which appeared in the transcript. When the session ended, we all adjourned to a nearby restaurant, where I had my second dinner at the expense of the FBI; the first, it will be recalled, was on the occasion of the visit I made to the San Francisco FBI office in 1944, after my banishment from the Manhattan Project.

A few days later, I was in New York, about to take off for Israel, when federal agents invaded the travel agency processing my ticket and confiscated my passport. No explanations were forthcoming, and when I wrote to Mrs. Ruth B. Shipley, the chief of the Passport Division, I got in reply only the noncommittal statement that "it is not in the best interests of the United States that you go abroad at this time." This statement, or variations thereof, was to be repeated ad nauseum for the next seven years.

I was still not too concerned about the passport impasse. During a visit from the distinguished organic chemist Ernst Bergmann, a major mover in the establishment of the Weizmann Institute, he assured me, as had the representatives of the Australian and French governments about the same time, that official proceedings would be instituted to get my passport back. Meanwhile, other matters filled my time and dulled my perceptions of federal harassment. I finished an article for the *Scientific American* on tracers[7] and attended a meeting on metabolic aspects of convalescence, sponsored by the Macy Foundation in New York, where I renewed contacts with the Columbia group, including Dave Rittenberg, De Witt Stetten, and Al Marshak. I also gave the Harrison Howe lectures at the University of Buffalo, and presented seminars and taught on the Washington University campus. My dossier was growing bloated in various favorable environments, meanwhile, as it was handed about from one federal agency to another. Events soon to come indicated that it could already be found in the Departments of State, Defense, and Justice, as well as in the hands of an indeterminate number of former Manhattan Project security types.

In early August I arrived in Berkeley to hear exciting news. Howard, who had been in Pacific Grove for a month, had achieved a most astounding result in his first trials of a new synthetic medium he had contrived after reading an article by Seymour Hutner relating to microbial nutrition. Using this medium, which employed salts of glutamic acid—an amino acid—as a nitrogen source in place of the usually preferred, and simpler, inorganic ammonium salts, he had prepared inocula of *R. rubrum*, as customary, in glass-stoppered bottles filled to the top to exclude air and illuminated in a thermostated incubator. Some similar cultures in bottles covered by aluminum foil were included to act as dark "controls." After a few days, he found to his amazement that there had been considerable evolution of a gas, insoluble in the slightly alkaline medium, therefore definitely not CO_2. No such gas production was seen in the dark controls. No precedent existed for this unexpected finding, even in the exhaustive studies by van Niel. Howard tested the gas produced, which still remained in some of the bottles (the stoppers had not all blown out), and found that it was combustible. He concluded that hydrogen gas somehow had been made photosynthetically. Van Niel, just as much astonished, expressed incredulity, and suggested that there might have been accidental contamination by gas-forming nonphotosynthetic bacteria in the illuminated bottles. Howard stubbornly persisted, however, and demonstrated that he could obtain the hydrogen repeatedly with illuminated bottles, but not in the dark controls. He was encouraged to hold his ground by I. C. ("Gunny") Gunsalus, an eminent microbiologist who was spending the summer at the lab, as well as by another investigator, Jack Stokes. It was on this occasion that I first met "Gunny" and began a lifelong, mutually fruitful friendship.

This finding of photohydrogen in bacterial photometabolism had come about wholly accidentally because Howard had chanced to substitute glutamate for ammonia as the nitrogen requirement. Hutner's medium had apparently been designed for a glutamate-*requiring* mutant. We were to find that the usual nitrogen source, ammonia, inhibited photohydrogen production, explaining why the phenomenon had not been seen previously by the numerous investigators who had worked with these bacteria. They had all used ammonia, the simplest and most available nitrogen source, especially as the work of van Niel had shown conclusively that ammonia nitrogen was a suitable general source in bacterial growth.

It was a paradox that the bacteria would throw away energy in the form of hydrogen when they were already such poor energy converters compared to green plants. Howard and I were naturally anxious to probe further as soon as we returned to St. Louis, where we had the necessary facilities to take the next step of determining how much hydrogen was produced for each carbon substrate used up. Late in August, I returned to Berkeley for a round of parties and chamber music sessions with all my old friends as well as some new ones, including "Lilick" Schatz, a talented Israeli artist, who took me on a memorable visit to Henry Miller's eyrie atop a canyon in Big Sur one night. I passed an evening in stimulating conversation with Miller about the future of civilization in a nuclear world, and I was surprised to see how little there was in him of the rebel and iconoclast I had come to expect from his writings.

Only one small event cast a pall over the general ebullience. I still had many friends in the Radiation Laboratory and one of them invited me to dinner shortly after I arrived in Berkeley. The next day the invitation was hastily and lamely withdrawn. It was clear, as I had already suspected, that I was still persona non grata in the laboratory. Apparently, there was reason for the personnel not to be seen fraternizing with me. In view of what was to happen in September, it is likely that people there already knew trouble was brewing in Washington.

What this trouble was began to be evident when headlines, inspired by handouts and leaks from the House Un-American Activities Committee (HUAC), appeared in the press all over the nation claiming that certain scientists on the Manhattan Project had been part of espionage rings working for the Soviet Union. I was soon identified as one of these "links" to Communist Party organizers. The first story, leaked to the *St.*

Louis Post-Dispatch, and appearing in various editions on September 2, was written by Joseph Hanlon, a *Post-Dispatch* Washington correspondent. It was no coincidence that my hometown newspaper had been chosen. The preparation for this gambit had been a sensational story in a number of newspapers quoting Congressman R. B. Vail, a member of HUAC, as being "certain, on the basis of testimony by a 'high Army officer'"—quickly identified as Lt. Gen. Leslie R. Groves, the former director of the Manhattan Project—that Russian spies had obtained atom bomb data. Closed hearings probing this allegation were promised for September 15. My projected role as a star witness in these hearings had been revealed to Hanlon by a transparently careless trick. He had "found" testimony relating to me given by one Harold Zindle, an agent who had allegedly been in attendance at the Bernstein Fish Grotto dinner. It had been bound "inadvertently" in a book (probably of the Chambers-Hiss investigation) containing testimony of Louis Budenz, the former managing editor of the Communist paper, the *Daily Worker.* According to Hanlon, Budenz's testimony had been taken in closed (executive) session, and reporters later saw a transcript of Zindle's testimony "although they weren't supposed to at that time." So Hanlon got his scoop where it should do the most damage—in my hometown. The headline read, "Washington U. Scientist Linked in Atomic 'Leak.'" A caption below it stated, "Martin D. Kamen Fired From Army Project at California U. After Talking to Reds." There was a smear of the United Nations thrown in for good measure, as the Hanlon story mentioned that the Russians I had talked to were in San Francisco attending the conference where the United Nations was formed.

For the next two weeks, the committee leaked further information to the press establishing that I was the atom scientist involved, HUAC keeping on with its leak strategy, naming various suspects lumped together as though they were all part of an atom spy ring that would be under investigation. The smear was deepened by claims that I had been overheard by government agents discussing "classified" information with Russians and dismissed from the Manhattan Project for "indiscretion." In this way, my reputation was badly damaged in the public mind before I had a chance to appear at a hearing to defend myself.[8]

I was not without defenders, however. The home forces began rallying, notably Chancellor Compton. He released a statement supporting me and denying the HUAC allegations. One story, appearing in the *New York Times,* datelined from St. Louis September 4, was headlined

"Compton Backs Teacher," with the subheading "Finds No Evidence D. H. [sic] Kamen Gave Information to Russians." The same statement also appeared in the St. Louis newspapers. In the *Globe-Democrat,* for example, a story headlined "Compton Clears Kamen of Giving Reds Atomic Data" was printed in the September 4 issues. The most heartening reaction came from the *St. Louis Star-Times,* the "other" afternoon paper. Unlike the *Post-Dispatch* and the *Globe-Democrat,* which had adopted a wait-and-see attitude, the *Star-Times* launched a strong attack on HUAC. Robert Blakely, the chief editorial writer on the *Star-Times* staff, wrote two scathing editorials under the headings "Close-Up of a Smear " and "Persecuting our Scientists" in the issues of September 3 and 4. He mentioned particularly the "cheap theatrical trick," as he characterized it, whereby Hanlon had been "inadvertently" given the "secret testimony" about me. Alongside the editorial, there was a cartoon by the celebrated Daniel Bishop, showing me being inundated in tar and mud by a giant hand labeled "Un-American Activities Committee" wielding a huge brush. The cartoon was captioned "Smear First, Ask Later." (I still have the original, which was given me by Bishop.) This no-holds-barred action by the *Star-Times* contrasted starkly with the timid behavior of the *Post-Dispatch,* supposedly a great muckraking paper. Nevertheless, it was clear that HUAC could not count on seeing its smear tactics amplified by my hometown press or effective in pressuring for termination of my employment. The support I received from Washington University was not typical, unfortunately, of the position taken by other universities in the Cold War years to come.

I was finally subpoenaed to appear in Washington. The committee underwrote my travel expenses, obviously anxious that nothing impede my scheduled appearance as one of their star exhibits. Aware that I needed help, I called Abe Fortas of the law firm Arnold, Fortas, and Porter, because I had heard that it was heavily involved in defending security cases in Washington. I also talked with Edward U. Condon, a distinguished American nuclear physicist, who was at odds with HUAC because of smears they were directing at him.[9] Arriving in Washington, I was put up by HUAC in modest quarters at the Stratford Hotel, then at 25 E Street, N. W. It was September 12, and I had a day or two to plan strategy before my hearings, scheduled for Tuesday morning, September 14. I immediately called Milton Freeman, a lawyer with Arnold, Fortas, and Porter and recommended by Fortas. He arranged some meetings with Condon and Dr. William Higginbotham, who was working with the much beleaguered Federation of Atomic Scientists, an organization started soon after the end of the war to contend with lobbies intent on vesting control of atomic energy in the military, as spelled out in the May-Johnson Bill, and to disseminate information to the public about the facts of nuclear physics and the hazards of nuclear war. The association was particularly concerned about the threat of public reaction against scientists, fed by the propaganda from HUAC and much of the press.

Our sessions were gloomy enough. For the first time I comprehended not only that my continuing career was endangered, but that my life itself might be threatened if HUAC could press its charges of espionage, however false, to the point of an indictment. The tactics of the committee were to build up a case by headlines first, then call its intended victim into a closed session where there was no legal protection. There it could browbeat the witness and, with any luck, pressure him to make statements that, lifted out of context, could be very damaging. It was not clear what hard evidence the committee had, but I was sure from my recent experience with the FBI that they would be reading appropriate canards culled from my dossier.

The atmosphere in Washington was frighteningly favorable to these HUAC operations. The press eagerly received handouts and printed headlines feeding the anti-Soviet hysteria that had been mounting ever since the war's end and the beginning of the Cold War. Although the American public had been favorably disposed toward Russia during World War II, fears of Soviet expansion had changed the public perception. Attempts were being made to show FDR as a dupe of the Russians and the United States as the victim of a vast communist conspiracy to take over the world. The "loss" of China was cited as but one result of the influence exerted on the Democrats by "fellow travelers" and "crypto-Communists." With these sentiments growing, the political malaise to be known as McCarthyism was beginning to make its appearance.

The major effort of this Cold War ambience was a realization on the part of some Republicans that they could use anti-Soviet feeling to regain power, which they had been deprived of all through the Roosevelt years. They only needed to discredit Roosevelt and his less charismatic successor, Harry Truman, as tools of leftist Soviet sympathizers. HUAC was an appropriate sounding board. As soon as the Republicans seized control of the House in the elections of 1946, J. Parnell Thomas became chairman of the committee. There was every indication that the Republicans would take over the White House in 1948, so Thomas and his cronies on the committee felt encouraged in pressing the defamation campaign already launched in the Whittaker Chambers–Alger Hiss hearings. These were just winding up when I and others appeared on the scene to start the next show (there had already been a hearing on September 9 involving a Clarence Hiskey).

To add to my troubles, I now heard from Beka, who was present at our strategy meeting, that a hatchet job was to be done on me at *Time*. A story vilifying me and others as active Soviet spies, or at least collaborators, was planned to coincide with the eventual HUAC release of my testimony. It took no imagination to see what would happen to my career and my reputation if such a story, published with all the authority of *Time* and distributed through its immense circulation, were to appear. Beka had the brilliant idea of arranging an unprecedented interview for me with Ben Williamson, the editor assigned to do the story, noting that Williamson was basically a decent sort and would find it difficult to carry through his assignment once he had seen his subject. *Time* policy was not to conduct such interviews, as they might inhibit editorial comment. Nevertheless, she managed it. I was to see Williamson in New York as soon as I could go there. For the immediate present, it was decided that I try to blanket the HUAC handouts by some of my own, getting up early enough to have them in the hands of reporters before the HUAC operatives were stirring. Finally, I was to demand a public hearing so as to prevent their manhandling me in executive session. The effect of such a demand would also bring a horde of reporters to the hearings, where they would be milling about in the corridors, available for statements from me if I emerged to make them.

On the morning of the hearing, reporters—knowing of my demand—were in the corridors in force outside the hearing room in the Old House Office Building. I handed out a statement for release at 9 A.M. (an hour before the hearing) written for me by Freeman. It read:

> I have come to Washington from California at public expense under subpoena of the House Committee on Un-American Activities. This committee has already released to the press derogatory innuendoes about me. I have nothing to conceal. I wish to cooperate fully with the professed objectives of the Committee's work and am prepared to answer in *public*

> hearing any and all questions the Committee may ask. In the light of what has already happened, however, I consider that this investigation is one about which the American public is entitled to receive full information. Therefore I shall refuse to cooperate with any *secret* proceedings and shall not answer any questions in closed session.
>
> The Committee may object that the hearing must be secret because it might involve the presentation of restricted atomic energy data. This simply is not so. I would not in any case reveal such data to the members of this committee without being officially notified—as I have not been—by the Atomic Energy Commission that the members of the committee have been cleared for access to such restricted data.
>
> I am fully aware of the regulations applying to atomic energy information and I solemnly declare that I have never revealed in the past, nor do I intend in the future, to reveal such information to unauthorized persons.

While waiting for the hearing to begin, I had a small conversation with Steve Nelson, who was scheduled to appear after me. I did not know him, but had read the HUAC handouts in which he was represented as active in the alleged Berkeley "atom-spy ring." Although he had been stated to be closely associated with scientists such as Frank Oppenheimer and Joseph Weinberg, I certainly had no recollection of having seen him around the lab.

On entering the hearing room, I found Thomas and two of his Republican party henchmen, Congressmen McDowell and Vail waiting. Ironically, Vail came from my home district of Hyde Park in Chicago. No Democratic members were in attendance. Some staff members, led by Robert E. Stripling, the chief inquisitor, were present. We began by sparring over the question of why I wanted a public hearing. Clearly the committee was upset by my demand. I stated in part: "I have no protection with regard to what kind of quotes come out from the committee to the press. . . . It seems to me there is less chance of the whole truth not being told in a public hearing than in a private hearing." A little later, Stripling pointed out that the committee was not responsible for every story somebody might print or "dream up."

As a good example of how the committee reports were doctored (indeed they were labeled "Excerpts from Hearings," but this reservation was not usually emphasized or even noted in the press summaries), there is no mention of a discussion off the record in which the chairman agreed to give me the opportunity to go public if I felt I was being browbeaten or mishandled. The hearing started, to no one's surprise, by taking up my associations among liberal circles in San Francisco and the East Bay. Soon we were into the Bernstein Fish Grotto banquet. I was read portions of testimony by committee investigators. It was evident that the committee had a copy of the Manhattan Project transcript, which was supposed to be classified and not available to them. From their questions, it was clear that the same farrago of misquotations, fabrications, and plain lies that had intrigued the FBI was providing them with a basis for what appeared to be a fishing expedition.

At one point the matter of leaks about Oak Ridge and Hanford as sites for the construction of the atom bomb arose. I was alleged to have been asked by the Russians about a news story they had read concerning the sudden appearance of a large community in Tennessee where research on nuclear energy was rumored to be proceeding. I attempted to supply some perspective by noting that there had been an even worse, documented security breach when a story appeared about Los Alamos at the end of 1944, written by a reporter for a Cleveland newspaper. Congressman McDowell entered the quiz at this juncture. "It is my belief as a member of the American Editorial Association that there was not one violation [of requests to eliminate references to Oak Ridge or Los Alamos] in several thousands of publications that were entrusted with that information during the war," he pompously asserted. I listened incredulously, as there had been innumerable instances of newspaper breaches of security, including the infamous revelations by the *Chicago Tribune* in its "expose" of U.S. war plans in November 1941. McDowell was displaying a convenient lack of memory, or else deep ignorance of the facts.

The committee attitude became more wary as the hearing continued. This seemed especially so after another interchange, in which I elicited from Stripling the statement that the text they were using came from testimony of the intelligence agents and not from any authorized document. All of the remarks cited, including some stated by Stripling to be gibberish, were supposed to have been remembered by the agents. When I pointed out that hours of such testimony could hardly have been "remembered," Stripling hastily changed the course of the questioning to cover other aspects of the Fish Grotto meeting. It was clear that the committee had been given a transcript in violation of the president's directive forbidding transmittal of classified files to the congressional investigators. (This exchange also cannot be found anywhere in the "excerpts" of my testimony.)

Finally, after nearly three hours of questioning, the ordeal ended. There had been a break during which I had emerged to talk freely with the reporters, who had maintained their vigil patiently waiting for the public hearing. Disappointed in the end by our failure to go public, they left to write stories with headlines such as, "Kamen Changes His Mind, Gives Secret Testimony" (*St. Louis Post-Dispatch,* September 14) and "Atomic Scientist Reverses Stand and Talks Freely" (*Washington Star,* September 14). I fared better than Steve Nelson, whose refusal to testify, relying on his right not to incriminate himself, earned him a contempt citation and feature billing in a story headlined "Spy Investigators Act to Cite Balky Witness for Contempt" (*New York Times,* September 14). I was mentioned in this story as having been a "frank" witness who had decided to cooperate in a closed hearing, where I had denied allegations that I had told the Russians anything in violation of security regulations at the Fish Grotto meeting.

I had my defenders in the press, too—notably Albert Deutsch and I. F. Stone in the *New York Star,* who dissected the HUAC reports and demonstrated that they proved nothing except that the committee was adept at framing unsubstantiated charges.[10]

In reviewing my testimony, edited as it was, some years later, I must admit that it was not wholly admirable as a textbook example of proper legal caution, but it did serve its purpose of blunting the HUAC attempt to make a mountain out of a molehill in my case. I was treated well as regards the expense of staying on nearby to remain under subpoena, as ordered at the end of the hearing. While still in town, I took the opportunity to make a quick journey to New York, where I met with Ben Williamson at *Time* headquarters. Beka's strategy worked well. Williamson did not write his story, or at least not one that served *Time*'s purpose. The Luce cabal decided to feature the Hiss case and ignored the atom spy hearings completely. So I was successful in spiking the guns of *Time,* as well as in minimizing damage from the HUAC hearings. I left Washington on September 26. Two days later the committee issued its report calling for indictment of everybody except me. The report nevertheless repeated the allegation of the committee that I had conceded "gross indiscretion," which I heatedly denied as a misquotation.

I returned to St. Louis wounded, but not fatally. Chancellor Compton personally reassured me that the trauma in Washington had not weakened my position at the university, and even expressed the hope that there would be no further harassment from the feds. Neither of us really believed that this would be the case, but it was a relief to have the immediate troubles over. I had been immobilized only a month, so the work at the lab had not been delayed inordinately. I hastened to join Howard and Jack and press on with our research, particularly to pursue the new findings on light-induced production of hydrogen.

Jack had experienced a setback in his investigations. He found that neither Foster's original strain of *Rhodopseudomonas* nor his own enrichment cultures exhibited the simple removal of hydrogen from isopropanol that had so excited us. He had the less sensational task of continuing with the strains he had isolated that would grow in isopropanol, even if they oxidized it beyond the state of acetone. Howard and I busied ourselves determining the "stoichiometry" of the hydrogen production reaction—that is, finding out whether the hydrogen evolved from compounds such as malic and succinic acids was a primary product. If it was made in only small quantities, the reaction had no great significance.

The course of research is never smooth. Jack's reverses were matched by ours with the hydrogen evolution process. Using media made in St. Louis, Howard and I had found that light did not cause *R. rubrum* to make hydrogen. We were crestfallen momentarily, as the process seemed not to be reproducible. We performed a number of experiments to test what components might be inhibiting the reaction and discovered that our yeast extract, which we added to make up the trace vitamin requirements, was the culprit. If we diluted it, the hydrogen reappeared. It turned out that the commercial extract we were using contained trace amounts of ammonia. The extract used in Pacific Grove, being handmade and much more dilute, contained too little ammonia to inhibit the hydrogen production. Again, an accidental factor had been at work. The yeast extracts in Pacific Grove were produced by hand to save money. The discovery of photohydrogen evolution had come about partly because there was a need to economize.

Happy at having apparently resolved our dilemma, we set about assaying the gas production quantitatively. We used manometers consisting of test flasks attached by closures to liquid-filled columns of glass to measure changes in pressure. To make these measurements, we needed something inert to fill the gas space of the manometers. This had not been a problem in the bottles, where all the space was filled by liquid. We decided to use nitrogen as the inert gas.

Again, we found no hydrogen, although the carbon compounds

disappeared at a rapid rate. There was only CO_2 production. Puzzled, I suggested that perhaps oxygen was present as a contaminant and was inhibiting the reaction. This was reasonable, because the bottle cultures were maintained free of oxygen. Furthermore, oxygen was known to inhibit the action of the enzyme hydrogenase, which we assumed was involved in the mechanism of molecular hydrogen formation by the cells.[11] We spent some days trying to build a manometric system free of oxygen—no simple task. One day Howard, Jack, and I were conferring about our various problems when Herta Bregoff, a new graduate student who was just beginning her research, breathlessly broke in to tell us about an experiment she had just performed.

Hoping to help in the search for the lost hydrogen and also to improve her skills with our manometric setup, Herta had repeated one of our routine experiments. On her first attempt, she had succeeded where we had failed. She showed us flasks in which light was producing vigorous evolution of CO_2 and H_2. The experiments were simple enough. A few milligrams of washed *R. rubrum* cells were suspended in buffer, to which some sodium malate had been added from which the cells would make CO_2 and H_2. The suspensions were pipetted into flasks equipped with small center wells filled with appropriate gas absorbents (strong alkali for CO_2 and methylene blue dye on palladinized asbestos for H_2). Control flasks were prepared in the same way, but covered with aluminum foil to exclude light.

The secret Herta had stumbled on was to flush out the air in the flasks and manometers with argon, instead of nitrogen. Luckily, she had not been with us long enough for me to impress on her that she was not to use argon because it was expensive and reserved for use as a gas filling in our homemade Geiger counters. The fact that argon worked where nitrogen did not was all the more astonishing because we knew that our argon, being a product of air liquefaction, was not pure. It contained at least 5 percent oxygen contamination.

The famous lines of Keats in "On First Looking into Chapman's Homer" flashed through my mind:

> Then felt I like some watcher of the skies
> When a new planet swims into his ken.
> Or like stout Cortez when with eagle eyes
> He stared at the Pacific and all his men
> Look'd at each other with a wild surmise
> Silent, upon a peak in Darien.

"Wild surmise" was a perfect expression of our state of mind, confronted with photohydrogen evolution in the presence of argon, but not of nitrogen. It was clear that nitrogen was doing something; it was not inert. Perhaps the bacteria were actually fixing it.

To appreciate our excitement, it should be remembered that in those times, only a few very special microbes were known to fix nitrogen. The possibility that a photosynthetic bacterium was such an agent and that nitrogen fixation might require light in the case of such bacteria threw us into feverish activity to probe the phenomenon further.

First, it was conceivable that the action of gaseous nitrogen was merely catalytic—that is, it was blocking hydrogen evolution just by adsorbing onto a hydrogenase system responsible for producing hydrogen, and not acting chemically. There was no evidence that nitrogen was actually disappearing in the flasks. However, the rate of photohydrogen evolution observed in argon could be used to calculate how much nitrogen should have been taken up in place of the hydrogen expected to be evolved. We could assume ammonia, the inhibitor par excellence of hydrogen production, might be made by reaction of nitrogen with hydrogen, so that the failure to make hydrogen came about as a result of this ammonia production from nitrogen in the light. Calculations on this basis were comforting in that they showed that the very small amount of nitrogen that would have been taken up was too little for us to have observed.

We therefore set up experiments to settle the matter. Three vessels were used. In one, labeled *A*, we put washed cells (equivalent to about thirty milligrams of nitrogen as assayed by a routine Kjeldahl digestion of cellular nitrogen to ammonia), and added alkali to the center well to absorb CO_2 and not hydrogen. The medium, as usual, contained malate, salts, and a trace of biotin as a needed vitamin. Another vessel, labeled *B*, was identical, but was covered with aluminum foil as a "dark control." A third, labeled *C*, contained boiled cells to provide a "dead cell control." All vessels were flushed with helium and left at atmospheric pressure. Three flushings were carried out. Then we filled one-tenth of the gas space with nitrogen labeled with ^{15}N made from ^{15}N-ammonium nitrate. This nitrogen contained almost one hundred times as much ^{15}N as normal nitrogen gas. The remaining gas was helium, so that the total pressure was one atmosphere. We allowed these vessels to remain in the thermostated water both at 30° C. for 120 hours, then removed them, took off the cells by centrifugation, and prepared ammonia samples by the usual Kjeldahl digestion. We had so much labeled nitrogen in the gas phase, compared to what was present as unlabeled nitrogen in the cells, that even traces of labeled nitrogen incorporation would be seen if they existed.

I sent the samples to Irving Sucher at Columbia University. He was the technical associate of David Rittenberg in charge of their mass spectrometer assay system. A few days later, he phoned me with the results. With much trepidation, I awaited his findings. He began by telling me that sample *A* showed ten times the normal ^{15}N content. This result I knew could only be possible if there were nitrogen fixation. Breathlessly, I inquired about *B* and *C*. He said there was negligible ^{15}N—barely above normal in the former and insignificant amounts in the latter. I was stunned—there was no question but that nitrogen fixation had occurred in the illuminated sample. As I remained silent for some seconds, trying to assimilate the news, Irving inquired worriedly if there was something wrong. I hastily assured him that all was well.

We now needed not only to confirm fixation by actually showing that *R. rubrum* could grow using only molecular nitrogen as a nitrogen source, but also to convince the experts in the field that this was true. So I phoned P. W. ("Perry") Wilson at the University of Wisconsin at Madison. Perry was an acknowledged authority on nitrogen fixation. He had had to spend much time disproving the many claims made through the years for the existence of nitrogen fixers other than the few such agents that had been established. He was accustomed to getting such calls and was understandably pessimistic when he got mine. I offered to bring our cultures to Madison so that he could test them in the ^{15}N-growth apparatus set up by Robert Burris, a faculty associate just beginning his distinguished research career in the study of nitrogen fixation.

Howard and I built an octagonal wooden box to transport the cultures, with holes for ventilation and a central socket for a bulb to provide light. We could insert a thermometer to check that the temperature did not rise above approximately 35° C., so that the microbes would not get overheated during our journey from St. Louis to Madison. When in operation, a red glow emanated from the holes in the box as the white light passed through the culture flasks. Howard and I took turns through the night in our roomette checking that the bacterial cultures did not overheat and that they kept their characteristic purple color, while continuing to make hydrogen. Arriving in Chicago, we transferred to a first-class lounge coach because, like the roomette, it had the appropriate electrical outlets. We settled on a bench with our box, understandably drawing suspicious glances from our fellow passengers. The sight of two strangely intense, wide-eyed young men hovering anxiously over a box from which emanated a red glow must have been alarming.

In Madison we were met by Wilson and Burris, with whom we had been in correspondence through the early part of 1949. Wilson had cautioned us that we were probably dealing with some microbial contaminant, a suggestion we appreciated not at all. We saw that no one expected our observations would survive critical reexamination. The experiments were set up using the Burris apparatus. Growth medium based on malate, omitting ammonia, had been prepared according to our mailed instructions.

After a few days had elapsed to allow sufficient time for incorporation of the ^{15}N-enriched N_2, Howard expressed some worry that the cultures did not "look right." They had an unhealthy peach color, instead of their usual deep purple tint. So we sampled a bit of the culture medium and found to our horror that it was very acidic. Apparently, the medium had not been neutralized before use. (I should explain that, to save money, we used malic acid rather than sodium malate, its neutral salt. We would bring the medium to neutrality before sterilization in an autoclave by titrating it with sodium hydroxide.) Apparently, the hapless graduate student who had prepared the solution had forgotten this step. We feared the experiment had been ruined. Unhappily, we watched Burris remove the cells after a quick remedial titration. To our amazement, and his as well, there was no doubt that there had been nitrogen fixation, despite the maltreatment to which the cells had been subjected.

Quickly, the Wisconsin group confirmed and extended our results, showing that nitrogen fixation was a general phenomenon in the photosynthetic bacteria and was not restricted to *R. rubrum*. We had thus made it possible to increase the known catalogue of nitrogen fixers at least tenfold or more. Later we demonstrated actual growth of the bacteria in light at the expense of molecular nitrogen, after repeating our isotope experiments under a variety of experimental conditions, thanks to the help we received from Mildred Cohn, who had set up a mass spectrometer in Carl Cori's Department of Biochemistry at the Washington University Medical School.[12]

The history of these two remarkable discoveries—photohydrogen evolution and nitrogen fixation in the photosynthetic bacteria—illustrates dramatically the chance, or serendipitous, character of true basic research. By a succession of lucky mishaps, we had made these discoveries while pursuing wholly unrelated objectives.[13] The failure of such noted investigators as C. B. van Niel and Seymour Hutner to observe these phenomena previously had occurred because van Niel had never tried molecular nitrogen in place of ammonia and Hutner, assuming the gas evolution (he had actually seen it, too) was due to CO_2, had never checked the acidity of his media during bacterial growth.

An amusing sidelight is the historical circumstance that M. W. Beijerinck, the great predecessor of A. J. Kluyver and C. B. van Niel in the Dutch school of microbiologists, had been struck by the morphological and nutritional similarities of a nitrogen-fixing microbe he had observed to those of *Chromatium,* a photosynthetic bacterium. Beijerinck had toyed with the idea of calling the microbe *Para-chromatium,* but had discarded this notion because it fixed nitrogen, a highly specialized nutritional characteristic. He did not know that it would be demonstrated for *Chromatium* itself half a century later. Instead he called the microbe *Azotobacter* ("nitrogen bacterium," from the French *azote*, nitrogen). This is an excellent example of the remarkable intuition great scientists sometimes display and how scientific discoveries rarely fail to cast their shadows before them.[14]

In the midst of all this excitement, I did not respond happily to the realization that some residual reek from the sewage of September in Washington persisted. The summary reports of HUAC, which contained gratuitous slurs based on my alleged involvement in their fabrications of atomic spy "disclosures," appeared at the end of September 1948. They received front-page treatment in the press around the nation. One of these outbursts had been particularly offensive. I was made aware of it by a clipping someone sent me in October. Ironically it had appeared in the *Chicago Sun-Times,* the "liberal" daily financed by Marshall Field and editorially at the opposite pole from the McCormick publication, the *Chicago Tribune.* It carried a highly uncomplimentary picture of me, unshaven with a cigarette dangling out of a corner of my mouth, looking like a gangster. The picture was captioned "Martin D. Kamen" and placed over an article headed "Red Faces Contempt Action." The story dealt not with me, however, but with Steve Nelson. It was a flagrant and wholly indefensible libel, perpetrated largely because of the incompetence of someone on the paper's editorial staff.

By this time, my fuse was short, and I reacted sharply. I rushed up to Chicago, consulted with a lawyer there, Stanley Kaplan. He agreed that the *Sun-Times* was guilty of technical libel and that it was very likely that substantial fines would be automatically levied. He estimated the penalty could amount to hundreds of thousands of dollars, as the libel had appeared in practically all the early editions of that day's issue.

I had no trouble obtaining an interview with the chief editor, a Mr. Finnegan, who expressed dismay and offered profuse apologies for the egregious errors of his staff. I was unimpressed, but found myself persuaded not to press the suit by a phone call from Harrison Brown at the University of Chicago, who had heard of my intention to sue the *Sun-Times*. Brown, acting on behalf of the Federation of Atomic Scientists, implored me to call off the action on the grounds that the *Sun-Times,* with its liberal attitude toward the atomic scientists and its willingness to publish the federation's handouts, had to be kept friendly. So I settled grudgingly for an apology written by the *Sun-Times*'s featured columnist, Herb Graffis, under the heading "Guilty of What?"[15] This probably made me the federation's most generous contributor ever. Later, however, I had good reason to be thankful I had done the federation this good turn.

A typical distortion by HUAC of what I had said at the hearings appeared in a column titled "On the Line," by Bob Considine, a columnist for the International News Service with wide circulation. In an article headed "Russia's Long Race for A-Bomb," dated September 26, he repeated the committee's assertion that I had conceded a "gross indiscretion" in the course of my conversation with the Russians in the Fish Grotto. I wrote him an angry letter, accusing him of having swallowed the HUAC version whole without really checking the facts. He answered somewhat contritely, requesting I drop him a note so he could print my version. This particular canard appeared in so many news stories, however, that I felt it useless to pursue the matter further.

My preoccupation with *R. rubrum* had distracted me from the gathering storm in Washington. A Republican clique was determined to discredit Roosevelt and the Democrats by playing on public fears of atomic horrors and a growing Russian menace. They were pressing their campaign of vilification through HUAC more vigorously than ever. President Truman, no tyro as a politician himself, was keenly aware of the threat. He had sprung into action from the first hint of the HUAC attacks, calling the charge that he had protected subversives a lie.[16] At the centennial meeting of the American Association for the Advancement of Science in late September 1948, coincident with HUAC's pillorying of atomic scientists, he had appeared on the platform with Ed Condon, one of HUAC's prime targets, to make the opening address. Truman had roundly condemned HUAC and defended scientists, who, he said, needed to work in "an atmosphere free from suspicion, personal insult, or politically-motivated attacks." He was further quoted as saying, "Continuous research by our best scientists is the key to American scientific leadership and true national security. This indispensable work may be made impossible by the creation of an atmosphere in which no man feels safe against the public airing of unfounded rumors, gossip and vilification."[17]

These were brave words, but as history was soon to show they failed to stem the tide, which was running against "intellectuals" and would engulf J. Robert Oppenheimer in a few short years. The vilifiers received a momentary setback with Truman's unexpected victory in the national elections of November 1948. The chairmanship of HUAC returned to the Democrats in the person of Congressman John S. Wood, who was much less addicted to headline hunting than his Republican predecessor, J. Parnell Thomas. The latter had gambled heavily on beating back a charge of receiving kickbacks, expecting a Republican takeover of the White House. Instead he found himself naked to the attacks of his gleeful Democratic opponents. In short order, he was indicted, proven guilty and sent to brood on the vagaries of fate in a jail where, so I am told, his job was raising the prison's chickens. (Appropriately enough, claimed some wags, they were Rhode Island Reds!)

For the moment, all seemed under control and the antics of HUAC did not look particularly frightening to me. I took solace in the belief, shared by Chancellor Compton, that HUAC had done its worst in my case and that I could put the whole sorry spectacle of September in Washington behind me. Indeed, the only letter I received in all this affair that seemed critical came from a Mr. Bishop, representing the "Human Engineering Foundation," who expressed sorrow that a "top-notch" person such as myself should be pictured smoking a cigarette, thus furthering the interests of the tobacco industry.

Events were crowding together so fast in those years that I find it difficult to record them in any coherent fashion. At the Mallinckrodt Institute, Sherwood Moore was retiring, to be replaced by Hugh Wilson from the Yale School of Medicine. In several letters we exchanged, the latter made it clear that he wanted me to stay on at the institute and promised to continue Dr. Moore's generous support in providing both research space and money. Wilson believed that the educational experience of radiologists should include fundamental instruction in both radiation physics and chemistry. He intended to add to the staff a competent biophysicist to expand the institute's offerings in basic science related to radiology. This post was soon occupied by Dr. Michel Ter-Pogossian, a choice that turned out be a most happy one in later years when the institute built a program based among other things on visualization of isotopically imaged organs using short-lived positron and gamma ray emitters.

I was of two minds about staying at the medical school. Although I had been successful in attracting the three gifted graduate students working with me in this period, it was unlikely that the Chemistry Department would continue to be a reliable source of support, for reasons I have already mentioned. There was, moreover, the assurance of space and money from the institute that I could not expect to obtain from the campus. Wilson's offer of a stable research budget was decisive. I would have available money for a technician, a secretary, and a lab helper, as well as funds to cover purchase of supplies and equipment, and travel money amounting to the then generous total of some $10,000. I could still expect help in the form of postdoctoral associates. So the die was cast, and I remained in radiology full time.

An important and happy development, bringing with it added financial support, was my sponsorship by the C. F. Kettering Foundation. This was the enterprise founded by Charles F. Kettering—one of America's great inventors—to probe the mystery of photosynthesis (see chapter 6). The laboratory group there was no longer under the aegis of O. L. Inman, but had been set on a new course led by a perceptive young engineer, Edward Redding, who was aware that the foundation's research policy needed new directions. At a meeting in my office, he told me that Kettering (also known as "The Boss" or "Ket") wanted to set up grants to help promising young investigators, particularly those not part of "big teams," feeling that real progress in solving photosynthesis problems would most probably come from the efforts of "loners" unimpeded by big budgets and organizations. Kettering's own life experiences were excellent examples of such successes. Redding had been sent to interview me as a possible grantee. Shortly after our meeting, I received a letter informing me that I could spend up to $5,000 each year our agreement lasted. There were no conditions on how the money was to be spent. It was to be kept at the foundation and could be drawn on as needed.

This was the beginning of a wonderful association with Kettering and his colleagues at their laboratory at Yellow Springs, Ohio. It was to last for nearly two decades, continuing long after Kettering's death, and providing me with an irreplaceable source of support that made possible much research that I would have found it difficult, or even impossible, to pursue with funds subject to the usual restrictions.

In this connection, I recall a story Kettering told us at one of the yearly meetings the foundation scheduled at his palatial home near Dayton. As Ket told it, he was approached by Alfred P. Sloan, the president of General Motors, who asked whether he would not like to have a laboratory subsidy to further his interests in seeking the solution to the mystery of photosynthesis. Ket said he would be happy to have help, whereupon Sloan inquired whether something like ten million dollars would be enough. Ket thanked him but said he thought that was too much. They settled on a mere million.

However, Ket knew that despite the backing of Sloan, there would be trouble sooner or later from the accountants, who had to worry about the stockholders and fiscal responsibility. Sure enough, one day the chief accountant waited on him and asked in what category the expenditure of this large sum was to be entered. Ket said, "Call it 'insurance'!" This was a word the accountant understood and he went away, mollified for the moment. But it was not long before he was back inquiring, "Insurance against what?" Ket, in no way surprised or unprepared, replied, "Call it 'Insurance against surprise'!" I think this statement still stands as the best definition I have ever heard of industrial research.

Meanwhile, the effects of the bad publicity I had experienced were beginning to wear off. Prospects of offers from Stanford, New York University, and Harvard were materializing, in addition to renewed pressures to consider posts at the Weizmann Institute and at the Australian National University. I was also in correspondence with Phil Morrison at Cornell about some intriguing phenomena we had uncovered at Washington University, starting with a phone conversation I had had one day with Joe Kennedy. Calling from his office in the Chemistry Department, Joe had asked whether anyone knew how viruses multiplied. When I said no one did, he asked slyly whether anyone would like to know. Sure that Joe had something up his sleeve, I assured him that anyone lucky enough to learn how viruses reproduced could certainly be expected to be wined and dined by the Nobel committee in Stockholm. Joe then came up with his proposal. Suppose a virus molecule, highly specifically labeled with ^{32}P in its phosphate groups, reproduced. Would the progeny contain any of the labeled phosphate, or would all the label remain with the parent molecule? If the latter were true, this would mean the parent acted as a simple template on which progeny were manufactured and split off; if not, one could infer that material from the parent was distributed among the progeny by fragmentation and reincorporation. He suggested that we try tobacco mosaic virus, but I thought a bacteriophage would be better from the standpoint of quantitation, especially as Alfred Hershey on our staff in microbiology was an expert in handling it.

When I told Hershey about Kennedy's suggestion and my modification, he enthusiastically set to work devising protocols to test the scheme. The first set of experiments were, not surprisingly, inconclusive. We found that about 30 percent of the parent labeled phosphate appeared in the progeny. Kennedy then came up with his second suggestion: would not the radioactive decay of the labeled phosphate inactivate the phage by breaking phosphate-oxygen bonds, thereby causing disruption of the whole molecule, or at least enough damage to inactivate the virus molecules?

We did find that viral progeny grown in media of highly radioactive phosphate were markedly unstable. This raised the question of whether the energy released was concentrated at the site of the phosphate bond or was dissipated in vibrations and rotations distributed through the whole big virus molecule. We had calculated that the recoil energy of a sulfur atom leaving the site of the ^{32}P beta decay should certainly be enough to disrupt the original bond between P and O. Furthermore, it seemed certain that the turmoil induced in the participating atoms by the electron rearrangement after transformation of phosphorus to sulfur would certainly disrupt the bonding. Phil Morrison at Cornell made some more sophisticated calculations and concluded that our expectation that ^{32}P decay could kill phage with a higher efficiency than would be the case merely with beta decay was essentially correct.

The experiments we carried out in the course of the next few months demonstrated conclusively that the rate of phage inactivation could be related quantitatively to the amount of ^{32}P originally incorporated in the phage. Howard Gest had joined these efforts. His participation was especially helpful in that there were times when I was away and ^{32}P assays were needed. The research had to be discontinued after our initial findings because with the retirement of Professor Bronfenbrenner, Al Hershey, unsure about the policies of his successor, decided to leave and take up an appointment at the Carnegie Institution's Cold Spring Harbor laboratory. Our article—in which we used the term "suicide" to describe the self-destruction of the virus that had ingested the ^{32}P—was published shortly thereafter and became the starting point for a whole new area of phage research.[18] Al went on, helped by this start in isotopy, to win a Nobel Prize in 1969.

In connection with this research, and because of my interest in the general mechanisms whereby radiation affected living tissue, I proposed an informal symposium to bring together the best minds in radiation physics and physical chemistry to interact with radiologists. In this way the insights of both kinds of scientists might be brought to bear on the basic problems in radiobiology. Ideas would be emphasized rather than work in progress, minimizing compartmentation of the various disciplines. The conference was held in Highland Park in the spring of 1951, and the proceedings were later published.[19] They explored the best ideas then available on the nature of mechanisms involved in radiobiology. Somehow, I also found time to completely rewrite and revise a new edition of *Radioactive Tracers in Biology* for the Academic Press. The sensational success of the first edition and the rapid obsolescence of portions of the text made this task obligatory.

In the laboratory, research was moving well. Howard had finished his thesis and obtained an appointment as instructor in the Department of Microbiology at Case–Western Reserve University. Jack Siegel had recovered brilliantly from the setback he had experienced with Foster's strains. He had isolated several new strains that carried out the metabolism of isopropanol and acetone. Supported by a U.S. Public Health Service fellowship, his research had finally established, using isotopic labeling, that isopropanol could be converted directly to acetone by simple removal of hydrogen. The further conversion of acetone to cell material proceeded by condensation with CO_2 to form acetoacetate. Siegel helped demonstrate that photoproduction of hydrogen was a general phenomenon, not restricted to a particular bacterium, such as *R. rubrum,* or to a particular carbon source. His results on the relationship between amounts of H_2 and CO_2 produced and amounts of carbon source (acetate, malate, or oxaloacetate) assimilated were in accord with the view that two-carbon fragments formed by degradation of the carbon sources were incorporated directly into cell material.

After her momentous breakthrough, Herta was successful in carrying forward her experiments on quantitation of photohydrogen production as influenced by nitrogen sources. She demonstrated that when one mole of malate disappeared, one mole each of H_2 and CO_2 were produced in the absence of N_2 or ammonia. In their presence, only the CO_2 appeared. She offered exhaustive data on ammonia uptake and inhibition of photohydrogen production, showing that there was no definite (stoichiometric) relation between the amounts of ammonia disappearing and the H_2 expected to be produced from the quantities of malate that disappeared. This encouraged the view that N_2 (or NH_3) interacted with the enzyme systems involved in some catalytic manner. Her thesis was accepted a year after Jack had left, and was the last such doctoral publication I would supervise for almost a decade.

Over the next several years, my research effort went forward, aided by the highly competent and hard-working collaborators who came from Europe to work with me on photometabolism. These included Stanley Ranson from Newcastle, John Glover from Liverpool, Raymond Collet from Geneva, and H. Van Genderen from Utrecht. Funds from the Commonwealth Foundation and the World Health Organization supported their stays. The next decade saw more than thirty major publications on light and dark metabolism in photosynthetic bacteria, based largely on the pioneer efforts of these co-workers in my laboratory. Coming from the Department of Microbiology was the young Samuel Ajl, whose collaboration with our group provided additional momentum.

Summarizing our findings in a review paper at a Federation of American Biological Societies meeting in 1950,[20] I noted—prophetically as it turned out—that the activity associated with photohydrogen production was dependent on nitrogen metabolism and was different from the hydrogen activating ("hydrogenase") systems previously found in bacteria. One major difference was the fact that while non-photosynthetic evolution of hydrogen could be reversibly inhibited by H_2 gas as hydrogen pressure was increased, this was not the case with photohydrogen evolution. In later years, it developed that the same enzyme complex that catalyzed N_2 uptake and fixation was also responsible for H_2 evolution, both in nitrogen-fixing photosynthetic and nonphotosynthetic microorganisms. Our early findings predated an upsurge of interest in the commercial exploitation of photosynthetic systems for bioenergy conversion some two decades later.

Among those recruited to the study of non-sulfur photosynthetic bacteria was the eminent biochemist and future Nobelist Severo Ochoa, who received a starting culture of our strain of *R. rubrum,* in return for one of his, a bacterium that metabolized a dismutation of pyruvate to lactate, acetate, and CO_2. We were interested in its use as a means of determining amounts of malate quantitatively. The friendly interchange of ideas which thus began lasted many years.

One might suppose that with all this activity I had little time for a private life. This was not at all the case. I was as active as ever in chamber music groups. But, more significantly, there was considerable progress in my pursuit of Beka. She had begun to experience extreme disillusionment with the life of a journalist in New York. The atmosphere at *Time* was depressing, despite the unlimited expense accounts, unrivaled library resources, and prestige associated with being on its staff. Women researchers were kept subordinate to male colleagues, who rewrote material the women provided. Although researchers such as Beka had the ultimate veto over articles based on their findings, there was no assurance that the stories would survive the bias and news slanting constantly attempted by the senior editors. The distortion of fact in *Time* articles was notorious, and the magazine was held up as a model of irresponsible journalism by schools of journalism. In addition, there was a cabal of informers organized by Whittaker Chambers, who held a high position at *Time,* to identify and root out "leftists."[21]

The repression of truth was shown dramatically in one instance involving T. S. Mathews, the managing editor. Beka and George Burns, her editorial associate, had written a masterful and factually correct account of the operation and philosophy of the Menninger Clinic in Topeka. Mathews, who regarded all psychologically disturbed people as essentially "malingerers," rewrote the story, putting in unacceptable slurs on psychiatry reflecting his personal bias. Beka was able to veto the insertions, paying the penalty of incurring his ire. Feeling sure that her dismissal was only a matter of time, she decided to venture into the free-lance field.

Her first such effort, as I have already mentioned, was a book for the layman on cancer in collaborative effort with C. P. Rhoads, the head of the Sloan-Kettering Institute. It was a brilliantly conceived and written effort, almost wholly Beka's, as Rhoads did little except lend his name and make available the resources of the institute for gathering data and information. As soon as the book was finished in early 1949, however, Rhoads demanded that Beka's name be removed and he be listed as sole author. He pointed out that sales would be promoted by use of his name alone.

This professional arrogance revolted Beka's editor at Random House, Saxe Commins, and publisher Bennett Cerf, who courageously sided with Beka. Rhoads thereupon withdrew his name completely, and Random House was left to publish the book with Beka as the sole author. As Rhoads predicted, sales were disappointing, although the book was certainly a "succès d'estime."

This experience, together with the traumas at *Time,* helped convince Beka that she should leave New York and succumb to the temptation of marriage and continuation of her writing career in the less profitable, but more restful, ambience of St. Louis. There was no ignoring the fact that we were very much in love, and the separation forced on us by her remaining in New York had become intolerable. We were married at the Ethical Society headquarters in New York on March 16, 1949, with the Reverend Jerome Nathanson presiding. Beka's sister, Anne, was the bridesmaid, and Albert Deutsch, the noted social commentator, the best man. A small and motley group of witnesses, mostly friends from *Time,* gathered later at the home of George Burns. We crept away, unnoticed, and slept away the night in a kind of exhausted, roseate haze.

With life at home in St. Louis and in the laboratory fruitful and happy, I was not worried by the frequent reminders that I was still under a cloud, as evidenced by reports reaching me that the FBI was continuing its questioning of my friends and acquaintances. I was chiefly concerned about my continuing failure to obtain a passport, despite representations from high officials in the Australian, French, and Israeli government offices in Washington, inquiries by the National Academy of Sciences through its president, Dr. A. N. Richards, requests by Chancellor Compton, and the continuing efforts of Milton Freeman at the office of Arnold, Fortas and Porter in Washington, who convinced me that I should seek legal help. Freeman and Ed Condon recommended that I write to a young and brilliant lawyer, Nathan H. David, who practised in Washington and had been the prime mover in collecting and organizing material prepared for publication as a manual on the laws, regulations, and procedures involved in government security and loyalty cases.

I wrote to David on October 5, 1950 and received a heartening answer immediately. He knew something about my troubles from talks with Ed Condon and others, and replied to the effect that he very much wanted to help me, not only because of the "extreme frustration you personally have suffered but also because of the truly basic questions of principle that are involved."[22] He thought something could be done and invited me to visit him in Washington, bringing all the relevant papers (of which there were a great pile) so that he could get started on the case.

David's conviction that basic questions were involved, over and above whatever personal injuries were being inflicted on me, was eminently justified. The State Department's passport policy was essentially government by fiat and not law. There was a legislative no-man's land in the matter of passports, because Congress had enacted no law defining the authority of the secretary of state as to his discretion in their issuance. Any such law might well encounter sharp scrutiny from the courts, because freedom of travel could be argued to be a constitutional right, and not merely a privilege awarded at the pleasure of the secretary of state. Denial of a passport—an absolute requirement for travel abroad in modern times—could only legally take place, David argued, after a proper judicial hearing with all the safeguards of the process, such as a clear statement of charges, examination and cross-questioning of witnesses, and a decision arrived at by fair evaluation of evidence presented. We had precedents for the contention that freedom to travel was a right going back to the Magna Carta of A.D. 1215, Articles 41 and 42, which stated that anyone should be able to leave and return from the kingdom safely and securely by land or water. They mentioned "merchants, both English and foreign" and restricted the right of free exit and entrance only in the cases of prisoners, outlaws, and natives of countries at war with England. In the absence of any pronouncements by Congress, recourse could traditionally be made to common law of which there was no more weighty source than the Magna Carta.

As detailed in the many briefs and articles generated by suits against the State Department then and later, Mrs. Shipley's actions were on exceedingly shaky ground, even though they were based on a series of executive orders written to fill the policy vacuum created by the inaction of the Congress.[23] One of these quite typically read: "The Secretary of State is authorized in his discretion to refuse to issue a passport, to restrict a passport for use only in certain countries, to restrict it against use in certain countries, to withdraw or cancel a passport already issued, and to withdraw a passport for the purpose of restricting its validity or use in certain countries."[24] In other words, there was no need to consider due process as far as the State Department was concerned. It gave itself the power to act as arbitrarily or capriciously as it wished on grounds of its mandate to conduct foreign policy.[25]

To successfully challenge the position Mrs. Shipley was taking, we first had to press for a hearing in the State Department at which all charges could be heard and rebutted, bringing witnesses to uphold our case. David expected that such a hearing would probably force disgorgement of a passport. If it did not, he was prepared to carry on the fight in the courts. We had no illusions about the possibility that there would be a tremendous drain on time and resources.

I needed a passport to pursue my scientific career, however. I needed to visit the laboratories of colleagues wherever they might be, as well as to receive visits from them, so that the free interchange of data and ideas essential to scientific research could go forward.

13

War with the Establishment As Science Moves Forward

(1951–1953)

NATE DAVID began his campaign by obtaining an audience with Mrs. Shipley. He discovered that she had developed a strong bias in the case, claiming that I had "deceived" her. She felt I had downplayed my alleged importance on the Manhattan Project. She apparently regarded the HUAC hearings as making a substantial case against my loyalty to the country. When I had talked to her the one time I had been in her office a few years previously, she had not seemed so hostile. It was evident that she had been completely taken in by material from my security file.

Obviously, Mrs. Shipley had become our main worry. At Nate's suggestion, I wrote her again, rehearsing all the reasons why I needed a passport, citing the invitations to lecture at universities in Australia and other places abroad, and the prejudice to my career in limiting my freedom to travel. This effort brought only a restatement of her position. Nate then spent some time trying to work through his friends in the State Department. During the course of these efforts, I suggested we see what Thomas Hennings, the senior senator from Missouri, could do.

Meanwhile, Mrs. Shipley sent Nate a letter, dated March 13, 1951, in which she cited the usual claim that the secretary of state (and therefore she) had the right to act as seen fit in the issuance of passports as part of the conduct of foreign policy. She threw us a small morsel of hope by saying that she might change her mind if warranted by new facts in any future application. We then tried the strategy of Senator Hennings's intervention. This resulted only in a contemptuous brush-off of the senator's queries. Paradoxically, the State Department was seeking at this time to give the impression that it actually wanted to expedite travel for scientists. In the April 6 issue of *Science,* there was an official announcement of the establishment of the office of science advisor in the State Department, ironically enough alongside an article describing the new Australian National University, written by its vice-chancellor—whose invitation was one of the reasons I was seeking a passport.

Still grasping at straws, I reported to Nate that there had been an article in the August 1950 *Reader's Digest,* written by one Sidney Shallet, on the trial of some communists the previous year. A witness named "Balmes Hidalgo" testified that he had penetrated the Communist Party as an informant for the FBI and had heard someone say at a meeting than an important Party official was named Marty Kamen. Later, at a Party convention in June 1948, this same Marty Kamen had presided at a meeting where he had been heard to advocate using Hitler-like tactics to overthrow the government. I thought it possible that I was being confused with this character. Nate spent some of my scarce money and his time having a lawyer in New York track down this testimony, as it appeared in the records of the New York Circuit Court of Appeals. However, this inquiry led nowhere, as no one seemed to have heard of the communist Marty Kamen subsequently.

I was beginning to feel the pinch financially. My take-home pay was only about $500 a month and my expenses were nearly as much. Even the $42.43 we had to give the New York lawyer to run down the dead lead was a considerable strain. Though Nate was not charging for his time, there could, moreover, be expected to be a constant drain in the shape of various fees for preparing and filing legal documents, telephone calls, and other incidentals. I was regularly sending small sums to my father, who was finding it increasingly difficult to make ends meet. Even with the help of occasional earnings Beka contributed and my book royalties, we were just managing to keep up with the minimal demands our struggle was generating by practicing severe economies at home.

Despite living so marginally, 1950 had been a wonderful year for the family. Beka's health had improved markedly, and she was busily engaged as a writer and mother (our son David having been born on April 24). I was devoting much of my attention to the developing research program at the laboratory, where the metabolic versatility of the photosynthetic bacteria posed an infinity of exciting challenges. I had decided to tool up for an attempt to sort out the effect of photoactive (actinic) light on the dark (nonphotochemical) processes in *R. rubrum*. The early observations in the middle thirties by the Japanese investigator H. Nakamura were of particular interest. He had shown that respiration (uptake of oxygen) was inhibited by light in a photosynthetic bacterium, *Rhodopseudomonas palustris*, of the same family as *R. rubrum* (that is, a member of the so-called non-sulfur purples). If a culture of these organisms was incubated in the dark with any of a variety of carbon compounds on which they could grow, they would show the expected uptake of oxygen typical of a respiring organism. In the light, the carbon compounds continued to disappear at more or less the same rate, but the oxygen uptake dropped off, ceasing altogether at a high enough light intensity. It was reasonable to explain this by assuming that photosynthesis was accompanied by the production of hydrogen acceptor(s), which competed with oxygen for the hydrogen coming from the carbon compound. This might be the long-sought precursor of oxygen that produced oxygen in green plant photosynthesis but failed to do so in the case of bacteria. (Since bacteria cannot evolve oxygen photosynthetically, they would need a hydrogen donor as an alternative to react with the oxygen precursor.)

We set out to sharpen our skills in various forms of chromatography, which we combined with isotopic labeling techniques to help sort out the products of metabolism in light and dark. The technique best suited to our purposes was paper chromatography. This involved pipetting a drop of an extract from an experiment, in which all its labeled components produced were contained, onto a defined spot on the paper and drying it. Then the paper was placed in a closed chamber with one edge dipping into a solvent, or special mixture of solvents, which flowed through it by capillary action, eventually traversing the whole paper. Each component of the extract moved along at a characteristic rate, depending on its solubility in the solvent mixture. The result was a one dimensional chromatogram—a series of compounds spread out as spots in a linear array. The paper was then dried, and replaced with its edge at right angles to the chromatogram of the original mixture in a new and different solvent system. A second wetting then took place, with each compound moving away from the spot it had reached previously in a direction at right angles to that of its original motion. Each compound moved, as before, at a rate characteristic of its solubility. A two-dimensional chromatogram thus resulted, with the various labeled compounds separated out in different spots on the paper. After drying, the spots could be visualized by pressing a photographic film on the paper and exploiting the radioactivity of each spot to form a photographic impression on the film. After removal of the film, chemical tests could be applied to the paper to locate known compounds, the placement of which could be compared with that of the radioactive spots by juxtaposing the developed film with the chemically treated paper.[1] This technique, combining isotopy and chromatography, was coming into use all over the world in biochemical laboratories where experiments were being performed to establish the pathways of metabolism in living systems. It was the basis, for example, of the success of Calvin, Benson, and their associates in establishing the pathway of carbon from CO_2 as photosynthesis proceeded in green plants. Other types of chromatography, involving columns of absorbent rather than paper, were also widely used. This technology was to lead to a veritable explosion of new information and help create the new science of intermediary metabolism.

By the spring of 1951, we had established that the distribution of labeled carbon in carbonate and cell material in nongrowing cells, produced during short periods of dark respiration on acetate, was the same whether the acetate was labeled in its methyl carbon (C^*H_3COOH) or in its carboxyl carbon (CH_3C^*OOH). This was not seen in the light assimilation of labeled acetate in the absence of air, where the methyl-labeled acetate gave activity practically entirely in cell material and very little in water-extractable fractions or carbonate, whereas practically all the activity appeared in carbonate when carboxyl-labeled acetate was used.[2]

We could conclude that the operation of a cyclic mechanism that mixed the two carbons of acetate, as in the well-established Krebs citric acid cycle, worked in dark aerobic respiration, but not in anaerobic photosynthesis. Moreover, acetate carbon was assimilated directly into cellular compounds without first being oxidized to carbonate, a finding in accord with reports from other laboratories. Extension of such experiments to growing cells, using labeling times ranging from a few seconds up to several hours, established that when labeled carbon was supplied as carbonate with only molecular hydrogen as reductant, instead of organic compounds (such as acetate), the intermediates formed were like those found in green plant photosynthesis—namely, phosphoglycerate and phosphorylated sugars. Adding unlabeled acetate when labeled carbon dioxide was present lowered incorporation of labeled carbon into these compounds. A very large fraction of labeled acetate went into lipids even at 0° C. These results and others with nongrowing cells differed from those with growing cells (supplied with growth factor and a nitrogen source). In these cells the water soluble fractions were mainly the usual products of green plant CO_2 fixation. In dark aerobic metabolism, labeling patterns more or less similar to those obtained in the light anaerobic photometabolism were seen. Experiments on acetate metabolism with the nonphotosynthetic bacterium *Micrococcus lysodeikticus,* carried out with Sam Ajl's group, provided some evidence that a Krebs cycle was occurring in its dark aerobic metabolism, as was seen in the dark with the photosynthetic *R. rubrum.*[3]

In the meantime, Jack Siegel had finished his investigations on the metabolism of isopropanol,[4] with the results I have described previously, and Herta Bregoff had demonstrated that photohydrogen production could also be found in other photosynthetic bacteria e.g., *Chromatium* species.[5] Her thesis work was summarized in a paper, the twelfth in our series of studies on photosynthetic bacteria, in which we reported the quantitative relations between disappearance of malate, photoproduction of hydrogen, and nitrogen metabolism.[6]

These results showed progress made in a preliminary exploration of photometabolism in *R. rubrum.* To proceed further, however, I would have to devise a long-range program, using a variety of growth compounds, dissecting the various enzymatic steps in their overall metabolic pathways and trying a great variety of different conditions. Experiments would involve control of different phases of growth, possibly synchronizing culture development (a trick not yet reported as accomplished by any of the microbiologists who had studied the physiology of photosynthetic bacteria). To even consider such a program, I would need a greatly expanded corps of associates and added resources in equipment and supplies. Support for such expansion was unlikely. Moreover, I would have no research fellows or graduate students in residence in the fall.

Alternatively, there was another problem I could tackle, along the line of resolving Nakamura's phenomenon—the interaction of light anaerobic with dark aerobic metabolism. As yet, research in this direction required more ideas than manpower. I wanted to keep my objectives sufficiently restricted to be able to make reasonable headway working by myself with one postdoctoral fellow and a small laboratory staff, which was already in place, funded by my radiobiology budget. I therefore reoriented research toward an analysis of the enzymic composition of the bacterial chromatophores, as compared with the chloroplasts of green plants. I kept in mind the possibility that phosphorylation reactions would turn out to be important parts of the energy conversion mechanisms in both micro-organelles.

To obtain a subsidy for support of a postdoctoral fellow, I approached the Public Health Service for a "grant-in-aid." I already had a candidate in mind to fill the fellowship position. He was Leo P. Vernon, whom I had first met when Arthur Hughes and I had visited Harland Wood and Les Krampitz at Ames. Since then, Leo had spent some time at the University of Wisconsin, working on enzymes involved in biological oxidation. The particular enzyme Leo had worked with was one that reduced cytochrome *c* in mitochondria. According to prevalent dogma, cytochromes (the respiratory proteins, basic to reduction of oxygen, discovered by David Keilin decades earlier), and especially cytochrome *c,* were not expected to participate in any anaerobic type of metabolism such as we expected to encounter in the photosystems of chromatophores and chloroplasts. The former did not use oxygen in the light and the latter actually had a mechanism for eliminating it. Although he had worked only with mitochondrial enzymes, Leo had a good grounding in general enzymology and could be expected to adapt to whatever course of research might develop. He promised to come the following September if I got the money from the National Institutes.

Of great importance in improving my background in metabolism and enzymology was the arrival of Arthur Kornberg at this time to take over as chairman of the Microbiology Department at the medical school. He quickly built up one of the leading centers of research on nucleotide synthesis and metabolism, work that led to his Nobel award some years later. Arthur inaugurated a "brown-bag" lunch seminar in which we met at regular intervals to discuss papers of interest. These sessions were great learning experiences for me. Whoever was assigned to lead the discussion would prepare a short typed summary of relevant data provided in the paper. We would examine these without prior consideration of the author's summary and contentions and decide what they showed. Then we compared our conclusions with the author's. It was surprising how often authors of great prestige in the biochemical community failed this test.

Our dreams of a wonderful summer vacation on the West Coast and in Wyoming were abruptly shattered in the early morning hours of the day we were scheduled to leave. Around 2 A.M. the phone rang. Answering it sleepily, I was jarred awake by a reporter from the United Press saying he had what might be bad news for me. There was a story in the *Chicago Tribune* identifying me as the "atom spy" denounced on the floor of the Senate a week earlier by Bourke B. Hickenlooper, the Republican senator from Iowa.[7] The reporter explained that there had been a story on this speech the week before.[8] The senator had named no names but had claimed that he had a photo showing "one of the most prominent persons" working on the Manhattan Project walking with Soviet agents to whom he had given "secret papers." This person, he alleged, was a "traitor and spy" and was "selling secrets." Further, he charged, "administratively higher authority" had prevented General Groves and the Manhattan Project from prosecuting.[9] As I sat up in bed, dazed by this incredible libel and attempting to assimilate its impact, Beka took the phone and sharply questioned my informant as to his identity. Satisfied that the report was authentic, she said I would have no comment until I had seen the texts of the reports.

We slept little more that night. The phone rang again around 8:30 A.M. This time it was Richard Dudman, a reporter from the *St. Louis Post-Dispatch*, asking if I had heard about the *Tribune* story. I said I had, and that if the story was a reprise of the HUAC affair it probably concerned me—in which case there was nothing new, as I had been cleared in 1948 of any imputation of disloyalty. When Dudman remarked that Hickenlooper's speech had been provoked by an attack on General Groves by Senator Brian McMahon (Democrat, Conn.), I hazarded the opinion that I seemed to have gotten into the middle of a senatorial row that had resulted in dredging up the old canards. Soon thereafter, I had a call from Chancellor Compton to tell me that he had been awakened by the U.P. reporter on the phone about an hour before I had gotten my call. Compton had told the reporter that there was no substance to the story, and that he and the university would support me if necessary. Dudman's story appeared later in that Sunday's edition of the *Post-Dispatch.*

Beka and I had an intense and anguished discussion of what to do about our trip with our departure due in an hour. We concluded that there was no point in not going—indeed, we might need to get away from the hubbub until sufficient time had elapsed for us to better judge the situation. Having made arrangements to have our infant son stay with his nurse, a motherly and devoted woman, and leaving our hairy retainer, the long-haired dachshund Eno, at a kennel, we left as passengers in a car driven by Dr. John Frerichs, who was going west to take up a post at Fort Miley in San Francisco. We started off on a journey that took us through woodsy western Missouri, the parched and dusty plains of the Oklahoma and Texas Panhandles, the Colorado Rockies, and the grim wastes of the Humboldt Sink in Nevada to San Francisco.

It was a shock, however, to come back two weeks later to the grim reality of the situation we had left and fully appreciate the enormity of the slander the senator from Iowa had perpetrated. My father sent copies of the two *Tribune* stories, which provided not only flagrant distortions of the celebrated Fish Grotto episode but included a photograph purporting to show me walking with Russian consular officials, identified as Kheifetz and Kasperov. In Kheifetz's hand, the text went on to claim, there were "secret papers" I was supposed to have given him. This villainous piece had been published in my home town, where its effect on my family and friends would be most devastating.

Very soon, I became more and more paranoid. I was convinced that somewhere some faceless sociopath was dedicated to my annihilation, and that there was nothing I could do about it. My very survival, and that of my family, was at stake. As the pressure mounted, the relationship between Beka and myself became strained and we took out our frustration and despair on each other. I concluded that I had become an intolerable burden, not only to my family, but to Chancellor Compton and the university as well. One black night I reached bottom. In a suicidal fit, I made an abortive attempt to do away with myself. Beka discovered me lying on the bathroom floor, bleeding from numerous self-inflicted cuts on the face and throat. Fortunately, the knife I had seized had been dull. Terrified, she managed to reach Dr. George Saslow, a close friend of ours and an accomplished practicing psychiatrist as well as a member of the Washington University Department of Psychiatry. He hastened over and after some time brought me around to a more rational frame of mind.

The experience appears to have been salutary. After a month of shame and remorse, I started to climb out of the depths, developing a hardened resolve to fight back and not give up until I had won vindication. I wrote to Nate, enclosing copies of the stories and inquiring whether he thought they constituted an actionable case of libel. Nate, never one to go back on principle, answered that there was no doubt I had a case and that I ought to go ahead with it. He pointed out, however, as I was already dimly aware, that I might be badly damaged in bringing such a suit, as the *Tribune* would "throw the book" at me and I would probably lose. I might be so badly smeared that I might wind up completely unemployable. Nevertheless, he advised that we sue.

At Nate's suggestion, I went to see the chancellor to ask what his attitude and that of the university would be, in view of the mud that would be thrown not only at me but at the university. Compton was sympathetic, but warned me that one hardly ever "won" a libel suit. He agreed that I might be lucky and obtain a cash settlement, but said he needed to consult the university's lawyers as to the facts of the case before rendering an opinion. After several months, the university lawyer told him that if, in the chancellor's judgment the suit had a chance of winning, the university should not stand in the way of my seeking vindication, but on balance he still advised against proceeding. I sent a copy of the lawyer's letter to Nate, who wrote Compton explaining why he thought we should bring suit. Compton finally told me he would not oppose our effort to bring the *Tribune* to heel. This was a remarkable demonstration of his courageous insistence on seeing justice done, even if his own reputation might be damaged.

In the light of the Hickenlooper speech, Mrs. Shipley's behavior was not surprising. She had been taken in completely by the whole fabrication and was convinced that I had escaped conviction as a spy or traitor in some infernally clever way. At the least she was sure I was a subversive who had pulled the wool over the eyes of so eminent and irreproachable a patriot as Chancellor Compton. In her view, as she stated a few years later, our legal efforts were communist-inspired tricks to further subversion of the government. Quite obviously, it was her bounden duty to foil them. As a subversive, I was not entitled to consideration in a proceeding involving "due process"—a concept she never did come to comprehend. Lord Acton's observation that "Power tends to corrupt, and absolute power corrupts absolutely" might have grown hackneyed, but her behavior forcefully illustrated its truth. With Cold War fears increasing in intensity and so many like-minded congressmen and senators supporting her views, there was no one to say her nay in the State Department. For all practical purposes, as far as passport policy was concerned, Mrs. Shipley *was* the State Department. To remove her as a road block would require a long, expensive legal battle. Before we could even get to the courts, we would have to establish that whatever procedures existed for appeal from Mrs. Shipley's decisions in the State Department were either inoperative or failed to provide due process. There was also the problem of getting a definitive ruling from the courts that passports were a right and not a privilege, and hence protected by constitutional guarantees of due process. Eventually we hoped to convince the courts to take over adjudication of passport cases.

Nate could not possibly handle this burden as well as cope with the libel action, so he sought help. He found it, to our everlasting good luck, in Alexander E. Boskoff, whom he knew to be a remarkably astute and principled trial lawyer. Boskoff had an intimate knowledge of the workings of the federal establishment, having put in some time in the Department of Justice, and was actuated by the same strong desire for fairness and justice as Nate. He, too, was outraged by the way I had been treated.

Fighting on two fronts, the piles of paper mounted rapidly, as did the expenses. At first I met the bills by taking out loans on insurance and mortgaging everything I owned, but soon I had to seek added support. I had recourse to my cousin in Washington, D.C., Louis Bean, who generously provided some money from his own pocket and also mobilized some support from the family in Canada. In the next four years, this amounted in total to no more than a few thousand dollars, but it proved to be absolutely crucial in helping us to survive.

Nate's opinion as to the basis for libel was that although Hickenlooper's remarks, as reported in the first story, might not be libelous taken singly, the gratuitous publication by the *Tribune* of the second story identifying me and publishing the picture allegedly of me with the Russians was libel per se. Hickenlooper had carefully avoided mentioning me. Instead he had looked the other way while one of the henchmen in his office gave privileged information to Willard Edwards, the reporter who had written the story. The senator could even hide behind the cloak of congressional immunity, if need arose, and leave the *Tribune* holding the bag (as did actually happen when we brought our libel suit against the newspaper). Nate held that the only defense the *Tribune* would have would be to prove the truth of its allegations that I was a "spy" and "traitor," that "one of the Russians had in his pocket certain secret papers" that I "had transmitted to him," that I was "selling secrets," and so forth. It was hard to see how such fabrications could be proved to be true, especially as I had documentation and testimony completely refuting these allegations. Meanwhile, the case against the State Department could proceed. I continued to plague Mrs. Shipley with copies of invitations I kept receiving. She continued to turn down my requests for permission to accept them, or ignored them completely.

Nate fired the first shot against the *Tribune* in December 1951, attempting to establish that the paper was doing business in the District of Columbia, as it owned and administered an office there. It also ran the *Washington Times-Herald,* which we were told Colonel McCormick had taken over after a family spat with his niece, "Cissie" Patterson, just prior to the publication of the libel. It was obviously unwise to bring a suit against the *Tribune* on its home grounds in Chicago. Nate lost on this argument and for lack of adequate funds to appeal the decision, we had to be content at the time with the *Times-Herald* as the sole defendant. With expenses mounting, I asked Louis Bean if I might seek help from Abe Spanel, the president of the Playtex Corporation, who was a close family friend. However, Abe had just been through a wringer winning a libel suit he had brought against reporter Westbrook Pegler, and accordingly was doubtful I was on the right track. He felt that I should not proceed with the case in view of his experience, and especially the uncompensated expense incurred in winning mere vindication.

Nate, still determined to prove his point that the *Tribune* could be sued in the District of Columbia, subpoenaed Willard Edwards and Walter Trohan, the head of the *Tribune*'s Washington Bureau. He anticipated trouble in getting them to testify, as they would claim newspapermen's privilege, and expected that he would have to get a court order to proceed (again an expensive business). The deposition he finally obtained from Trohan was not particularly helpful, but that from Edwards provided some interesting information. From him we learned that HUAC was still very much in business. It had published several summaries of its investigations into Soviet espionage, in which it had gratuitously included a sideswipe at me entitled "Charges Concerning Martin David Kamen," under the category "Charges Not Proven in a Court of Law."[10] Edwards had looked up a HUAC investigator, one Donald Appel, who had given him the picture of me with the Russians. Edwards had also been given access to the HUAC files and seen material otherwise not publicly available. The picture had been obtained originally from a former captain working in Army intelligence on the Manhattan Project named David S. Teeple. This officer had been in charge of the security detail reporting on my movements. The picture was claimed to have been taken by agents following me under his direction. Teeple had become convinced I was involved in espionage and, frustrated because he could not offer proof, had done the next best thing in supplying the picture to the committee. Edwards also claimed that Teeple had bragged in his interview that he had actually thought of having me done away with for the security of the United States![11]

The *Tribune,* hoping to get some support for its case, forced an appearance by Teeple. An Army representative was present to make sure Teeple divulged nothing of significance, however, and the deposition proved unfruitful. At this juncture I suggested we approach John Lansdale, now a lawyer in Cleveland, who had been the colonel in overall charge of Army intelligence on the Manhattan Project, and thus Teeple's superior. It was Lansdale who, on being questioned by Compton about my being hired at Washington University, had assured the chancellor that there were no grounds for questioning my loyalty to the United States.

To provide us with some diversion on the Shipley front, there was a laudatory piece in *Time* at the beginning of the year that called her the "actual" secretary of state in passport matters, applying terms such as "most respected, feared, unfireable, admired" and quoting Franklin D. Roosevelt, who had called her his "marvelous ogre." We did not find this enlightening.

In February I wrote Lansdale asking for an appointment to see him at his office in Cleveland when I would be in the vicinity in April giving lectures at Western Reserve University Medical School. Lansdale was cordial in his answer and agreed to meet me. In the meantime, more notes from HUAC files were made available to us in the Edwards testimony. In them, Lansdale, appearing as a witness at a committee hearing in July 1948, stated that a "Colonel Park" had supervision of the duo following me—two agents named Wagener and Zindle. Wagener had later been hired to oversee security at the Union Terminal in St. Louis. I even talked with him by phone but found him, not unexpectedly, wholly uncooperative. Lansdale's testimony as to how the picture was taken was at odds with that of the other agents and with Hickenlooper's story that it had been taken by a "telephoto" camera. Lansdale said that a street photographer had taken it and, when I refused it, had given it to the agents following me.

When I saw Lansdale in Cleveland, he assured me that he wanted to help me straighten out the mess, but would not appear as a witness. He reiterated that while he thought I had been extremely careless, I had done nothing to warrant the extreme position taken by Teeple, whom he characterized as an excellent investigator, but fanatically anti-Communist. Lansdale reinforced my impression that Teeple had focused his frustrations on me, with resultant amplification of my dossier. I now had the satisfaction of knowing, rather than feeling, that the whole case against me was a fabrication.

During the summer, the State Department at last issued specific regulations regarding appeals against refusals of passport applications. In response to court action by the American Civil Liberties Union on behalf of a writer, Anne Bauer, whose passport had been arbitrarily revoked by Mrs. Shipley without explanation, a federal court ruled that the constitutional guarantee of due process had been violated and that the State Department procedures required revision. The secretary of state, Dean Acheson thereupon issued new regulations, providing for an appeals board that would conduct hearings in accord, presumably, with some regard for due process. The Truman Administration was entering its final days, however, and it was left to the incoming secretary, John Foster Dulles, to implement the decision and create the board. No test of the appeals procedure had yet been made, because Miss Bauer was unwilling to take the chance of being prohibited from leaving the United States once she returned. Instead, she renounced her citizenship and stayed in France, where she married.

Nate suggested I reapply to go to Australia so that we could test the procedure if another refusal came through. He also noted that there would be some action on the *Tribune* case, as one of the lawyers retained by the *Tribune,* Perry Patterson, would be taking a deposition from me some time during the year. Nate wanted to keep it from becoming a fishing expedition by limiting the procedure to written interrogation. He was worried by my tendency to talk too much. I accordingly submitted an affidavit claiming that teaching and lecture commitments interfered with the dates Patterson wanted to set, as well as generating travel expenses. (I was feeling a real pinch at the time because I had to find $1,000 to enable my father to be eligible for Social Security.) This argument, not unexpectedly, came to naught. I was ordered to appear for deposition, either in Washington or Chicago, by the end of the year. The date was finally set for December 20, 1952. It would be my first test under fire.

I was spending less time in the laboratory, as I had to meet a mass of lecture and teaching commitments and to finish writing papers on the work our group had done through 1951 and the early part of 1952. In addition, I attended meetings of various government research committees. Two important symposia also took place in this time—one on radiobiology sponsored by the National Research Council at Oberlin, Ohio, and another on phosphorus metabolism at Johns Hopkins University. At the first, I presented a general report on localization of radiation effects in biologically important molecules, mostly dealing with the experiments Hershey and our group had done on effects of ^{32}P recoil and radioactive decay in inactivating bacteriophage. At the phosphorus symposium in Baltimore, I presented a paper, with Howard Gest as co-author, dealing with the remarkable phenomena we had uncovered in *R. rubrum* relating to photohydrogen production and nitrogen fixation. (I used the word "serendipic" in the title; apparently, it was unknown to some of my auditors and aroused much comment.)

In February, I received a letter from the Public Health Service informing me that Leo Vernon's fellowship would be funded by a grant from the surgeon general ample to cover his salary for a year, with a commitment to continue the grant at the same level for two more years. In preparation for Leo's arrival, I spent much time mulling over the approach we might use in establishing whether a process analogous to the Hill reaction in green plant chloroplasts might occur in bacterial chromatophores. There was an obvious difficulty. If the primary reaction in photosynthesis was a "splitting" of water to produce oxidants and reductants simultaneously, how could we supply corresponding reagents that would not react with each other instead of with the photochemically produced systems?[12] To get around this, we would most likely have to exploit the remarkable specificity of enzymes—but which?

We began by looking at the chromatophores themselves to see what enzymes were present in them that might be useful. In sonic disruption we had a simple and effective way to break up the *R. rubrum* cells and make viable chromatophores. This procedure had been successfully applied by others in making cell-free extracts in cases where alternative procedures, such as grinding with abrasives or treatment with detergents, had inactivated enzyme systems. Leo's previous experience caused us to begin by looking at enzymes associated with cytochromes even though we did not expect to find any such enzymes in the airless photosystems of chromatophores. Leo also looked for oxidation-reduction ("redox") enzyme systems, which were known to be involved in oxidation of Krebs cycle organic acids, such as succinic, malic, and lactic acids.[13]

All the enzyme activities associated with cytochrome *c* were present in the chromatophores. We were surprised, as we had thought that these enzymes, the basis of the physiological ability of *R. rubrum* to grow aerobically, as is the case with yeasts, would not be found in the chromatophores. In plants and algae, they are found in the mitochondria. Of course, what bodies might correspond to the latter in a bacterium, itself only about the size of a mitochondrion, remained for speculation.

Another surprise came when we found that chromatophores under illumination showed a substantial increase in their ability to oxidize cytochrome *c*. In the dark, cytochrome *c* remained essentially unoxidized. It seemed to be foregone conclusion that cytochrome *c*, or its bacterial analogue, was not normally present in chromatophores. If it were, there would be no way to explain why illumination of the intact organism *repressed* uptake of oxygen rather than stimulated it.[14]

The only thing wrong with this rationalization was that when Leo actually looked at acid extracts of chromatophores, treated as one would to isolate cytochrome *c* from mitochondria, he found first by spectroscopic analysis and then by isolation and purification that *R. rubrum* chromatophores contained what looked like cytochrome in amounts even greater than in mitochondria.[15] Furthermore, the physico-chemical properties of this chromatophore cytochrome were, as far as we tested them, identical to those of mitochondrial cytochromic *c*. Enzymically, it reacted with the mitochondrial cytochrome *c*–reducing enzyme equally as well as did normal mitochondrial cytochrome *c*.

We were stunned, because we could not imagine that in all the years *R. rubrum* had been around as a major object of interest to microbiologists, no one had looked at extracts, or even whole cell cultures, spectroscopically à la Keilin. In retrospect, it could be understood that prevailing dogma assigning cytochromes *c* to strictly aerobic metabolism would not suggest its presence in the strictly anaerobic chromatophore. More likely, few microbiologists possessed the skills, or even the simple visual spectroscopes, used by Keilin to demonstrate cytochrome redox reactions in yeasts and animal tissues. The many surveys of cytochrome composition in bacteria in the wake of Keilin's initial findings had not included any of the chemosynthetic or photosynthetic bacteria. The finding by Robin Hill and Robert Scarisbrick that chloroplasts had a *c*-type cytochrome (called "cytochrome *f*") was still too new to have alerted the microbiologists.[16] In 1951 the business of tracing a precise absorption spectrum was not a trivial matter, as it was later when fast recording spectrophotometers became routine laboratory instruments.[17]

The discovery of the bacterial cytochrome *c* was to exert a profound effect on the rest of my research career. Gradually, I would be drawn into a study of cytochrome chemistry and away from the pursuit of problems relating to photosynthesis. The observation crucial in effecting this diversion was yet to come, however. Meanwhile, devising protocols with Leo to isolate cytochromes and their associated redox enzymes from various fractions of *R. rubrum* extracts, I was learning cytochrome chemistry.

In December 1952, an event took place that accelerated my progress. Robin Hill visited Bob Emerson at Urbana, and we were invited to come up and meet him. While there we showed him our data on photo-oxidation of mammalian and bacterial cytochromes *c* and with his collaboration performed some experiments on the quantitative relations between amounts of cytochrome *c* oxidized and amounts of oxygen used. We published our findings shortly thereafter, including a speculation that bacterial photosynthesis could proceed through reactions in which compounds other than water that could donate hydrogen and were required for growth might be photo-oxidized directly rather than via photosynthetic fission of water.[18] This had been suggested earlier by E. I. Rabinowitch.[19]

In the midst of all this exciting activity, the trumpets blared, and I was called back into action on the front in Washington. There were two battles proceeding—one in the wake of the appearance of the new regulations by the State Department on issuance of passports (Code number 686, dated September 2, 1952) and the other the confrontation with the *Tribune* lawyers, in which I was to give a deposition beginning December 1952. To prepare for the latter, I went East in late December and went over the material we expected to be covered. A deposition is a scary thing because the only protection the witness has is that supplied by his own counsel, and there is practically no way of telling what can happen in such an adversary proceeding. No judge is present to see that normal protection against undue harassment is provided. Perry Patterson, the lawyer representing the *Tribune,* turned out to be firm but not brutal. Nate did a good job of keeping the proceedings within bounds as we wound our way tortuously through two days of questioning, almost entirely devoted to an exhaustive probe of my testimony to HUAC in 1948. Patterson attempted to pin me down on precisely how I thought I had been damaged by the *Tribune* story. He also fished around in many private aspects of my life, including the question of whether I had been a corespondent in, or responsible in any way for, Beka's divorce. He explored in detail all my associations going back to my college career. As predicted, the *Tribune* had done extensive research on my past. One of Patterson's aims was to show that damages to me had been minimal and in any case not confined to what the *Tribune* story had occasioned. He showed us copies of equally scurrilous stories that had appeared in publications other than the *Tribune* and *Times-Herald,* which had escaped our notice. His investigators had done a thorough job of probing all my contacts, whether or not they might be related to Communist Party activities, and, as Nate had predicted, had piled up a mass of circumstantial evidence that would be claimed to lend substance, however tenuous, to the innuendoes contained in the stories the two McCormick papers had printed. Studying this evidence later was of some help in our preparations for the final battle.

Beka had some windfalls about this time that slightly helped alleviate our worsening fiscal position. She had been commissioned to collaborate with Dr. Lena Levine on a book about menopause. This was finished and published before the year was out, receiving high praise.[20]

The war on the passport front bogged down soon after the hopeful break of the appearance of the new regulations. To continue our campaign, I had sent in an affidavit to counter the bill of particulars brought against me by Mrs. Shipley, as required in the new procedures. Her contentions remained a superficial rehash of a few items in my dossier, nothing new or substantial being presented. We then awaited her decision on my new passport application to visit Australia and give my invited lectures. We kept being put off. We came to see that a fatal flaw existed in the regulations. The appeals board, presumably to be appointed to hear our complaints and rebuttals, could only act if there was a denial. However, there was no time limit set on how long Mrs. Shipley could take in considering my application. In the meantime, a new administration had been sworn in office. Eisenhower had swept into the presidency, putting the Republicans in complete control of the executive and legislative branches. McCarthyism was on the march. The guillotine was being readied for Oppenheimer—although he did not know it.

It was apparent that the new political atmosphere encouraged Mrs. Shipley to think that she could forestall our attempts to force an eventual showdown in court, if only by sheer attrition. It was expensive to keep up the legal pressures on her, and might prove prohibitively so. In fact this was a reasonable supposition. Beka's free-lance writing was bringing in less and less. Her brief flirtation with Educational Television proved abortive, and our legal costs were increasing. Sadly, we found ourselves forced to leave our spacious apartment in the high rent district of Maryland Avenue near the medical school for a lower-middle-class suburb in the far western reaches of St. Louis County. All in all, our morale had fallen to an all-time low in contrast to the happy days of only a year before. This must have shown in my letters to Nate, because he wrote at one point that he wanted to see no further evidence of feelings of futility.

The only rays to pierce the gloom were coming from the laboratory where Leo and I were still reaping rewards from *R. rubrum.* We had found two workable redox systems to detect reactions in chromatophores analogous to the Hill reactions in chloroplasts. We could conclude that the bacterial chromotophores had the capacity to carry out Hill-type reaction just as did chloroplasts.[21]

Evidence that I was still being systematically pursued and avenues of opportunity closed by the FBI continued to come to my attention.

Early that spring, Ric Skahen applied for a job at the Oakland, California, Veterans Administration Hospital. Later that spring, Dr. B. Gerstel, the hospital laboratory chief, was visited by the FBI, who informed him that Ric was a friend of mine and that I was alleged to be a communist and had been cited by HUAC for security violations on the Manhattan Project. Gerstel persisted in hiring Skahen, but he had to sign a statement saying that he had been warned. All this came out later that summer, when Ric arranged to have me lecture at the hospital and the hospital administrator, a Dr. Lake, called him in and ordered him to withdraw the invitation. There were other incidents, all reminiscent of actions preventing my employment in 1945. The feds continued their underhanded persecution. Nothing had changed in their attitude in the decade since I had left Berkeley.

A welcome respite from the heat of the St. Louis summer came thanks to the generosity of relatives in New Hampshire. A cousin of mine on my mother's side had a cottage on the shores of Lake Winnepesaukee, where we spent a wonderful, relaxed month far from the worries at home and in Washington. Coming back to St. Louis in late August, I found a letter from Dr. Sidney Elsden, a noted microbiologist in Manchester, England, who had read Leo's note announcing the existence of the new cytochrome *c* in *R. rubrum* with disbelief. Like many established investigators, he found this claim incredible and had resolved to repeat the experiment.[22] He had found the observations Leo reported were reproducible, but noted that the bacterial cytochrome differed from the mitochondrial in two important ways. First, its "isoelectric point" (the *p*H at which its inherent electrical charge was zero) was close to neutrality in contrast to that of the mammalian cytochrome, which was well known as a "basic" protein with an isoelectric point near *p*H 10. Secondly, it did not react with the mitochondrial cytochrome *c* oxidizing enzyme, whereas it did with the reducing enzyme. We had not checked the isoelectric point or the reactivity with the oxidizing enzyme (oxidase) because we had expected to do so immediately on resuming work in the fall. The result of this exchange of letters was a short note with joint authorship announcing "a new soluble cytochrome."[23] Robin Hill had told David Keilin about our new cytochrome. I do not remember now whether the suggestion that it be called cytochrome c_2 came from Hill or Keilin.

This finding that *R. rubrum* cytochrome c_2 differed in structure and function from mammalian cytochrome *c* despite its close spectroscopic similarity was the crucial observation in firing my interest to pursue further the comparative biochemistry of the cytochromes and the beginning of research that would occupy my attention almost exclusively for the rest of my investigative career. Leo and I decided to look at other microorganisms, and in short order found that a strict anaerobe, the green photosynthetic bacterium *Chlorobium limicola*, which we obtained from Roger Stanier and grew in sufficient quantity for extraction and isolation of cellular proteins also contained a *c*-type cytochrome.[24] In this case there was no possibility of an aerobic alternative mechanism to explain the presence of a cytochrome *c* as in *R. rubrum*. Almost simultaneously in places as far removed as they could be on this globe from our laboratory—England and Japan—a cytochrome *c* provisionally termed "cytochrome c_3" was reported to occur in strains of another strict anaerobe, the sulfate-reducing nonphotosynthetic bacterium *Desulfovibrio*.[25]

Thus, in the short span of a year, the dogma that cytochrome *c* was uniquely involved in the reduction of oxygen by mitochondria, and not in metabolism in the absence of oxygen, was decisively disproved. The particular cytochrome *c* that occurred in air-breathing organisms and was localized in their energy-transducing organelles, the mitochondria, was a critically important catalyst for biological oxidation. It could now be seen, however, that it was only one of a large family of similar heme proteins with the same iron porphyrin as part of the active catalyst, reactive generally in redox metabolism not only in aerobes but also in anaerobes. I would spend the next quarter century prosecuting research with many associates, collaborators, and students at home and abroad elaborating this generality, thereby establishing a new comparative biochemistry of iron-containing proteins.

As the year's end approached, I was beginning to feel a guarded upsurge of optimism, based not only on the progress we were making in the laboratory, with the new vistas so unexpectedly revealed by our cytochrome work, but also by signs of movement on the passport front. We had been waiting since April for action by the State Department to appoint an appeals board as stipulated in their new regulations, issued with such fanfare at the end of 1952. Despite continued needling by Nate, nothing had happened. It appeared that no one in the department was taking these regulations seriously—at least, no board had been appointed. It was obvious that the department hoped we would give up and leave Mrs. Shipley to savor the triumph of her delaying tactics. Nate disabused the department of this notion by filing a suit requesting that the secretary of state, John Foster Dulles, and Mrs. Shipley be directed to make a final decision, which could be taken on appeal in the event (as we expected) that it proved to be negative. In accord with the regulations, a bill of particulars on the basis of which the decision had been made would have to be supplied us.

The suit brought quick action. Mrs. Shipley sent me the usual denial, together with an appropriate selection of charges from the dossier.[26] (It transpired she had some sixty such applications backed up but was acting only on mine.) Nate got his copy and immediately wrote that he would amend his complaint already on file in court to require Dulles to move on appointing an appeals board. In the meantime, I was to provide answers to the charges and get testimony in the form of affidavits about my character, background, professional competence, contributions to knowledge, and the importance to me of travel, in the expectation that there would soon be a full-scale hearing. He also addressed a letter to the State Department's "Board of Passport Appeals" asking for quick action. A week later, the board was appointed. The *St. Louis Post-Dispatch* December 29 issue carried a story headed "Dulles Appoints Passport Board to Hear Appeals. Had Delayed for More Than Year—Washington U. Professor Threatened Suit."

Happy now that we had some actual allegations to disprove, I went about the business of getting the affidavits and making preparations for a personal appearance in front of the board. I did entertain the suspicion that it would merely rubberstamp Mrs. Shipley's action. The massive effort I would have to make might go nowhere. Nevertheless, it was crucial to go through with the departmental hearing, especially as we could never hope to get court action without showing the judges that we had exhausted all other recourse. Meanwhile, I was already getting a fantastic response from everyone and everywhere. Offers to provide affidavits and testimonials were flooding in. For example, I had a letter from my old friend Dean Cowie, at the Carnegie Institution in Washington, volunteering not only to give personal testimony but to arrange a lecture for me at his laboratory, providing a fee that would help meet expenses for my trip to appear at the hearing.

On the *Tribune* front, Perry Patterson suggested that Chancellor Compton provide a deposition. This indicated to the ever hopeful Nate that they might be thinking of a settlement. However, the chancellor was leaving his post, in a few days' time, and would be on a world tour for nine months. His place was being taken by Ethan A. Shepley, who could be counted on to continue the strong support I was enjoying from the university. As for the damage done me by the libel, I had testimony about losing appointments specifically offered at the University of Illinois and at the University of Colorado in the wake of the appearance of the *Tribune* and *Times-Herald* publicity.

A new and heartening development was taking place. Dr. Geoffrey Chew, the chairman of the passport committee of the Federation of American Scientists, had written to us about our suit against the secretary of state. His interest had probably resulted from the efforts of old friends such as Charley Coryell, Jules Halpern, and others on the executive board. He had written a report that would appear in an FAS newsletter in which he reviewed the history of appeals on passports. He pointed out that my case was important because it would be the first on record in which a final decision had not been made by Mrs. Shipley, and in which the applicant would be represented by counsel in a formal hearing at which the usual rules of evidence applied. This included cross-examination of witnesses bringing the charges, whatever they might be, and a decision based on an impartial evaluation, it was to be hoped in good faith, of all the evidence so obtained. The fact that the FAS was beginning to take an active interest in my case was tremendously encouraging, in that it showed that the community of scientists was at last stirring and considering organized action against the many attacks being made on its members.[27] A letter to Chew from H. M. Levy, staff counsel for the ACLU, indicated that it would participate as amicus curiae in our case.

The makeup of the board also seemed encouraging. There was no indication, as I had darkly suspected, that it would be merely a rubber stamp for the secretary of state and Mrs. Shipley. Its chairman was Thruston B. Morton, the assistant secretary of state for congressional relations. The counsel for the board was John W. Sipes. Both of these men were known to be officials of unquestioned integrity.

In a letter to Secretary Dulles, Nate submitted my affidavit and petition to appeal the Shipley denial. I denied all the allegations she had made and requested that the board schedule a hearing at an early date. Just a few days previously, the ACLU had filed a brief challenging the State Department procedures in another case, that of William Clark, who had served as chief justice of U.S. courts in Germany during the occupation. Apparently, his passport had been revoked after he had opposed jailing of American citizens by German police without right to bail or speedy trial. The ACLU spelled out, as the basis for its complaint, what Nate had already contended—that English Common Law and the Magna Carta established freedom to travel as a right, and that its abridgment was in violation of the Fifth Amendment. The question was whether denial of a passport by the secretary of state merely through exercise of his "discretion," without any hearing of a type familiar to due process of law, was constitutional. The Constitution definitively supplied an answer in the negative by virtue of the principle that no person could be deprived of liberty without due process of law. In the Clark case (as well as countless others), the secretary of state had acted under no express law or regulation by Congress, but merely under his own assumed power to deny exit at whim. The ACLU brief concluded by stating that, regardless of circumstances depending on the variety of interests that might be at issue, nothing less than a full and fair hearing would serve to satisfy the demands of constitutional due process.

It looked as though 1954 would be a year Mrs. Shipley would definitely not find boring.

14

The War Is Won

(1953–1955)

THE NEW year began with feverish activity both inside and outside the laboratory. Nate and Alex were busy fencing with the legal staff of the *Tribune* in Washington. I was assembling the documentation needed to continue the struggle on the passport front—when I was not in the laboratory engrossed in the implications of our discovery that large amounts of a cytochrome *c* existed in chromatophores.

The riddle of how adenosine triphosphate (ATP) might fit into the scheme of photosynthesis still nagged me. Largely through the work of Fritz Lipmann and his associates, it was beginning to be clear that ATP is an all-purpose reservoir of chemical energy that supports all the metabolic activities of cells. An ATP molecule has five parts: adenine (an organic base), ribose (a sugar), and three phosphate groups. Adenosine diphosphate (ADP) and adenosine monophosphate (AMP) are the two-phosphate and single phosphate versions of the molecule. Energy for cell metabolism is stored in the bonds between two of the phosphates and the rest of the molecule. When one phosphate is stripped from ATP, yielding ADP and a free phosphate, ten to fifteen kilocalories of energy are released as heat. The removal of another phosphate, leaving AMP and a free phosphate, again yields ten to fifteen kilocalories. To reverse the process and add phosphorus requires an energy input of the same amount. It is also possible to take both terminal phosphates off together as pyrophosphate, obtaining just the energy of removal of one phosphate.

The energy stored in ATP can be used to drive synthetic reactions.

Instead of phosphate loss with a release of energy, the phosphate groups can be transferred to other molecules (symbolized RH). These transfers are facilitated by enzymes called kinases. In this ingenious way, nature has contrived to store and then use energy in the highly regulated and specific manner needed to sustain life.[1] (See also chapter 9 n. 2.)

It took no great leap of imagination to hypothesize that the transfer of phosphate from ATP to the organic compounds RH by the action of the appropriate kinases provided the ten to fifteen kilocalories per molecule needed for the uptake of CO_2 to make the primary fixation product in photosynthesis. Sam Ruben's suggestion to this effect seemed right on the mark.

There remained the original question of how light energy was harnessed to form ATP from AMP and ADP, and in what subcellular sites this occurred. A clue lay in the demonstration that in the cells of respiring organisms using the energy available from the burning of carbon compounds, the mitochondria contained the chemical apparatus for the manufacture of ATP. In these subcellular organelles there was a system of enzymes and compounds structured and organized like a bucket brigade to carry electrons from the electron-rich carbon compounds supplied in food (the reductant) in small steps energetically to the ultimate electron acceptor, molecular oxygen (the oxidant). This resulted in the eventual removal of the valence electrons from each carbon of the food to produce fully oxidized carbon, as CO_2. At the other end of the chain, electrons were soaked up by the oxygen, along with protons, to form water.

It seemed plausible that a similar process took place in chloroplasts and chromatophores, with the difference that oxygen was not the eventual acceptor of electrons. It could be carbon dioxide acting as an electron acceptor, which thereby was reduced to cellular material. The conceptual jump involved was to assume that an electron transfer chain such as that in mitochondria existed in the photosynthetic organelles. At this time Robin Hill was having similar thoughts about chloroplasts, in which he and Scarisbrick had found a kind of cytochrome *c* that they called "cytochrome *f*," analogous to another kind of cytochrome *c* we had found that we called "cytochrome c_2" in chromatophores. In short, mitochondria, chloroplasts, and chromatophores all very likely shared the same ground plan of energy-transducing catalytic chains, differing only at the ends. In mitochondria, a specific cytochrome oxidase was present to direct electrons from cytochrome *c* to molecular oxygen. In chloroplasts and chromatophores, a special light-activated mechanism existed to direct electrons to carbon dioxide. All of the energetic mechanisms of respiration and photosynthesis could thus be contained in this one neat package.

Reasoning further, one could explain the "Nakamura phenomenon"—inhibition of respiration in *R. rubrum* and other non-sulfur photosynthetic bacteria—as the result of competition for electrons coming down the transfer chains. The electrons had to traverse a path common to both respiration and photosynthesis, but at some point, presumably at the terminal end, it bifurcated, the electrons going either to CO_2 in the absence of oxygen as an acceptor or to oxygen if aerobic conditions were established. It was a beautiful scheme, but simplistic, as became evident in later work. Nevertheless, it was a good working hypothesis.

It could be imagined on this basis that chromatophores, as well as chloroplasts, carried out a light-induced synthesis of ATP from ADP and inorganic phosphate, storing the energy generated by the movement of electrons down their respective oxidation chains in a manner analogous to that employed by mitochondria. It remained to find such chains. There were already vague indications that they existed. At this particular stage of research, we knew little about the specific enzyme systems in chromatophores or chloroplasts, especially those that might correspond to the terminal mitochondrial cytochrome oxidase.

It would have been a simple matter to do an experiment testing whether the predicted photophosphorylation occurred in chromatophores in the absence of air. However, I reasoned that we would first have to separate the oxidizing and reducing systems made by light absorption in chromatophores, as Hill had done for chloroplasts (see chapter 13 n. 12). ATP production could not occur unless this happened, because electrons had to flow from one end of the chain to the other and this required that one end of the chain be supplied with its reductant and the other with its oxidant. It seemed to me, and I managed to convince Leo, that we would have to get the chromatophores to undergo a Hill reaction in the absence of oxygen before attempting to see if photophosphorylation occurred. There was no such impediment to the reaction with chloroplasts where the separation of the two systems already existed, but we were preoccupied with the chromatophore problem, and did not try the experiment.

Around this time I received a letter from Albert Frenkel. Albert had worked with Sam Ruben and me back in Berkeley, where he had been a graduate student in plant physiology, and had published an article with us on metal exchanges in plant material. He had since taken an appointment in the Middle West, and now wrote that when he did the simple experiment incubating chromatophores from *R. rubrum* with ADP and inorganic phosphate, he found a vigorous production of ATP when they were illuminated. I reacted just as all the experts had to us in the past when we had described our results on photoproduction of hydrogen and nitrogen fixation, or the existence of cytochrome c_2. Like van Niel, Wilson, and Elsden, I was expertly skeptical and wrote back to Albert that he had probably had some oxygen present in his preparations, so that what he was seeing was a light stimulation of the ordinary kind of phosphorylation, such as occurred in mitochondria.

Albert replied that he had checked this point and found there was no uptake of oxygen during the light-induced synthesis of ATP in chromatophores. In fact, he could set an upper limit of less than one mole of oxygen used for every ninety or a hundred moles of ATP made. The energy needed to make so much ATP was thousands of times greater than might be available from the minimal amounts of oxygen reduction that I had suggested might be occurring. This result decisively established that he was seeing the predicted phosphorylation induced by light (subsequently to be called "photophosphorylation").

Later, I learned that Albert had been set on the track leading to his discovery of bacterial photophosphorylation by Fritz Lipmann, in whose laboratory he was doing postdoctoral research. Lipmann told me he had suggested to Albert that he repeat some of the experiments Leo and I had reported, remarking something to the effect that "whatever Kamen is doing always has some importance."

Essentially simultaneously with the discovery of photophosphorylation in chromatophores by Frenkel came the announcement of the same process in green plant chloroplasts. Daniel Arnon, working with a young English researcher, F. R. Whatley, and none other than Mary Belle Allen in Berkeley, had done the simple experiment with chloroplasts. And so the basic mechanism of energy transduction in photosynthesis—an anaerobic photophosphorylation—was established. I had missed finding it by failing to remember that it is best to do the experiment first and think later. (When someone once asked Röntgen what he had thought when he first saw x-rays, he replied, "I didn't think, I experimented!")

There were some consolations, to be sure. *R. rubrum*, like Al Capp's proverbial Shmoo (the fabulous little bird that supplies all good things to the Dogpatchers in *Li'l Abner*), continued ever bountiful. Our isolation procedure for cytochrome c_2 was based on successive fractionations of soluble extracts by graded additions of ammonium sulfate. (The development of various column chromatographies, involving absorption chromatography and "molecular sieving" was still in the future.) Proteins "salted out"—that is, precipitated at a given pH—depending on their becoming insoluble at characteristic concentrations of salt in solvent. We would begin with our extract, clarified by centrifugation. Following procedures well established in the literature by other cytochromologists, such as David Keilin and E. F. Hartley, we isolated precipitates formed as different fractions, obtained with increasing amounts of added ammonium sulfate. There would be "cuts" for 20, 40, 60, and 100 percent salt saturation. We noted that cytochrome c_2, like the mammalian cytochrome c in Keilin-Hartley procedures, came down at salt saturations of 60 percent or greater. In the fraction between 40 and 60 percent, we kept seeing brownish precipitates, which on spectroscopic examination appeared to be very much like animal myoglobin or hemoglobin. In fact, we at first called it "pseudo-hemoglobin" because while it showed absorption spectra like hemoglobin, it did not form the characteristic oxygen-carrying complex that authentic hemoglobin did. It had the same iron porphyrin active grouping as heme proteins such as cytochromes, catalases, and peroxidases, but it differed from each of these in some fundamental way or other. For instance, it bound heme to protein as in cytochrome c, but it was rapidly oxidized by air without the need for a special enzyme catalyst like an oxidase. This property also distinguished it from myoglobins and hemoglobins, which functioned by not being oxidized while combining with oxygen. To make matters even more confusing, it did combine with small neutral molecules such as nitric oxide and carbon monoxide, but with none of the many small electrically charged "ligands" bound by hemoglobins, such as cyanide, hydrosulfide, isocyanate, and so on.

This strange new protein was to remain an object of intense research for years as we struggled to rationalize its behavior at the structural level. The publication of our observations[2] was met with much disbelief by the authorities. Keilin thought we were observing some artifact produced in our procedures, as he had never heard of so strange a heme protein. One of the great thrills of my career was the day I spent with him some years later when I brought the new protein for his examination. With Robin Hill also in attendance, Keilin exhaustively covered all aspects of its combination with a large range of compounds. Using his amazing facility with a visual spectroscope, he quickly confirmed our original observations and extended them. It was a great course in cytochrome chemistry for me.

The problem of how to classify the new protein, or even name it, was to prove troublesome for years. We changed its name almost every time we published a new paper on its properties. The day I spent with Keilin resulted in a suggestion by Robin Hill that we call it RHP, standing for *rubrum* heme protein. Of course, we knew by then that it also occurred in another non-sulfur purple photosynthetic bacterium, a *Chromatium* species. Later, it was found to be one of the most ubiquitous heme proteins in bacteria.

Finally, a special meeting of an international committee on cytochrome nomenclature was needed to straighten out the mess into which the classification of cytochromes in general had fallen trying to retain the classical system based on the properties of cytochromes as they occurred in mitochondria. The Pandora's box we had opened produced a vast proliferation of new forms of cytochromes and heme proteins as research progressed and new structures and functions appeared based on the common theme of combination between heme and protein. The final designation for RHP was cytochrome c', the prime denoting that it has its heme group bound like cytochrome *c* but unlike the latter does not show the typical absorption spectrum of a cytochrome in the reduced form with sharp maxima for absorption bands in the green part of the spectrum.

We had been preoccupied with this work all through 1953 and into the new year. Our results had attracted much attention, and we were getting an enthusiastically favorable response from our funding agency at the National Institutes of Health. The NIH officials wanted to see the work continue. It was indicated that we could count on support for an indefinite time even after the termination of the initial three-year period. I had also heard from Hugo Theorell, the noted Swedish authority on cytochrome *c,* who commented favorably on our work. (Theorell would be awarded a Nobel Prize in medicine the next year. He was to become a close friend and collaborator in future years. Another bond between us was his ability as a violinist of professional caliber.)

With everything so upbeat on the research front, I made plans to spend the summer with Leo at van Niel's laboratory extending our work on bacterial cytochromes to the great variety of bacteria in cultures available there. Meanwhile, I had completed the preparation of our case for the hearing by the Passport Appeals Board on March 10. The results of my solicitations for testimonials were truly impressive. I had documents attesting to my character, career, and accomplishments from people in all walks of life, decisively refuting the charges that I had been a communist, or a communist sympathizer. The letters disposed point by point of the fabrications and distortions in the specific allegations of the Passport Division. I got a real surprise from Don Cooksey, who wrote offering to send a supporting statement. (I had formed the impression that the upper hierarchy at the Radiation Laboratory, while setting no store by the allegations against me, nevertheless thought it best to keep its distance.)

Waldo Cohn and Charley Coryell had been especially active in organizing the campaign to get the letters written and sent. In addition, three old friends were able to appear and give oral testimony at the hearing, as well as to provide painstaking and thorough documentation of their convictions as to my loyalty and integrity. One of these, Dean Cowie, had been an associate at the Rad Lab. Another was George Boyd, one of the old gang from my graduate student days in Chicago. George, whose respectability was beyond question, wrote a masterly defense of my activities in those days and gave impressive personal testimony.

The third supporter was my old roommate in Berkeley, Paul Aebersold. He deserves special mention as a devoted friend whose loyalty swept aside all thought of personal hazard. Paul had defended me fiercely at the time of my troubles on the Manhattan Project and through the HUAC confrontations. Now he was in a vulnerable position himself. As developer and director of the new isotopes division at Oak Ridge, he was in direct opposition to Lewis Strauss on the policy of making isotopic materials freely available at home and abroad. This was a very sensitive topic. During congressional hearings on the subject, Oppenheimer had held Strauss up to ridicule.[3] Now Strauss was chairman of the AEC and busily involved in the campaign to discredit and crucify Oppenheimer. Proficient character assassins, such as my would-be nemesis, Teeple, were working in his office. (Another, to surface later, was one R. L. Borden, who actually put together a fantastic accusation that Oppenheimer was a Soviet spy.) Despite this, Paul still insisted on coming to testify, and submitted an impassioned and documented case refuting the charges against me. In fact his affidavit had to be toned down a bit by Nate.

We kept the new chancellor at Washington University, E. A. Shepley, informed of events and the progress of the hearings. He was a most remarkable person—thoughtful, anxious to see justice done, and unflappable. He had been asked to function as acting chancellor until a replacement could be found for Compton, who had retired; but soon it was apparent that he was ideally suited for the position, and he was asked to stay on permanently, even though he was a lawyer and not an educator. He had written an affidavit strongly supportive of our case to add to Compton's testimony.

Though confident of victory, we could not help noticing some serious flaws in the appeals procedure. The regulations left the nature of the charges and conduct of cross-examination up to the discretion of the security people. The proceedings would still be essentially a security hearing, rather than a true judicial process. Only charges the security people were willing to supply would be presented, and decisions would not necessarily be based on answers to those particular charges. The makeup of the board, as I have already remarked, was to consist wholly of officials from within the State Department. In other words, the department was to act both as judge and defendant. However, we found no reason to complain about the membership of the board. We had researched all of its members, and found them not likely to be pawns for Mrs. Shipley.

The hearing began at ten o'clock in the morning and lasted until six in the evening. The whole massive case, which we had taken three months to prepare, was presented. This was the first formal procedure in the history of the country involving presentation of evidence before a special State Department Appeals Board in a passport application. The board was receptive and seemed to be convinced in our favor. Indeed, at the end of the hearing, the chairman, Thruston Morton, declared that he was most impressed with my patience and forebearance in prosecuting the appeal and even remarked that he felt it a privilege to have met me. He then arose, came around the table and shook my hand, as did some other members of the board.

A few weeks later, the storm broke around Oppenheimer, and

shortly thereafter we received a shattering letter from the undersecretary of state, Walter Bedell Smith, curtly informing us that my passport was still being withheld. To us this could only mean that the appeal procedure was meaningless. The power of Mrs. Shipley to make decisions and have them stick, however capricious, willful, and arbitrary, remained as firm as ever. The widespread impression of the board's integrity remained, but there were indications—at least Nate thought there were—that it had been overruled because the State Department administration lacked the will to override Mrs. Shipley, whose friends in Congress were numerous and powerful.[4]

Our only hope now was the judiciary. In the meantime, Nate kept up the battle, submitting a strongly worded critique of the decision based on his careful rereading of the hearing transcript. He petitioned the secretary of state for a reconsideration of the decision. He also wrote Chancellor Shepley at Washington University explaining in detail what had happened and outlining what our future course of action might be.[5] He felt that any undisclosed material that might have been influential in causing the board's decision to be overridden would be required to be made available at least to the judge in a court proceeding, and that on the basis of his intimate knowledge of legal actions in security cases, there was every reason to bring a suit against the government. The question of how to fund such an action remained to be tackled and suggestions were welcomed. The petition for a reconsideration was filed in June, but there was little reason to hope it would receive any attention or anything other than a perfunctory reply. The general gloom was immeasurably deepened by the tragic outcome of the Oppenheimer hearings. He was deprived of security clearance, branded a security risk, and excluded from all contact with federal atomic research.

The failure of our efforts to bring Mrs. Shipley to heel was shocking to me and Nate, but it had a devastating effect on Beka. Somehow, I had the resilience to recover and consider with Nate how to continue the battle, but she went into a complete depression. I began to sense for the first time the desperate psychological state in which she was trapped. She had no religious beliefs to help her in adversity, having renounced her faith as well as her family years before. She also lived in constant fear that when the telephone rang, it would be some dire message from her family.

Little seemed to have gone right for her since she had left Scranton, her hometown. In a nationwide contest, she had won a much coveted scholarship given by the French government to attend the Sorbonne, with full remission of all expenses for four years, only to have the outbreak of World War II deprive her of the opportunity to accept it. Poor and with no available resources, she had gone to New York and managed to work her way up to a responsible job in the public health field. Although her education had been wholly without science content, she had acquired a strong working knowledge of how science was done and a basis for evaluating the significance of research projects in medicine as well as general science. She had a natural aptitude for probing and interviewing and had become a first-rate journalist.

Her career had not, however, taken wing. Women's liberation was still far in the future, and preference always went to the men with whom she was required to collaborate. She had had a taste of this at *Time.* The struggles we were engaged in with the government and the press had been particularly hard on her, as she had to sit by helplessly on the periphery. (That she could accomplish much when given the opportunity was evident from the way she had choked off the story *Time* had planned to run on me in the 1948 fiasco.) A final blow, timed especially badly, came from a television interview of Lena Levine, her co-author on the book on the menopause. The entire interview, intended to provide publicity for the book, contained no reference whatever to the fact Beka was a co-author. The impression given was that Dr. Levine was the sole author. To Beka this was not an oversight but a deliberate effort to deny her credit.

All these traumas cumulatively produced a severe depression. Beka had, moreover, been having trouble with sciatica and had resorted to increasing doses of alcohol to get some relief from her physical and psychic traumas. Probably as a result of inbreeding in her family for centuries, she bore a heavy burden of genetic vulnerability, easily visible in such signs as pathological sensitivity to bruising, areas of red blotches on her hands, and an almost transparent skin. Her metabolism could not cope with an alcohol intake that was rapidly increasing as her basic depression deepened. The crushing outcome of our passport appeal was the last straw. Beka became convinced that neither she nor I had any future.

As in previous years, we arranged to stay for the whole summer in Pacific Grove on Monterey Bay, where Leo and I would continue our research at van Niel's laboratory and we could all enjoy the pleasant climate. I hoped that Beka would rally and come home renewed in health and spirits. We had been in California only a few weeks, however, when a new blow fell. Suddenly in early August I was sent a curt notice from the chief of the Division of Research—who had previously been sure I had nothing to worry about—informing me that our grant funding Leo's fellowship fell in a "category which can not be supported by the Public Health Service." This was only a month before the scheduled renewal of the grant. This decision was not only morally indefensible because of the commitment made by the PHS when the grant was originally awarded, but unwarranted on the basis of the quality of the research the grant had made possible. Moreover, Leo was given no time to look for a position elsewhere.

The action of the security people in forcing the PHS (now under Oveta Culp Hobby, the new director in the Eisenhower administration) to withdraw support and renege on its commitments was to penalize Leo and not me. The last person with whom to associate the idea of subversion was Leo, and yet he was being made to pay merely for working with me. Happily, he soon had an offer of a fine position at Brigham Young University in Provo, Utah, his hometown, so he was able to avoid unemployment. This gave us more time to complete a survey of bacterial cytochromes and their associated redox enzyme systems in a large selection of bacteria, both nonphotosynthetic and photosynthetic.[6] The summer's activity also began a new friendship of much importance to me in later years. I met an Australian, Dr. Malcolm Winfield, whose help in the laboratory aided greatly in prosecuting our researches.

We concluded that all photosynthetic bacteria contained a *c*-type cytochrome and exhibited associated enzymic activities, such as those catalyzing reductions, oxidations, and peroxidations. This generalization held for strict anaerobes as well as for bacteria that metabolized using oxygen. We also confirmed the existence of the *c*-type cytochrome in the strictly anaerobic sulfate-reducing bacteria, reported by others previously. Moreover, there was evidence for the occurrence of other cytochromes, notably of the *b* type, as well as the new heme protein we were still calling pseudo-hemoglobin. We had examined two species of aerobic bacteria that lived primarily using the energy obtained by reducing nitrate and nitrogen oxides, rather than oxygen. Again, we found a cytochrome *c* that closely resembled the cytochrome *c* of mitochrondria—in fact, more so than did the cytochromes c_2 of the photosynthetic bacteria. (One of these—the cytochrome *c* of *Micrococcus denitrificans*—was later the object of extensive structural studies in a number of laboratories.) We also measured the electrochemical oxidation behavior as a function of pH, the first studies of this kind ever attempted, and established that each microorganism had a cytochrome *c* unique to itself.

Our paper was the first to describe a system of comparative biochemistry for cytochromes and the starting point for development of a whole new area of bioenergetics. We finished our discussion of its significance by remarking that light appeared to affect metabolism primarily through the cytochrome system in analogy with aerobic nonphotosynthetic organisms using the chemical energy of respiration. Speculations about how cytochromes might be involved in photometabolism had already appeared in a paper by Robin Hill and H. E. Davenport.[7] I was now well on the way to a complete commitment to cytochrome research.

Arriving back in St. Louis at the end of the summer, I was rescued by the Kettering Foundation, which picked up the tab for my research expenses for the year. In the meantime, I had the good fortune to meet some representatives of the newly created National Science Foundation—Bill Consolazio and Estelle ("Kepie") Engel. Through their efforts, I received a grant for five years totaling $32,000. Somehow, the security madness had not affected the foundation, although it had taken over the PHS completely. It appeared that what was treason in one area of government-supported science was patriotism in another. The backing of the National Science Foundation was to continue for the rest of my research career, providing absolutely crucial support in all those years and a reliable backup source when the National Institutes of Health came out from under the security cloud and were able to take over a major share of the funding I needed.

During the summer, Nate received a letter from Geoffrey Chew inquiring further about the status of our case and expressing interest in providing support for it through the Federation of American Scientists. As he would be away through the summer, he referred Nate to Dorothy Higginbotham who worked in the Washington office of the FAS. This was the start of what eventually became a full-fledged fundraising effort on the part of the federation.

Chew submitted a memorandum he proposed to circulate to the FAS membership outlining the status and nature of our case as related to the general problem of passport policy. Nate replied encouraging him to proceed and including the hope that sufficient funds could be raised in a preliminary fashion to get started on court action. At about the same time, he wrote H. M. Levy, the chief counsel for the ACLU in New York, bringing him up to date on what had transpired at the passport hearing and emphasizing the strength of our case for possible support by the ACLU. Levy's reply was cautious and noncommittal. The FAS reaction was much more positive and indicated an inclination to go it alone even if the ACLU dragged its feet. The federation would start raising funds, a step it was taking somewhat hesitantly, as it never had been involved in a legal action before. A clarification of our thesis as to the issues raised—whether they addressed the general question of freedom of travel or only my special case—and a rough idea of how funds would be spent were requested.

While Nate worked on the negotiations with the FAS, the *Tribune* case was beginning to heat up, bringing with it further strains on my meager resources. I received my first directive from Alex Boskoff, who was taking field command of the battle with Colonel McCormick's minions. He had gone over all the material that would enter into the trial and had a mass of specific assignments for me to undertake. Among other items, he asked me to communicate with about a dozen people whose testimony might be especially helpful to us. He enclosed a memorandum to give Dr. Compton, who was back from his world tour and who had volunteered to give testimony at the trial. In the meantime, he was asked to provide written answers to a questionnaire included in the memorandum. Later, Alex expected to talk with him in Washington to prepare him as to the testimony he might give.

Nate tried to get a copy of the transcript of our passport appeals hearing, but the department refused to cooperate. As the year ended, the FAS formally decided to enter the case, its council passing the following resolutions:

1. The recent record of the State Department suggests that due process is not being observed in procedures leading to refusal or revocation of passports. Because this situation has obstructed travel by American scientists to an extent which damages both our country and science, the Federation of American Scientists has long been concerned. Our organization regards as one of its specific objectives a set of passport procedures which fulfills the constitutional guarantee of due process.

In particular the F.A.S. will lend support to the legal action against the State Department, on the due process issue, by Dr. Martin Kamen. We are concerned that the procedures followed in denying a passport to Dr. Kamen were in violation of his constitutional rights.

2. The passport committee is instructed to lend support to the legal action by Dr. Kamen against the State Department and is empowered to raise a special fund for appropriate expenses. The Passport Committee is urged to enlist the support and co-operation of the American Civil Liberties Union in this case to the fullest extent practicable.

It was noted in the printed announcement of these resolutions that the ACLU would probably enter the case as amicus curiae, but that complete control of the case would be in Nate's hands.[8]

Earlier in the year I had had an opportunity to see the face of the enemy—at least one of them—and a bizarre experience it was. In Chicago one day, Harold Kupper, an old friend from the time of my musical career there offered to take me to a performance of "Theatre of the Air," the big evening radio show of the *Tribune* station, WGN (for World's Greatest Newspaper). Featuring a gigantic display of talent presented at prime time in the evening, it included a full complement of vocal soloists, a chorus, and a symphony orchestra, all under the harried direction of a large production staff. I say "harried" because the centerpiece of the evening was the regular appearance of the *Tribune* publisher and owner, Colonel R. R. McCormick himself, whose presence and talk had to be accommodated in whatever time and at whatever length his whim dictated. He might arrive at any time during the complex hour's proceedings and the show would have to stop dead while he was shoehorned in. The evening Harold and I (we were the only audience present) sat in the *Tribune*'s auditorium, he made his entry after the show had been on the air a few minutes. Shambling in unexpectedly, the colonel was a grotesque sight. He wore a lavender evening jacket with a white frilled collar and lavender spats. Advancing with a slow and unsteady gait, he came toward the microphone speaking in a dull monotone, while the great crowd of performers strained to be as quiet as possible. His speech gave no indication of where punctuation marks might be, but droned on, reciting a written text that made little sense. I think these orations were also published as a regular column on the editorial page of the *Tribune*. In them, the colonel was wont to preach his gospel against the East Coast Establishment, Wall Street, and other interests dominated by what he perceived as servile obeisance to the British. He usually included observations on the conduct of military affairs, interspersed with claims that he had invented some of the modern weapons of war.

McCormick's reading had gone on for about five minutes when suddenly, in the middle of a sentence apparently, he stopped, turned abruptly, and shambled off the stage. Thereupon the show began again, a frenzy of activity seizing the whole multitude. I had the impression that the colonel, really one of the great newspaper men of his time, had reached an advanced stage of senile decay. More relevantly for me, it seemed unlikely that such a massive ego would permit the troops in Washington to sue for peace or a settlement. I could understand why Willard Edwards was nervous about having embroiled his paper in a libel suit. Undoubtedly, he had not endeared himself to the colonel or the management, and would even less if there were a cash settlement. Nate occasionally suggested that Patterson seemed to be hinting at something along such lines, but I held little hope that there would be any cash capitulation. The *Tribune*'s strategy was likely to continue to be based on mudslinging and attrition.

The trial was set for January 10 in district court, and all interlocutory matters, motions, depositions, and other preliminary maneuvers were ordered completed by January 3. We had compiled a whole new set of depositions and affidavits from appropriate sources, dealing with all aspects of the HUAC record. We also had the transcript of the deposition I had given Patterson, which we could use to develop a strategy to counter the inquisition we expected the *Tribune* lawyers to subject me to on the stand.

We had tried to involve Colonel Lansdale again, but he was determined to keep away from the whole matter of security administration on the Manhattan Project. He had had his fill of it in being subpoenaed to give testimony at the Oppenheimer hearings where he had a run-in with the prosecutor. We did have a letter Lansdale had written me early in 1954, confirming Arthur Compton's testimony that he had been advised by Lansdale that there was no reason not to hire me in 1945 when I was approached by Washington University. In this letter, Lansdale had also noted that the security situation had changed drastically since the days of the Manhattan Project. New criteria had developed that it was unwarranted to apply retroactively to cases of a decade earlier. Rather disturbingly, though, Lansdale referred to the possibility there might be "new" facts that he did not know of. In that case, his testimony might not be relevant. (Later, however, he told me he had uncovered nothing new in looking at the files again.)

The date set for my appearance in Washington was unfortunate in that it fell only a few weeks after a new postdoctoral fellow was to succeed Leo. After an interim period of some six months between Leo's departure in the summer of 1954 and the beginning of the year, I had found time to look at the heme protein content of the strictly anaerobic purple sulfur bacterium, a photosynthetic *Chromatium* species, and had detected a cytochrome of the *c* type as well as evidence of RHP. The new fellow, Jack W. Newton, had come after I was bailed out by the Kettering Foundation from the poverty-stricken state into which I had been cast by the Public Health Service. Jack, whose recruitment was another stroke of good luck for me, had worked with the group on nitrogen fixation at Wisconsin and had an excellent background to pursue collaborative research on the metabolism of photosynthetic bacteria. I left him to continue the work on *Chromatium*. Later, when I returned to rejoin him, he contributed important work on the composition of chromatophores, and their photophosphorylation capabilities. He also began novel studies on their development in cells, using immunological techniques.

In the weeks before the trial began (the date had now been set for January 18, a Tuesday) I haunted courtrooms observing the various judges, one of whom we might draw to preside over our case. At other times I worked in the Library of Congress as a kind of clerk for Nate and Alex, looking up various references for them and saving us some of the cost of hiring clerical help. Just before the trial date, Alex put me through a rough grilling, asking the kind of questions he expected I would encounter on the witness stand. The problem as he saw it was to overcome my lifelong conditioning to hide emotion and give qualified answers rather than positive replies. As a scientist, I looked at everything in terms of probability, whereas in an adversary proceeding, such as a law suit, an appearance of certainty was needed.

Going through Alex's hardening process was a rough experience. For example, he asked how I would reply when the *Tribune* lawyer confronted me with the question: "How did you feel when you heard the story about you?" I gave some lukewarm reply such as "I was startled, shocked, disbelieving." Whereupon Alex roared, "No! you're not going to say that! You're going to say what you really felt. Say, 'I was sick to my stomach!'"

Witnesses we expected to bring forward to bolster our case had to go through a similarly thorough preparation. By now, we knew that the chief tiger on the *Tribune* home legal staff, J. B. Martineau, was coming to conduct the defense personally. He would be seconded by Perry Patterson and a new young recruit to the firm, Herbert J. Miller, Jr.[9] A jury had been chosen consisting of both whites and blacks. The *Tribune* was happy with the inclusion of the blacks, because it was assumed that they would be less likely than whites to think in terms of big money if damages were assessed. Here, as in many other aspects of the trial, they were to be proven mistaken.

We learned that our judge would be Thomas Jennings Bailey, a very old and distinguished jurist, said to be the son of a Confederate general. At the advanced age of eighty-seven and already seventeen years retired from the bench, he was still working to help clear the docket of district court cases. We soon discovered that a better choice could not have been made. Judge Bailey was beyond the reach of mortal temptation. He was responsible to no one but his Maker and impervious to pressures the *Tribune* might hope to exert, such as bringing influence to bear based on the prospect of possible promotion. He would dispense judgment evenhandedly and from a unique background of learning and experience. All of this we came to understand as the trial progressed.

We had heard from Arthur Compton that he would be available to give testimony during the first week of the scheduled trial, around January 12–15. The date kept being shoved backward, however. We suspected that the *Tribune* was delaying the trial to avoid exposing the jury to Compton, which was highly undesirable from their standpoint. Finally, on Saturday, January 15, we were reduced to taking a deposition, as he could not continue to wait. Compton submitted to this exhausting ordeal, much of it badgering by Martineau, the whole day. Martineau's constant objections and attempts to muddy the waters, tactics quite to be expected in a proceeding such as a deposition not monitored by a judge, were doggedly fought by Alex. Compton remained unruffled. The only indication he was feeling some pressure came in the middle of the hearing, when he mentioned that he had a headache. A coffee break restored him.

The deposition, running to 234 pages of doublespaced typed text, was later read in court. It excited the admiration of the judge so much that he halted the proceedings and called both counsel to the bench to remark that in over a half century of experience he had never heard better testimony. We never ceased to regret that the jury had not had the opportunity to hear Compton deliver his testimony in person. We were certain that the eventual award of damages to me would have been tenfold greater than it was.

Quite in contrast to Compton's performance was the experience we had with E. O. L. I learned that he was in town from Don Cooksey, who was accompanying him. Nate and Alex suggested that E. O. L. might be willing to testify. Certainly, outside of Compton and Lansdale, no one would be better able to set the record straight. Lawrence agreed to meet us, and so I saw him again for the first time since I had left Berkeley a decade before. Both he and Don greeted me warmly. I noticed, however, that Don seemed somewhat preoccupied and withdrawn, especially when we broached the subject of E. O. L.'s appearance in court. E. O. L. stunned us by agreeing to testify, but only if there were to be no cross-examination. We could not understand how he could suppose that he would be immune from due process in a courtroom. Unlike Compton, he was not prepared to undergo the trauma of an adversary confrontation. Nate and Alex drew some uncomplimentary conclusions about how real a friend E. O. L. was to me. I was baffled and unable to rationalize his behavior. Very probably, E. O. L. had been advised that he would do well to stay out of the trial, because one never knew where cross-examination would lead. Looking back on the affair decades later, I still have no clue as to why he was so resistant about making an appearance.

Two other witnesses who were to prove crucially helpful to us were Dean Cowie and Joe Kennedy. Dean had appeared at the Passport Appeals Board hearing, and the *Tribune* had done the necessary homework to blunt his testimony if possible. Dean could talk authoritatively about the non-classified nature of the conversation I was alleged to have had with the Russians concerning radiation hazards connected with cyclotron operation. He himself was a living example. A tall, attractive, and athletic figure, Dean had suffered considerable loss of eyesight because of cataracts resulting from accidental exposure to the cyclotron beam while building the cyclotron at the Bartol Foundation. He could be expected to make a profound impression on the jury.

The other witness, unknown to the *Tribune,* was Joseph W. ("Joe") Kennedy, chairman of the Chemistry Department at Washington University and the co-discoverer of plutonium. Joe had been a section chief at Los Alamos. He, too, was an impressive figure—a tall blue-eyed Texan with short cropped hair, the epitome of an upstanding young American. Joe happened to be in town and eagerly agreed to appear. He showed no wariness whatever of a courtroom appearance.

In the middle of the trial, Isaac Stern came to Washington, as I recall to play the Brahms Double Concerto with Gregor Piatigorsky and the National Symphony. Alex and Nate wondered whether he might not be a big help as a witness. After all, the whole trouble had started with the party at his house back in San Francisco. I did not want to involve Isaac, who stood a good chance of ruining his career by running afoul of the *Tribune*, which was capable of organizing a campaign to brand him a Red. Nevertheless, he agreed to help us. Fortunately, we did not need his testimony.

The manner in which these old friends rallied to our cause, volunteering to face the withering fire from the defense was, and remains, one of the most heartening things that ever happened to me. Years later, I wrote an appreciation of Dr. Compton's display of moral courage in my case and his continued espousal of my efforts to obtain justice.[10]

The courtroom was laid out for the convenience of the judge, in view of his age and infirmities. He was hard of hearing, so he used a hearing aid connected to an amplifier mounted on his table. He could leave the bench at will and use a portable microphone attached to his hearing aid to prowl about the courtroom. Because of his advanced age, he tired quickly, so court sessions were confined to a few hours in the morning and afternoon, three times weekly. Sometimes he seemed to doze off, but in reality he was merely concentrating on the evidence and argumentation.

The trial lasted a month. The transcript, regrettably, no longer exists.[11] My recollections, based on memory and some fragmentary notes, are particularly sharp relating to events during the early part of the trial when I spent two days under stiff cross-examination by Martineau. I knew that I was the case. Everything depended on the impression the jury had of my character and credibility. The pressure was enormous, but somehow I did not feel nervous. I remember very little of the questioning, except the extraordinary insight of Alex in predicting the questions Martineau would ask. The depositions we had submitted included one from Dr. Saslow describing my near suicide. Martineau made fun of the claim that I had really been suicidal, asking how much blood there was, and so on. This was but one of many brutal and unfeeling asides that were to prove counterproductive for the defense. I believe he was so accustomed to trying cases in the favorable atmosphere of the Chicago courts that his reflexes had become dulled. All through the trial, he exhibited an obsessive need to rough up witnesses. He even patronized the judge on occasion. This was particularly foolish, as the judge was more learned on the various points of law than the sources Martineau cited. Once, arguing against our contention that the *Tribune* stories were libelous, Martineau quoted a certain authority. Bailey smiled and observed that he had known the man as a classmate and had occasionally found his briefs in error and his footnotes faulty. Sometimes, the judge would call all the lawyers to the bench at the end of a session and assign them what amounted to homework—that is, to prepare briefs on some point of law raised that day. It was very much like some senior professor running an examination. He always prefaced his requests, however, by complimenting the lawyers on the excellent arguments they were making and remarking that he felt it a privilege to be presiding over a case involving such learned counsel.

When my testimony was completed, it marked the end of the preliminary phase of the trial. At this juncture, the *Tribune* could move to have the case dismissed on the grounds that the plaintiff had not presented adequate grounds for continuance. To argue for the defense, young Miller had been chosen. We were surprised, as we would not have expected such an important argument to be entrusted to a relatively inexperienced member of the defense team. Of course, it was also possible that Martineau and Patterson did not really expect they could win and were just handing a hopeless job over to Miller. Nervously, Miller arose and began his speech. Apparently suffering from a bit of stage fright, he spoke too loudly, and the judge halted him in mid flight. Then, turning the knob on his amplifier and hearing aid down, he signalled Miller to proceed. This helped Miller's nervousness not at all. At the end of Miller's oration, we waited apprehensively for the judge's decision. Much to our relief, he ruled that the trial should proceed.

During one of the intermissions in this part of the trial, I met Willard Edwards. He walked up to me and expressed some regret that he had not known more about me when he wrote the story. He had heard I was a White Sox fan, which had somewhat shaken him. He could not reconcile this with the impression he had of me as a "longhair." He also mentioned that he had been assured by Senator Hickenlooper that in the event there was any trouble (such as a libel suit) the senator would be available to testify. The informants who had fed him his data had assured him, however, that I would not have the nerve to sue. Instead, I had, and the senator had turned tail and run. Edwards and the *Tribune* had been left to fend for themselves. They resented this betrayal deeply. Edwards hinted that the senator might have some trouble when he came up for reelection in Iowa.

One of the defense contentions at the beginning of the trial was that the stories in the *Tribune* and *Times-Herald* were not libelous per se—that is, when not read together. This was a slippery point. While we were certain of our ground, we nevertheless waited with some anxiety the reaction of the judge. He called for the papers and read through them, as though he had not already seen them numerous times. Finally, after what seemed an eternity to Nate, Alex, and me, he looked up and ruled that they were libelous per se.

During the presentation of our complaint, we brought Cowie and Kennedy in. Martineau made some egregious blunders in handling them. He began by asking Dean Cowie when had he first become aware of the *Times-Herald* story. Dean replied that somebody had read the account to him at the Cosmos Club. With a sneer, Martineau then asked if Dean did not think it important to have read it himself. Dean composedly looked at him and said, "Well you see, sir, I was blind at the time!" Having been given this opening, he described how he had suffered cataracts and went on from there to inform the jury about radiation hazards. Without this opening he might not have been allowed to bring up the important point that eye injuries could result from radiation (this was well known to radiologists). Later we heard that one of the jurors also had cataracts and had felt an immediate common cause with Dean and resentment of Martineau's treatment of him. He could hardly wait to ask Dean about his medical history as soon as the opportunity offered.

His own compulsive nature trapped Martineau more disastrously in his encounter with Kennedy. Not having had sufficient time to look up Joe's background, Martineau went up against him without knowing much about Joe's experience at Berkeley and Los Alamos. To Martineau, Joe may have looked like a naive, not particularly bright, country bumpkin. The situation promised to give Alex, as an onlooker who knew better, some great moments. Martineau started as usual by asking what the witness would think of someone the House Un-American Activities Committee said had given classified information to the enemy. Joe favored him with a steady stare and replied that he would think nothing of it. Startled at this unexpected answer, Martineau made the cardinal error of asking a question on cross-examination to which he did not know the answer. He blurted out, "What do you mean [you would think nothing of it]?" Joe, with a slight smile, replied, "The question is meaningless because I would have to know how the material was classified, whether secret or non-secret. If non-secret, I would have no reason to have an unfavorable reaction." (The wording may be imprecise but that was the gist of his reply.) At this, the jury sat up startled and a look of sudden enlightenment and anger showed in their eyes. No one had even thought to bring up such a point before. Everyone had assumed that "classified" meant "secret." Suddenly, the suspicion dawned among the jury that HUAC had used a trick to beguile the public.

I seem to recall that Patterson tried to signal Martineau to back off and cut his losses, but Martineau could not repress his desire to fight back. He charged back into the fray. Waving a thick book, he asked Joe if he was not aware that I had violated security regulations spelled out in it. Joe asked to see the book. Riffling through its pages, he looked up after a few seconds and said that he knew of the book—it was the general Army manual of security—but it had nothing to do with the Manhattan Project. In fact, there had been no such manual that applied to the project at the time of my conversation with the Russians. He knew this because he had helped draft one a year later.

Having been given this opening, Joe went on to discuss security on the Manhattan Project in general, explaining how criteria had changed as time passed. In 1945, he pointed out, no one would have paid attention to casual inquiries about nuclear energy—the topic was a staple of Sunday supplements in the newspapers. Even a year or so later, he would not have felt such inquiries worth bothering the security officials with. After all, if he and others had done so, Army intelligence would have been swamped by a mass of trivia. However, he would have been much more concerned if someone had appeared seriously interested in the nuclear project in 1949.

I think Joe's testimony was the most catastrophic in its effect on the defense case. Coming as it did from so obviously authoritative a figure, whose very appearance belied any suspicion of disloyalty, it must have impressed the jury deeply. Martineau must finally have realized this and reluctantly indicated he had no further questions. Alex, of course, quite happily indicated that he had none either.

I remember just one more of Martineau's maneuvers that backfired. In my testimony at the HUAC hearings and in my deposition for the trial, I had been asked about a joke I had made when the Russians expressed a desire to give me some token of appreciation for helping them reach John Lawrence in connection with the case of the leukemic Russian consular official. I had suggested jestingly that they send me some Russian vodka, saying I understood it was much better than the vodka available in America. The HUAC investigators had tried to find something sinister in this exchange, and so did Martineau. Ironically, the foreman of the jury, a liquor salesman, told me after the trial that he had decided at that point that I must be okay, because he agreed with me about Russian vodka.

All through the trial, we kept wondering about the possibility that the *Tribune,* in its burrowing among the files in various government agencies, to which it seemed to have unusual access, might unearth material that could be damaging if sprung on us unawares. We could not imagine what it might be, but there was no guarantee that some fabrication we could not demolish decisively on the spot did not exist. The speculations I have mentioned about Lansdale and E. O. L. not being willing to testify and the continued denial of my passport after the favorable reception of our case by the Passport Appeals Board suddenly welled up in me when we learned that the *Tribune,* through its informers proliferating through the Army and State Department, had unearthed the document dealing with my dismissal from the Manhattan Project and had obtained agreement from the Army for a photostatic copy to be provided to the court, despite its original classification as "Top Secret." Nate and Alex drafted an appeal to Secretary of Defense Charles Wilson protesting this favored and unprecedented treatment accorded the defense. They mentioned that President Eisenhower had issued a directive prohibiting such actions in the instance of the McCarthy committee attempting to obtain security material in May of the previous year. It was asked why the Army allowed the *Tribune* to see material we were not privileged to see.

The appeal got us nowhere. As we watched agonized, a bemedalled officer strode into the courtroom and handed Judge Bailey a two-page document adorned with ribbons and a seal, signed by the secretary of the army and its deputy administrative assistant. Martineau looked on with satisfaction as the judge examined it, while we stared at the judge in bleak anticipation of possible disaster. After what seemed hours, Judge Bailey looked up at Martineau, then at me, and ruled that the document was inadmissible, being apparently a collection of unsupported allegations and rumors mixed with facts of no relevance to the case. Martineau argued angrily, seeing his great coup about to fail, but the judge remained unmoved, except to remark that he would admit the papers in evidence if the plaintiff wished to do so, on the grounds that the matter it contained concerned only the plaintiff and he was entitled to dispose of it as he saw fit. Then he handed the papers to me to read.

Nate and Alex stared at me, fearful that I might not have the wit to avoid some trap, while Martineau glared. What I read was incredible. It was a communication to the provost marshal general from Lansdale, dated July 15, 1944. The first paragraph described how I had been overheard discussing "highly secret" matters in public places. It mentioned in particular a talk of about ten minutes in the Faculty Club at Berkeley between me and several colleagues on the Manhattan Project. (I recall that the conversation referred to was not on "highly secret" matters; the trouble as I have explained in previous chapters, lay in the security forces thinking that *any* reference to atoms was "highly secret.") The second paragraph gave facts about my life history such as might be found in any biographical sketch. The third paragraph claimed that I and my sister were members of the American Student Union and that "he has been reported to have stated he has been a member of the Communist Party all his life." (This was a complete lie, undoubtedly supplied by some unbalanced and malicious gossip.) There followed a noncommittal sentence or two about my addresses in Berkeley. Finally, the document briefly recited the history of my parents—the fact that they were Jewish and that my father had changed his name on naturalization.

This, then, was the official charge that had caused me so much trouble. There was no mention of the Fish Grotto and the Russians at all. Dazed, I stood there wondering how to reconcile this bland collection of nothings with the vicious persecution I had endured for a decade. Thanks to the efforts of the *Tribune* and its libel, I had finally seen a document I would not otherwise have known existed. Nothing could be clearer than that this insubstantial piece of paper, with its mixture of a few bald lies and innocuous facts, had been built on by unprincipled scoundrels and misguided zealots to create the monstrous onus laid on me. Handing the papers back to the judge, I said that I would not agree to their being admitted. Martineau made vigorous objection, but the judge placed the paper in the custody of the court.

A recess was called and Alex and Nate immediately asked what I had seen. I told them. Stunned, they looked at each other. Finally, Alex asked, "But what happened to the Russians?" It was a shock to realize, after all the years I had thought the Fish Grotto affair so important, that it had not been the cause of my dismissal. The Army action had been solely preventive, on the grounds that I was so gregarious and my contacts among liberal circles so numerous that keeping me on the Manhattan Project constituted too much of a risk. In fact, Compton had said as much in his deposition explaining why the dismissal had occurred.

The *Tribune* presented a defense based on the hearsay and half-truths in the HUAC hearings and in the various reports HUAC and the newspapers had printed in 1948 and thereafter. They took several days doing this, and so the trial dragged on into the middle of February. I was staying in Nate's small apartment, sleeping on a couch in the living room. We passed much time after court sessions speculating on what the jury might be thinking. As is invariably the case, we were entirely wrong in our guesses. Eventually the *Tribune* finished, and the judge ruled there was no basis for their defense of accurate report. On hearing this, they amended their plea and entered a defense of justification by truth. We were outraged that they were permitted to do this, as now the jury was being asked to consider whether I was a spy and traitor with all the scandalous lies and innuendoes of the 1948 affair in front of them. The *Tribune,* while not claiming to have proved any intent on my part to commit espionage, tried to convince the jury that they had drawn a fair "inference" from the transcript of the HUAC report that I was really a spy and traitor.

The jury was out a long time. (We feared there might be a "hung" jury, especially as a unanimous decision was required.) For me it was a slow torture as I sat at the table watching the jury file in, not knowing whether or not I was being consigned to oblivion. For Beka back home it had been much worse. She had not had the distraction of being in the battle, but had been forced to sit a thousand miles away and wait, utterly helpless, while our fate was decided.

The jury brought in a verdict clearing me of all imputed charges and awarding me compensatory damages of $7,500. Ten members of the jury, some in tears, came over to shake my hand, and one apologized for the nominal sum awarded. One juryman we thought would be on our side had held out for no damages, saying that I had been wronged, but ought not to get rich as a result. To get a verdict, they had reached a compromise figure ten times less than they had felt I deserved.

It was a victory unprecedented in the annals of journalism. No one had ever sued Colonel McCormick's organization for libel and won. Of course, the *Tribune* had been removed as a defendant and we had only obtained damages from the *Times-Herald*. But now we could reopen the case against the *Tribune* with the verdict of the jury hanging over it.

The verdict was the more remarkable in that a majority of the jury were government employees and not expected to favor someone in trouble with the government. In this connection, we had had a stroke of luck before the trial began. One of the jurors had to be replaced because of illness. The first alternate was Miss Othello M. Nelson, a young black woman who was a college graduate. We learned that she had been a powerful force in influencing the blacks and led them in pressing for much higher damages than were finally awarded.

The *Tribune* could be expected to appeal the verdict against the *Times-Herald*. We had spent very little on a case that must have cost the newspaper much more. Amid the general rejoicing we noticed that there had been no fanfares announcing our triumph in the press. A few stories appeared, buried in the back pages of the Washington newspapers. An amusing development was an anxious query from the *Washington Post*, which had bought the *Times-Herald* in the middle of our court battles. The *Post* was worried about what action we might take as a result of the judgment against their new acquisition.

The *Tribune*, not conceding anything, took another sideswipe at me, publishing a self-serving squib repeating the history of my troubles with HUAC and Senator Hickenlooper, while playing up the fact that we had only received $7,500 compensatory damages, whereas we had sued for a total of $200,000 compensatory and punitive damages. Martineau stated that there would be an appeal. The unflattering photo the *Chicago Sun-Times* had used in their libel of me back in 1948 was resurrected and included. The *Washington Post–Times-Herald* printed a less objectionable notice, but again emphasized that there were no punitive damages. The only laudatory version to appear was in my little hometown newspaper, the *Hyde Park Herald*.

Meanwhile we moved forward on both fronts. Alex and Nate made several court appearances, arguing before Judge Bailey against motions of appeal, while Nate stepped up our campaign against the State Department. By April, the *Tribune* caved in and agreed to settle the case against it by paying an added $6,000. The money was sufficient to reimburse Nate and Alex and leave me and Beka with enough to make a down payment on a house of our own in a pleasant neighborhood near Washington University.

At last we had come out from under the cloud and could sit back savoring our long-awaited vindication. Letters congratulating us on our victory flooded in from everywhere. I even heard from the Radiation Laboratory. Luis Alvarez wrote telling us how happy everyone there was on hearing the news. Dr. Compton sent a letter in response to mine, expressing his pleasure at the thought that his testimony had helped. Chancellor Shepley joined with a warm letter of congratulation and praise for my legal counsel.

We could now turn our attention wholly to the passport battle. Nate had filed a suit against Secretary of State Dulles demanding he appear in court to show cause why I should not be issued a passport. Just prior to this, the department had sent us a defiant refusal to reopen the proceedings. This enabled us to assure the FAS that the department was unlikely to give up the battle before we obtained a definitive ruling in court on the question of whether passports were a privilege or a right. Dr. Chew wrote that the campaign for funds was moving well.

In the meantime, matters were not moving so well for the State Department. Our experience with the appeals procedure and the whole history of our maltreatment by Mrs. Shipley was available to support arguments by her other victims that a proper hearing could be obtained only in the courts. Lawyers for these applicants were writing to Nate asking for copies of our various briefs and motions. One of these came from the counsel for Dr. Otto Nathan, the executor of Albert Einstein's will, who had been prevented from going to Switzerland to deal with legal matters concerning the estate. Another suit was filed in behalf of Max Schachtman, whose livelihood (I think in import-export trading) had been seriously hindered by inability to obtain a passport. The pressures building up against the State Department were proving intolerable, as the courts kept ordering passports to be issued. As Nate and I held the door open, people were squeezing through and departing to do their long-delayed business. Conveniently enough, Mrs. Shipley was reaching her seventieth birthday and could be retired in a face-saving proceeding. Probably reasoning on the basis of "he that fights and runs away may live to fight another day," a retreat was set in motion, with appointment of a new director of the Passport Division, Mrs. Frances Knight.

Our case continued to work its way through the courts. Early that summer, I attended a symposium organized by the McCollum-Pratt Institute at Johns Hopkins University in Baltimore, and visited Washington to see how we were doing. Nate, in mock protest that I was distracting him from his efforts to earn a living, suggested I call on Mrs. Knight. He thought that if she saw me, she might be encouraged to take a more favorable view of our suit.

It was a hot day as I walked over to Mrs. Shipley's old fort, the Winder Building, and entered its sweltering confines. As I looked around, I noticed that many changes had taken place. There were numerous clerks and secretaries busily working at a profusion of desks where before there had been few. The place gave off an air of brisk efficiency. A security guard wanted to know my business. I told him I wanted to see Mrs. Knight. He asked if I had an appointment. I said I did not, but that I did not think one was necessary. I told him just to tell her "Dr. Kamen is here." Suspecting I might be some kind of crank, he backed off warily, picked up the phone, and talked to someone. I heard a crackling noise coming from the other end, apparently of sufficient import to galvanize the guard. He now regarded me with some awe and showed me the stairs to Mrs. Knight's office. As I made my way up, many of the workers on the ground floor stared in wonderment. At the head of the stairs, I was met by an elderly woman, somewhat out of breath. She had run down the hall on Mrs. Knight's orders to make sure I was brought to the right office. As we walked along the corridor, she apologized for the seedy aspect of the walls and rugs, explaining that they hoped to have the furnishings renovated soon. I remarked that everything looked much better to me than it had on my last visit.

Mrs. Knight turned out to be an attractive blond, probably in her forties—quite a contrast to Mrs. Shipley. Her legal counselor sat with her. Both looked worried. They had no notion why I had come, but sensed it was for no good. As it happened, I had received a letter from a cousin in Argentina, who held a high post in Peron's government and who had issued an invitation for me to lecture on nuclear science there. This was the branch of my father's family Isaac Stern had discovered when he had appeared in concert in Buenos Aires a few years before. I explained to Mrs. Knight that I did not know how to reply because I had no passport. I did not want to mention this, thus possibly creating embarrassment for our embassy in Argentina. It must have been a shock to her to find that I, an alleged communist, had such connections with the Peron regime.

Mrs. Knight expressed gratitude for my consideration. She regretted that she could not give me a definitive answer because the case was still in the courts. We then had a pleasant conversation about her efforts to modernize the office and get funds to renovate the antiquated accommodations left her by the previous chief. She invited suggestions from me as to how service on passports could be improved. On this note, I left the building, reflecting how times had changed.

In our action against the State Department earlier, we had actually issued a subpoena for Mrs. Shipley to testify. This move had created considerable hilarity in the press. Warren Unna wrote a humorous piece in the *Washington Post* under the heading "Chief of Passport Division Finally Gets Her Talking Papers."[12] It related how Hugh W. Duffy, a former deputy U.S. marshal, had penetrated her inner sanctum to serve the subpoena. Duffy had been hired by Nate because he was known to be one of the District of Columbia's most persistent and successful process servers. Three previous efforts to serve papers by another deputy had been unsuccessful. Mrs. Shipley had tried to evade Duffy, but he had conned his way in by telling her secretary he was an old friend. She had tried to refuse the subpoena and the check for $4.40 accompanying it as a witness fee, but he had told her, "Look, it's yours, lady. You've already got it."

"Some big fellow" had "eased over toward him," but he had faced down this character. The story continued: "He was bigger but he was older. He didn't touch me but if he had I would have knocked him right on his kisser." Unna had been unable to reach Mrs. Shipley for comment. The *Washington Evening Star* joined in the merriment with a similar report, entitled, "Subpoena Fails to Elate Mrs. Shipley, Deputy Says."[13]

The subpoena required Mrs. Shipley to submit to the taking of a deposition by April 25 to ascertain what grounds she had for denying me my passport. Chief Judge Bolitha J. Laws indicated that although Mrs. Shipley was retiring and scheduled to leave on a long-anticipated trip to Europe in May, court matters came first. If she did not cooperate, he might have to interfere with Mrs. Shipley's own passport arrangements. The tables certainly had turned! Mrs. Shipley was annoyed and quoted as considering the subpoena a "calculated annoyance." In a story under the byline of Selma Roosevelt in the *Star,* there appeared a laudatory notice about her retirement after forty years of service. There was a party attended by an impressive contingent from higher society and diplomatic strata. Asked about the subpoena, she replied, "He's only a man. Nothing like that's ever stopped me!" At another party she was given a plaque by the American Jewish League Against Communism, Inc., headed by George E. Sokolsky, in the presence of many notables including William Randolph Hearst, Jr., editor-in-chief of the Hearst papers, and Lieutenant General Albert Wedemeyer, retired from a distinguished Army career to become a business executive and fervent anticommunist. She took the occasion to make a speech explaining how lonely she had felt at times when she had been criticized for refusing passports to people she knew were subversives.

In late April, a notice in the form of an editorial finally appeared in the *St. Louis Post-Dispatch* headed "Passports: Right or Privilege."[14] Citing the precedent in English law, it declared "a citizen's right to a passport is as automatic as his right to vote." It went on to express the hope that the courts would make this the rule. This is just what the courts did a few months later. In the case brought on behalf of Max Schachtman against Secretary of State John Foster Dulles, the United States Court of Appeals of the District of Columbia held that passports could not be withheld solely at the discretion of the secretary of state. Nor was the issuance of passports to be confused with conduct of foreign affairs. Judge Edgerton, concurring with his fellow judges, wrote a separate opinion that spelled out the decision in the strongest terms.[15] The courts had gone as far as they could in decisively coming down on our side. There was no legal ground left for the State Department to continue holding out when our case came up in front of Judge Raymond Keech in district court. Earlier, Joseph Rafferty, the Department of Justice lawyer, had agreed to a stipulation stating that

> In the processing of the passport application of the plaintiff, Martin D. Kamen, the decision of the Department of State was based in part upon reasons and supporting information of which neither the plaintiff nor his counsel has ever been apprised by the Department of State, and that except for the information supplied by the plaintiff or his witness, neither the Passport Office nor the Board of Passport Appeals nor the

> Acting Secretary of State who ultimately denied the application were informed of the identity of the informers supplying any of the information upon which the decision was based.

Nate sent Chew at the FAS a copy of this shocking document. He explained that the government had been forced to reveal its willful and arbitrary behavior because Nate had told Rafferty he would file interrogations to be answered by Dulles that would certainly force disclosure.

The next step was to file a motion for summary judgment, based on the grounds that no significant facts were left in dispute and that I should have my passport as a matter of law. On August 6, the deputy undersecretary of state notified the security affairs chief that "in view of the position taken by the Department's Legal Advisor and representatives of the Department of Justice, the Department has no recourse than to grant a passport to Martin D. Kamen."[16] The same day, a scant half-hour before the case was to be argued, Nate was informed that a standard passport would be issued. This was done two days later on July 8—a great day!

Nate was not, however, happy about seeing the State Department cave in without some redress for the beating I had taken all through the years of our war with the Establishment. The judge agreed, and refused to dismiss the case without giving some consideration to the question of damages. In desperation, Rafferty agreed to a stipulation that the issuance of a passport to me constituted

> A ruling by the Secretary of State that under applicable law and regulations, Martin D. Kamen is entitled to such a passport and this ruling has the effect of superseding and reversing the contrary findings by the Passport Office of the Department of State, and by the Acting Secretary of State that the plaintiff was a person . . . engaged in activities which support the Communist movement and who was going abroad for the purpose, knowingly and wilfully, of advancing that movement.

The text was reported in the press, noting that a copy had been placed in the records of the State Department.[17] At first it seemed that this bland statement was hardly adequate compensation in lieu of damages, but sober consideration dictated that it be accepted, because otherwise there was every indication the Justice Department would begin a new series of maneuvers dragging the case on for an indefinite period.

Eventually the passport arrived. I could hardly believe it was real. The FAS issued a bulletin on July 25 welcoming the "breakthrough in passports." A total of $4,300 had been raised for the federation's Kamen Fund, most of which went to meet legal expenses. The publicity given my case by the FAS probably helped convince the government to abandon its defense. There was reason to think that the Justice Department lawyers were anxious to get out from under the load placed on them by Mrs. Shipley's high-handed behavior and the onus of having to argue the State Department's assertion that the secretary of state had absolute discretion in passport decisions.

An amusing sidelight on the effect of the FAS campaign should be mentioned. When Alex and Nate were receiving the check from Patterson in settlement of the libel suit, Patterson inquired how much money we actually had to continue fighting the case, probably because a factor in the decision to make such a settlement was the *Tribune*'s belief that the backing of the FAS would assure that a tactic of attrition would not work. Alex, making sure he had firm hold on the check, said we had practically nothing—only the money I had been able to raise from my family and my own meager resources. The FAS fund, which the *Tribune* had thought was in support of the libel action, was only available for the passport fight.

In November, as a bit of lagniappe, a letter came from Nate enclosing the original check for $4.40 that Duffy had forced on Mrs. Shipley as a witness fee. In the end she was not required to appear as a witness but only to answer a list of written questions.

After a euphoric summer spent savoring the freedom to live like normal American citizens, Beka and I returned to a new life. Eager to get back to probing the mysteries of chromatophores and cytochromes, I could presume, perhaps, to echo the sentiments of that "good man" Jake Fisher, who, as he lay dying, was asked if he did not want to forgive his enemies. "I ain't got none," he replied. "I licked 'em all!"[18]

Epilogue

The euphoria generated by our triumph did not allay the feeling of wary apprehension created in me by the trials we had undergone. The pattern of tragedy and transition reasserted itself once again. Beka's health, exacerbated by the stresses of our confrontations with the Establishment, worsened. Her illness had taken on an irreversible course, and she succumbed early in November 1963, leaving me and our son, David, bereaved and bitter at the fate that had cheated her out of fulfillment of a great literary talent. Although my scientific career burgeoned with new discoveries about cytochromes and bioenergetics and participation in exciting academic adventures, there was a pall overlaying it. It would be years before it dissipated.

Meanwhile, starting in 1957, I had a novel opportunity to help Nathan Kaplan found a unique academic institution—a graduate department of biochemistry at Brandeis University, offering instruction and research experience solely at the graduate and postdoctoral level on a university campus not connected with a medical school. The venture was a huge success. Later another remarkable project in academia took me to La Jolla, California, where I helped Roger Revelle and a handful of colleagues create a new campus of the University of California. In a short time, this campus achieved a leading position among the world's institutions of higher learning, not only in the natural sciences, but also in classics and behavioral science.

These successes and those of my own research slowly reduced my tendency to fear the worst. A positive factor in hastening the healing of old wounds was my marriage in 1967 to Virgina Swanson, a gifted pathologist and medical researcher, whose empathy and warm companionship restored the happiness of a family relationship to David and me.

Whatever doubt or insecurities remained were decisively ended by a tribute organized for me in the summer of 1978 on the occasion of my sixty-fifth birthday by Nathan Kaplan and a former graduate student, Arthur B. Robinson, then chief executive officer at the Linus Pauling Institute for Science and Medicine. Attended by a great throng of old and new friends, including many Nobel laureates and leaders in music, the celebration offered three days of symposia as well as chamber music sessions. The many papers included cover all aspects of my various research interests—isotopic tracer methodology, cytochrome chemistry and functions, biological oxidations, phosphorylation, photosynthesis, protein structure, and general metabolism.[1] A high point was the presentation by Dr. Harold Klein, chief of the Biology Division at the NASA Ames Space Center, of part of a ^{14}C detector of the kind used on the first Viking Mars mission to search for extraterrestrial life. Mounted handsomely in a lucite mold, it was one of only five such sensors produced, the others being on Mars or at the space center.[2]

The evil specters of the past had been exorcised.

Notes

1. Beginnings (1913–1929)

1. In the sciences reminiscences such as those of the biochemist Sarah Ratner ("A Long View of Nitrogen Metabolism," *Annual Reviews of Biochemistry* 46 [1977]: 1–24) and the geneticist Jack Schultz (see T. F. Anderson, "Jack Schultz," *Biographical Memoirs, National Academy of Sciences* 47 [1975]: 393–422) bear testimony to it, as do the recollections of the immunologist Michael Heidelberger ("A Pure Organic Chemist's Downward Path," *Annual Reviews of Microbiology* 31 [1977]: 1–12) to an earlier wave of immigrants from Germany. The progeny of highly cultured upper-middle-class Jewish families, as exemplified by the neurobiochemist David Nachmanson ("Biochemistry as Part of My Life," *Annual Reviews of Biochemistry* 41 [1972]: 1–28), subsequently further enriched American science.

2. College Years (1930–1933)

1. An excellent summary of the Hutchins era as well as the history of the university from the time of its founding under William Rainey Harper through its various presidents to the present may be found in A. Patner, "The Legacy," *University of Chicago Magazine* 72 (1980):14.

2. I had an opportunity to witness a typical performance by Adler because of the rare circumstance that Harold Laufman, the first violin in our university quartet, was also president of his fraternity, as well as an all-around star student athlete (he was on the conference championship water polo team and had top grades in medical school). He arranged a smoker and dinner at which Adler was the featured speaker. The quartet was pressed into service to provide background music. Such a highly intellectual gathering was not typical of most fraternity doings, but Laufman had unusual notions on these matters and insisted that there should at least occasionally be some reasonably cerebral activity. Thus, it came to pass that I was sitting with the quartet to one side of

the main gathering when Adler began his lecture with a tirade against the "pebble pickers" he claimed dominated modern science. After an attack on the Department of Education, he turned his guns on the medical school.

In rebuttal of Adler's call for a return to the principles of Galen and the philosophic precepts of the Greeks in medicine, Laufman observed that the dean of the medical school, himself a Greek scholar of some renown, had often discussed Galen and the Greeks and pointed out egregious errors Galen committed in speculating about disease without adequate experimental foundation. He recounted Galen's assertion that disease processes depended on color; and that "blue bloods" (the aristocracy) required blue medicines when ill, while "red bloods" (commoners) needed red medicines. This color theory of disease had been cited by the dean as an example of the pitfalls of pure deduction, so attractive to Thomists. Adler, surprised at hearing this unexpected display of erudition at a fraternity gathering, denied that such a proposition appeared anywhere in Galen's writings. He would certainly have known if it did, he claimed, because in his course on modern Western thought Galen was studied intensively. Laufman was made to appear a gullible know-nothing before his fraternity brothers.

The next day he angrily visited the dean and demanded to know why he had been misled. The dean grimaced on hearing that Adler was the source of the trouble, and took a well-thumbed volume of Galen in the original Greek off the shelf. Turning to the page about color theory, he asked Laufman if he cared to inspect it or have it read and translated. Many years later, I had a talk with Professor Max Radin, a well-known legal authority who was versed in the writings of Aristotle. The subject of Adler came up, and Radin remarked with some heat "If Adler has read the Aristotelian corpus, I'll eat it!"

3. Predoctoral Years: Tragedy and Transition (1933–1936)

1. See A. C. Lunn and J. K. Senior, "Isomerism and Configuration," *Journal of Physical Chemistry* 33 (1929):1027–39.

2. See C. Eckart, "Application of Group Theory to the Quantum Dynamics of Monatomic Systems," *Reviews of Modern Physics* 2 (1930):305–80.

3. An important member of our group at the postdoctoral level was another disciple of Lunn's, William Bender, whose paper "Properties of the Scale Co-ordinate" was eagerly studied as a formalization of the operational point of view. It was published in the *Journal of the Franklin Institute* 219 (1935):187–210.

4. I have a copy of a paper by Weissman and Freed, published eight years after Sam began work on his doctorate, in which the multiple nature of fluorescent emission lines from europium ions in solution is discussed (see "Multiple

Nature of Elementary Sources of Radiation-Wide-Angle Interference," *Physical Review* 60 (1941):440–42). Earlier papers beginning with the dissertation material started appearing in 1938 and continued after Sam had gone to Berkeley as a National Research Fellow to work with G. N. Lewis.

5. See "The Separation of Isotopes," *Journal of the American Chemical Society* 44 (1926):37–65.

6. "The Absence of Helium from the Gases Left After the Passage of Electrical Discharges . . . ," *Journal of the American Chemical Society* 46 (1924):814–24.

7. See "The Neutron Atom Building and a Nuclear Exclusion Principle," *Proceedings of the National Academy of Sciences 19* (1933):307–18.

8. He developed a general formula for nuclear composition based on the atomic number, or charge, *Z* and the "isotopic number," or excess of neutrons over protons, *I*. Using this relation, he derived his rules for nuclear stability. While the germ of the concept of magic numbers—so basic to the eventually successful shell theory of nuclei—could be said to exist in these empirical studies, there was no theory as such. (So-called magic numbers are those of protons or neutrons that belong to nuclei of great stability. They were first noted by W. Elsasser as occurring at values such as 50, 82, and 126. See "Pauli's Principle in the Nucleus," *Journal de Physique et le Radium 5* [1934]: 389–93.)

References that are helpful are the lectures by Eugene Feenberg, on the "Shell Theory of the Nucleus," in *Investigations in Physics,* no. 3 (Princeton, N. J.: Princeton University Press, 1955), and the classic text by Maria G. Mayer and J. H. D. Jensen, *Elementary Theory of Nuclear Shell Structures,* (New York: Wiley and Sons, 1955). Mayer and Jensen shared a Nobel Prize in physics with Eugene Wigner for their work on the shell model. The early paper by Mayer, "On Closed Shells in Nuclei," *Physical Review* 74 (1948):235–39, should also be consulted.

9. W. D. Harkins, D. M. Gans, and H. W. Newson, "Atomic Disintegrations by a Relatively Slow Neutron," *Physical Review* 43 (1933): 208–9.

10. E. Fermi et al., "Artificial Radioactivity Produced by Neutron Bombardment," *Proceedings of the Royal Society* (London) 146*A* (1934):483–500.

11. W. D. Harkins et al., "Scattering of Protons in Collisions with Neutrons," *Physical Review* 47 (1935):511–12. (My first publication!)

12. F. N. D. Kurie, "The Collisions of Neutrons with Protons" *Physical Review* 44 (1933):463–67.

13. For instance, benzene could be imagined as a composite of the two classic structures (in which the alternating single and double bonds between the six carbons changed places) and a few others in which there were different distributions of the "valence" electrons holding the atoms together. A special triumph was the demonstration that by mixing atomic orbitals of carbon in a reasonable manner one could calculate that the carbon valences comprised a

set of four directed to the corners of a tetrahedron, a finding that rationalized the empirical discovery of the eighteenth-century organic chemists that the structure of carbon compounds required a tetrahedral carbon. Using Pauling's system, one could calculate that the mixture of resonating structures for benzene gave a total binding energy to the molecule much greater than that of any single assignable structure, thus explaining the extraordinary stability of the benzene molecule. Pauling has provided an account of his experiences in developing resonance theory in *Daedalus* 99 (1980): 988–1014.

4. A Time and Place for Euphoria: Berkeley (1937–1938)

1. See H. Childs, *An American Genius* (New York: E. P. Dutton, 1968), p. 278.

2. F.N.D. Kurie and M.D. Kamen, "Disintegration of Nitrogen by Neutrons: Further Experiments in a Low Pressure Cloud Chamber," *Physical Review* 52 (1937):212.

3. In writing nuclear reactions, one shows nuclei of elements involved, in this case nitrogen, carbon, and hydrogen, with the corresponding chemical symbols. A superscript is added to indicate the mass to the nearest whole number ("mass number") and a subscript is used to specify nuclear charge. Nowadays, the mass number appears as a left superscript (e.g., ^{14}N, ^{14}C, etc.), but in the 1940's the superscript appeared on the right. Thus, nitrogen, carbon, and hydrogen nuclei would be written: N_7^{14}, C_6^{14}, and H_1^1, instead of in the current manner, e.g., $^{14}N_7$, $^{14}C_6$, and $^1H_1$. Neutrons, likewise, would be $^1n_0$ currently, and n_0^1 previously. The reaction for production of C^{14} using neutrons would be written in the 1940's as: $N_7^{14} + n_0^1 \rightarrow C_6^{14} + H_1^1$. Note that the sums of mass numbers and charges are equal for reagents and reactants. They total fifteen for mass number and seven for charge on both sides of the equation, as required by principles of mass and charge conservation. The reaction we were seeing in the cloud chamber eventually became the basis for industrial production of C^{14}. It had been reported previously by T. W. Bonner and W. M. Brubaker (see *Physical Review* 48[1935]: 469–70) and W. E. Burcham and M. Goldhaber (see *Proceedings of the Cambridge Philosophical Society,* 32 [1936]: 632–36).

4. J. M. Cork and E. O. Lawrence: "Transmutation of Platinum by Deuterons," *Physical Review* 49 (1936): 488–92.

5. The work at the Radiation Laboratory had become known nationwide. In 1937, *Time* published an extensive story on E.O.L. and the Lab, featuring a color photo of Lawrence at the target chamber of the 37-inch cyclotron on the cover, with the caption "He creates and destroys." The text was illustrated by a picture of Phil Abelson and Dean Cowie working inside the magnet coils of the machine.

6. E. McMillan, M. Kamen, and S. Ruben: "Neutron-induced Radioactivity of the Noble Metals," *Physical Review* 52 (1937):375–77.

7. Glenn Seaborg has related his involvement in discovering many highly useful radioisotopes, including those of iodine (I^{131}), cobalt ($Co^{59.60}$), technetium (Tc^{99m}), iron (Fe^{55}), caesium (Cs^{137}), and zinc (Zn^{65}). See his speech to the Society of Nuclear Medicine, July 1970.

8. Kurie published a description of the external target assembly he had designed in *Review of Scientific Instruments* 10 (1939):199–205.

9. P. R. Hahn et al., "Radioactive Iron and its Excretion in the Dog," *Journal of Experimental Medicine* 70 (1939): 443–51. This work was done at the University of Rochester in G. H. Whipple's laboratory.

10. For instance, L. Jackson Laslett had found a very long-lived sodium radioactivity. Assignment to ^{22}Na could be assured by the finding that all the activity (eventually shown to have a three-year half-life) acted chemically like sodium. The best production reaction was $^{24}Mg_{12} + {}^{2}H_{1} \rightarrow {}^{22}Na_{11} + {}^{4}He_{2}$. I estimated yields of about one milligram of radium equivalent for each five hundred microampere-hours of bombardment by 16 MEV deuterons. The activity was easily extracted by water, as expected for radioactivity of a sodium isotope. My own efforts, using alpha particles of about 32 MEV on thallium produced only a weak radioactivity arising from some contaminant. With Abelson, I did a slow neutron bombardment of thallium, as well as one with deuterons that produced only the radioactivity of the sodium isotope, ^{24}Na.

11. For an account of these beginnings of isotope tracer methodology, see M. D. Kamen, "The Rise of Tracer Methodology in Berkeley," *The Vortex* (publication of the American Chemical Society, Berkeley section) 24 (1963):406–16.

12. Schoenheimer's Dunham lectures summarizing the work at Columbia University were published as *The Dynamic State Of Body Constituents* (Cambridge: Cambridge University Press, 1940).

13. A partial list of compounds for which we developed assays, in addition to the classic guesses of a few simple carbohydrates (formaldehyde, glucose, and so on) may be reproduced from one of our notes: formate, oxalate, succinate, malate, glycollate, tartrate, lactate, maleate, fumarate, citrate, hexose phosphates, glucuronate, ascorbate, water-soluble polysaccharides, "polyhydric molecules," adipate, glutarate, glyoxalate, glycolaldehyde, glyceraldehyde, pyruvate, glycerate, glycerin, simple fatty acids, and amino acids such as glycine, alanine, aspartate, glutamate, serine, lysine, arginine, asparagine, and glucosamine.

14. See M. D. Kamen, *Radioactive Tracers in Biology* (New York: Academic Press, 1951), pp. 229 et seq.

15. See S. Ruben, W. Z. Hassid, and M. D. Kamen, "Radioactive Nitrogen in the Study of N_2 Fixation by Non-Leguminous Plants," *Science* 91 (1940):578–79.

5. New Vistas in Photosynthesis

1. For an account of the history of photosynthesis research in not too technical a vein see M. D. Kamen, *Primary Processes in Photosynthesis* (New York: Academic Press 1963, chapter 1.

2. Three letters in my files from Emerson, dated September 30, October 15, and November 26, 1948, chronicle his efforts to bring Warburg around to his viewpoint. "Warburg has resisted all efforts of Deans and Professors of Chemistry to persuade him to give bona fide lectures, but he has agreed to comment on other people's lectures," he writes in the first. "I can never foresee what his next impossible demand is going to be. But he is beginning to recognize that Emerson and Lewis found something funny about the method of Warburg and Negelein for measuring quantum yields. He does not accept our interpretation, but that may come later."

In the October letter, Emerson gives a list of the outside speakers who were invited to meet Warburg and present seminars, including Carl Vestling from the University of Illinois, I. C. Gunsalus from Indiana University, D. R. Goddard from the University of Pennsylvania, E. S. G. Barron from the University of Chicago, and Farrington Daniels from the University of Wisconsin. On December 8, he scheduled a "mammoth-all-day meeting on cancer at the University of Illinois School of Medicine in Chicago," following which there were plans for Warburg to visit Madison, Wisconsin, and Minneapolis.

The November letter outlines plans for Farrington Daniels to bring pieces of his apparatus for Warburg to see. "Also, Warburg has proposed a procedure for measuring quantum yields upon which both of us can agree, and I am reasonably certain the results will leave him holding the bag," Emerson says. Needless to say, he was too optimistic. A later letter from Hans Gaffron, writing from Wood's Hole in Massachusetts, June 25, 1949, indicates how clouded Emerson's crystal ball was. Gaffron states, "On June 22 we had another round between Warburg and Emerson. The new measurements of the Warburg group were presented by Dean Burk who said, or was it Warburg himself, that never have quantum yields been measured so accurately and definitively as at Bethesda (Burk's laboratory). . . . Not only have Warburg's results been confirmed as expected but it was seriously contended that quantum numbers of 3 have theoretical significance. Emerson, of course, showed numerous experiments evidently proving that no change in procedure will bring the quantum yield below the conventional 0.1 (quantum number must be 10 or greater). Most among the lay audience were inclined to believe Warburg who stated that the matter was settled and that Emerson's data were wrong."

3. Water (HOH) may be imagined as split into its two constituents, H and OH. Other participants in this process are symbolized as *x* and *y*; *x* accepts electrons from H, and so functions as oxidant; *y* is an electron donor, or reductant, that combines with the electron-deficient OH, itself an oxidant. The photosynthetic primary process can thus be written:

$$HOH + x + y \xrightarrow[\text{light}]{} Hx + y\,OH$$

In Hill's chloroplast reaction, he substituted an oxidant, such as potassium ferricyanide, for x. Other such substitutes could be certain dyes that changed color on being reduced, as did ferricyanide. The y OH remaining was postulated to rearrange, four molecules giving molecular oxygen and water, i.e., $4y\ OH \rightarrow O_2 + 2HOH + 4y$. Photosynthesis consists of repetitions of this process whereby the reducing power Hx transforms CO_2 to cell material.

4. Ironically, it was just at this juncture that Eugene Rabinowitch undertook the Herculean task of casting in concrete all the old knowledge. It was a thankless effort, especially as he could hardly edit out what might be dubious, or actually wrong, in the vast literature. As some long-forgotten American folk humorist put it, "It's not the things you know that kill you—it's the things you know that ain't so!" Rabinowitch labored and brought forth a massive monograph, replete with admirably ingenious, if irrelevant, reaction schemes devised to rationalize the many kinetic results published. Hardly any enzymology of the chloroplast was mentioned, except for the inclusion of catalase as the one recognizable enzyme present in the photosynthetic apparatus. Some tentative suggestions that liberation of oxygen might involve an organic peroxide were based on the presence of this enzyme, which was known to catalyze breakdown of inorganic hydrogen peroxide to oxygen and water. Nevertheless, the bringing together of the vast inchoate literature of photosynthesis was a boon to all researchers and Rabinowitch's monograph is still a classic.

5. K. V. Thimann, "The Absorption of Carbon Dioxide in Photosynthesis," *Science* 88 (1938):506–7.

6. H. Gaffron, "Metabolism of Sulfur-Free Purple Bacteria," *Biochemische Zeitschrift* 260 (1933): 1–17.

7. H. G. Wood and C. H. Werkman, "The Utilization of CO_2 by the Propionic Acid Bacteria in the Dissimilation of Glycerol," *Journal of Bacteriology* 30 (1935): 332; also, "The Utilization of Carbon Dioxide in the Dissimilation of Glycerol by Propionic Acid Bacteria," *Biochemical Journal* 30 (1936): 48–53.

8. D. D. Woods, "The Synthesis of Formic Acid by Bacteria," *Biochemical Journal* 30 (1936):515–27.

9. H. A. Krebs and K. Henseleit, *Zeitschrift für Physiologische Chemie* 210 (1932):33–66.

6. Happy Years: Berkeley (1938–1940)

1. The records are stored in the Bancroft Library at the University of California in Berkeley. As part of the History of Science project there, I have also provided some hours of oral testimony covering this period of my career.

2. S. Ruben and W. F. Libby, "Width of an Iodine Resonance Neutron Band," *Physical Review* 51 (1937):774.

3. We showed this in the following way. Assuming that the labeled carbon (C*) was in the carboxyl carbon of our organic acid (-C*OOH), where R stood for the molecular species attached to the carbon of carboxyl, it should have been possible to gradually remove the C* as CO_2 by evacuating the labeled solution so as to lower the pressure of CO_2 present with the RCOOH, according to the simple equilibrium: $RC^*OOH \rightleftharpoons RH + C^*O_2$. We evacuated the gas space over the RC*OOH solution to remove whatever C^*O_2 was there owing to the partial dissociation of RC*OOH, then replaced the gas with unlabeled CO_2. This CO_2 would then react with RH to form RCOOH, in which the original -C*OOH was thereby diluted. The process could be repeated many times, each time with a progressive loss of label from the solution. This even gave a basis for calculation of RH originally present.

4. I summarized our work some years later; see M. D. Kamen, "Some Remarks on Tracer Researches in Photosynthesis," in *Photosynthesis in Plants,* ed. J. Franck and W. E. Loomis (Ames, Iowa: State College Press, 1949), pp. 365–80. Original reports were: (a) S. Ruben, W. Z. Hassid and M. D. Kamen, "Radioactive Carbon in the Study of Photosynthesis," *Journal of the American Chemical Society* 61 (1939):661–63; (b) S. Ruben, M. D. Kamen, and W. Z. Hassid, "Photosynthesis with Radioactive Carbon. II. Chemical Properties of the Intermediates," *Journal of the American Chemical Society* 62 (1940):3443–50; (c) S. Ruben, M. D. Kamen, and L. H. Perry, "Photosynthesis with Radioactive Carbon. III. Ultracentrifugation of Intermediate Products," *Journal of the American Chemical Society* 62 (1940):3450–51; (d) S. Ruben and M. D. Kamen, "Photosynthesis with Radioactive Carbon. IV. Molecular Weight of the Intermediates and a Tentative Theory of Photosynthesis," *Journal of the American Chemical Society* 62 (1940):3451–55. Preparation of these reports usually began with my writing a preliminary draft, which Sam and I then revised until a final version was ready for typing.

5. S. Ruben and M. D. Kamen, "Radioactive Carbon in the Study of Respiration in Heterotrophic Systems," *Proceedings of the National Academy of Sciences,* n.s., 26 (1940):418–22.

6. H. G. Wood and C. H. Werkman, "The Utilization of CO_2 by the Propionic Acid Bacteria in the Dissimilation of Glycerol," *Journal of Bacteriology* 30 (1935):332; also, "The Utilization of Carbon Dioxide in the Dissimilation of Glycerol by Propionic Acid Bacteria," *Biochemical Journal* 30 (1936): 48–53.

7. A complete account of the history of CO_2 fixation will be found in *Molecular and Cellular Biochemistry* 5 (1973):79–101.

8. See the letter from H. G. Wood to J. T. Edsall and H. A. Krebs of May 24, 1974, in *Molecular and Cellular Biochemistry* 5 (1973):93.

9. Our conclusion that a carbon-carbon bond existed between the labeled carboxyl and an organic residue was based on well-established chemistry involving the reactions of barium salts of organic acids. When the doubly positively charged barium ions react with the singly negatively charged ions of carboxylic organic acids, a barium salt is formed which contains two of the negatively charged carboxylic ions for each barium ion. This salt is insoluble in a mixture of alcohol and water. It precipitates, so it can be filtered off, washed, dried, and then heated to temperatures of about 200° C. It then undergoes a process in which one of the carboxylic carbons is oxidized to carbonate and the other reduced to a ketone carbon. If the organic skeleton associated with the original carboxyl is large, the resultant ketone is not volatile after solution and acidification, whereas the carbonate comes off readily. We had shown previously that no small organic acids were formed with labeled carboxyl. We could expect then that our labeled material as the carboxylic acid of a large substance would come down with a mixture of carrier organic acids as the barium salt, but after the heat treatment would yield only half its label as carbonate (CO_2) upon acidification. This is precisely what we found. A completely conclusive demonstration required that we actually isolate and identify the organic moiety of our labeled material but the means were not at hand to do this at the time. The demonstration of incorporation of CO_2 into a known compound did not have to wait long. Just about the time we were beginning to try further experiments with our algal extracts, Evans and Slotin, working with the well-characterized Krebs Cycle system in animals (heterotrophes), showed conclusively that labeled carbon, administered as $^{11}CO_2$, appeared in the carboxyl group of alpha-ketoglutaric acid (see note 11, below).

10. In the letter cited in note 8, above, Wood refers to the contributions of Lebedev, published in 1921.

11. E. A. Evans and L. Slotin, "The Utilization of Carbon Dioxide in the Synthesis of Alpha-Ketoglutaric Acid," *Journal of Biological Chemistry* 136 (1940):301–2.

12. Quote from letter of B. Vennesland to H. A. Krebs of April 16, 1973, in *Molecular and Cellular Biochemistry* 5 (1973):86–7.

13. Schoenheimer prepared for the Dunham Lectures in 1941, but did not live to deliver them. They were finished and edited for publication by his associates and former students David Rittenberg and Sarah Ratner, aided by the then chairman of the Department of Biochemistry at Columbia University, Hans Clarke; see R. L. Schoenheimer *The Dynamic State of Body Constituents* (Cambridge, Mass: Harvard University Press, 1940). A reappraisal of his ideas on turno' · in the adult organism can be found in a paper by J. T.

Edsall in *Molecular and Cellular Biochemistry* 5 (1973): 108–10. A detailed account of his impact on tracer methodology and biochemistry is given by R.E. Kohler, Jr., *Rudolph Schoenheimer, Isotopic Tracers, and Biochemistry in the 1930s,* Historical Studies in the Physical Sciences, vol. 8 (Baltimore: Johns Hopkins University Press, 1977), pp. 257–97.

14. R.L. Schoenheimer, S. Ratner, and D. Rittenberg, "The Metabolic Activity of Body Proteins Investigated with l (-) Leucine Containing Two Isotopes," *Journal of Biological Chemistry* 130 (1939): 703–32.

15. See S. Ruben et al., "Photosynthesis with Radiocarbon," *Science* 90 (1939):570–71.

16. L. W. Alvarez: "Ernest Orlando Lawrence," *Biographical Memoirs, National Academy of Sciences* 41 (1970):251–94, quotation on p. 269.

17. Alvarez had a knack for ingenious experimentation and an unquenchable drive to do it. He had put together a massive apparatus to demonstrate the existence of structure in the neutron by showing it had a measurable separation of electron charge. He achieved a successful determination in collaboration with Felix Bloch, who came up from Stanford to provide the theoretical expertise. About the same time, Alvarez invented an apparatus for obtaining collimated beams of slow neutrons of controllable energies by use of the "time of flight" principle. Most importantly, he had made one of the few fundamental discoveries that could be credited up to then to the Rad Lab—the first experimental demonstration of the phenomenon of "K-capture." In the radioactive decay of isotopes to stable isotopes, caused by an excess of nuclear protons, the excess positive charge could be lost either by emission of a positive electron (positron) or by nuclear capture of a negative electron bound in the nearby atomic orbits of the so-called K-shell and L-shell. Using a positron emitter, produced in vanadium by deuteron bombardment of titanium, Alvarez found the expected soft gamma rays emitted when the hole left by the captured K electron was filled by one of the other electrons. In several papers he definitively established the process in this case and that of the gallium isotope (^{67}Ga) produced by bombardment of zinc. A large number of experiments establishing the general distribution of the K-capture process among radioisotopes were subsequently reported from the laboratory. An interest arose in the construction of x-ray spectrographs of all kinds, including the variety known as the "bent crystal" spectrograph, which played an important role later in Phil Abelson's demonstrations that rare earth isotopes were formed in nuclear fission.

18. R.R. Wilson, "A Vacuum-Tight Sliding Seal," *Review of Scientific Instruments* 12 (1941):91–93.

19. R.R. Wilson and M.D. Kamen, "Internal Targets in the Cyclotron," *Physical Review* 54 (1938): 1031–36.

20. An amusing and delightfully clear account of the way nuclear fission

was overlooked elsewhere as well as at the Rad Lab can be found in Joseph W. Kennedy, *Uranium, Fission and Transuranium Elements* (University Park: Pennsylvania State University, 1952).

7. The Carbon 14 Story: Bright Clouds with Dark Linings (1935–1941)

1. F. N. D. Kurie, "A New Mode of Disintegration Induced by Neutrons," *Physical Review* 45 (1934):904–5.

2. The nuclear reactions could be written: (a) $^{14}N + n \rightarrow {}^{14}C + {}^{1}H$, (b) $^{14}N + n \rightarrow {}^{13}C + {}^{2}H$, or (c) $^{14}N + n \rightarrow {}^{12}C + {}^{3}H$. Kurie suggested that reaction (c) producing tritons and ^{12}C was most likely because the most stable carbon isotope, ^{12}C, was formed.

3. T. W. Bonner and W. M. Brubaker, "The Disintegration of Nitrogen by Slow Neutrons," *Physical Review* 48 (1935):469–70; "The Disintegration of Nitrogen by Neutrons," *Physical Review* 49 (1936): 223–29.

4. J. Chadwick and M. Goldhaber, "Disintegration by Slow Neutrons," *Proceedings of the Cambridge Philosophical Society* 31 (1935): 612–16.

5. E. M. McMillan, "Artificial Radioactivity of Very Long Life," *Physical Review* 49 (1936): 875–76. McMillan suggested that ^{14}C was produced in the reaction: $^{13}C + {}^{2}H \rightarrow {}^{14}C + {}^{1}H$.

6. Later, I learned of at least one other abortive attempt, this time at Yale, where Ernest Pollard tried to detect radioactive carbon in boron and carbon targets that had received weak bombardments with 3 MEV deuterons, as well as alpha particles emitted by natural radioactive polonium and thorium C′. He had been studying the reactions,

$$^{13}C + {}^{2}H \rightarrow {}^{14}C + {}^{1}H, \text{ and } {}^{11}B + {}^{4}He \rightarrow {}^{14}C + {}^{1}H,$$

and seen the expected recoil protons. He could conclude that ^{14}C was a negative beta ray emitter with maximum energy of about 0.3 MEV. He used a thin-walled Geiger counter in an attempt to find these low-energy beta rays, but found none. His failure to see any meant that if ^{14}C had a mean life of several years, his bombardments totaling only about six microampere-hours, using 3 MEV deuterons, would not have been enough to make detectable amounts of ^{14}C; see E. Pollard, "Mass and Stability of ^{14}C," *Physical Review* 56 (1939):1168.

7. W. F. Libby, "Radioactivity of Neodymium and Samarium," *Physical Review* 46 (1934): 196–204; see also, W. F. Libby and D. D. Lee, "Energies of the Soft Beta-Radiations of Rubidium and Other Bodies: Method for Their Determination," *Physical Review* 55(1939):245–51.

8. Later I wrote a popular account of these happenings entitled "The Night Carbon-14 Was Discovered," which appeared in a rather obscure publi-

cation, *Environment Southwest* (November 1972). A detailed, more technical account is M. D. Kamen, "The Early History of Carbon-14," which appeared in both the *Journal of Chemical Education* 40 (1963): 234–42 and *Science* 140 (1963):584–90.

9. The reaction was $^{34}S + ^{2}H \rightarrow ^{35}S + ^{1}H$.

10. Nuel Pharr Davis, in *Lawrence and Oppenheimer* (New York: Simon & Schuster, 1969), mistakenly asserts that "Ruben and Kamen waited in the audience for him [Birge] to turn the limelight on them" (p. 80).

11. R. T. Birge, "Presentation of the Nobel Prize to Professor Ernest O. Lawrence, *Science* 91 (1940):323–30.

12. As an example, A. Brown and A. Frenkel stated, "The search for C^{14} was initiated by Ruben largely for use as a tracer in photosynthesis studies" (*Annual Reviews of Physiology* 4 [1953]:30).

13. Arthur Snell has written a short account of the history of the "Carbon-Fourteen Factory" that operated at Oak Ridge National Laboratory; see *Oak Ridge National Laboratory Review*, Fall 1976, pp. 45–47.

14. W. Bleakney et al. had published claims in 1935 that tritium was present in ordinary hydrogen in amounts of one part per billion. They based their estimates on the appearance of a faint line in the mass spectrogram of water hydrogen, which appeared to have a mass of three. See W. Bleakney and A. J. Gould, "The Relative Abundances of the Hydrogen Isotopes," *Physical Review* 44 (1933):265–68; also their article "The Concentration of H^3 and O^{18} in Heavy Water," *Physical Review* 5 (1934): 281–82.

15. L. Alvarez and R. Cornog, "He^3 in Helium," *Physical Review* 56 (1939):379, and "Helium and Hydrogen of Mass Three," *Physical Review* 56 (1939):613. Alvarez has recently written an authentic and detailed account of developments leading up to, and including, the discovery of tritium; see "The Early Days of Accelerator Mass Spectrometry," *Physics Today* 35 (1982): 25–32.

16. S. Ruben et al., "Heavy Oxygen (O^{18}) as a Tracer in the Study of Photosynthesis," *Journal of the American Chemical Society* 63 (1941): 877–79.

17. Warburg wrote a summary paper entitled "Photosynthesis," which first appeared in *Angewandte Chemie* 69 (1957):627, and was published in English translation in *Science* 128 (1958):68–73. Ruben and I were featured as perpetrators of experiments he dismissed as erroneously interpreted.

18. This latter area of research engaged our attention originally because we were attempting to locate the positions of labeled carbon in propionic and succinic acids formed during $^{11}CO_2$ uptake by propionic acid bacteria fermenting glycerol. Harland Wood and C. H. Werkman had shown that, as expected, only the carboxyl group of succinate formed in this process contained labeled carbon (^{13}C), indicating a simple addition of the CO_2 as a carboxyl moiety (carboxylation). We had expected a similar finding for the propionate formed. Using alkaline permanganate oxidation of the ^{11}C-labeled propio-

nate, we found instead that the labeled carbon seemed to be distributed randomly in all three carbons of the propionate; see S. F. Carson et al.,"Radioactive Carbon as a Tracer in the Synthesis of Propionic Acid from CO_2 by the Propionic Acid Bacteria,"*Science* 92 (1940): 433–34. Wood challenged this result, showing that the method of oxidative degradation we had used was unreliable; see H. G. Wood et al., "Position of the Carbon Dioxide Carbon in Propionic Acid Synthesized by Propionic Acid Bacteria," *Proceedings, The Society of Experimental Biology and Medicine* 46 (1941): 313–16. We, too, found on reexamining the reaction system that we had been misled by the alkaline permanganate oxidation degradative procedure; see S. F. Carson et al., "Radioactive Carbon as an Indicator of Carbon Dioxide Utilization. V. Studies on the Propionic Acid Bacteria," *Proceedings of the National Academy of Sciences (USA)* 27 (1941):229–35. It was evident that the propionate and succinate formed contained labeled carbon from CO_2 only in their carboxyl groups.

19. P. Nahinsky et al., "Tracer Studies with Radioactive Carbon: The Synthesis and Oxidation of Several Three-Carbon Acids," *Journal of the American Chemical Society* 64 (1942):2299–302.

20. T. H. Norris, S. Ruben, and M. B. Allen, "Tracer Studies with Radioactive Hydrogen: Some Experiments on Photosynthesis and Chlorophyll," *Journal of the American Chemical Society* 64 (1942):3037–40.

21. M. B. Allen and S. Ruben, "Tracer Studies with Radioactive Carbon and Hydrogen: The Synthesis and Oxidation of Fumaric Acid," *Journal of the American Chemical Society* 64 (1942):948–50.

22. S. Ruben, A. W. Frenkel, and M. D. Kamen, "Experiments on Chlorophyll and Photosynthesis Using Radioactive Tracers," *Journal of Physical Chemistry* 46 (1942):710–14.

23. We looked at numerous systems involving ferrous, ferric, cupric, and magnesium ions, exposed to chelating groups, such as tetraphemyl porphyrin, dipyridyl, orthophenanthrolene, and hydroxyquinolate, as well as biologically interesting chelating agents, such as heme, hemoglobin, and phaeophytin. We showed that structure was more decisive than the type of bond involved. For instance, so-called covalent bonding, in which electrons were shared between the nitrogen atoms of the organic chelator and the central atom, iron, as in iron dipyridyl, exchanged rapidly with free iron ions in water solutions, whereas there was no such exchange in the protoporphyrin complex of iron, which was held by only electrostatic forces to the nitrogen atoms of the protoporphyrin. See S. Ruben et al., "Some Exchange Experiments with Radioactive Tracers," *Journal of the American Chemical Society* 64 (1942):2297–98.

24. P. Nahinsky and S. Ruben, "Tracer Studies with Radioactive Carbon: The Oxidation of Propionic Acid," *Journal of the American Chemical Society* 63 (1941):2275–76.

25. S. Ruben and M. D. Kamen, "Long-Lived Radioactive Carbon: C^{14}," *Physical Review* 59 (1941):349–54.

26. M. D. Kamen and S. Ruben, "Production and Properties of Carbon 14," *Physical Review* 58 (1940):194.

27. See an early paper by R. Sherr et al., "Decay Rates of C^{14}, N^{14*}, and O^{14}," *Physical Review* 100 (1955):945–46.

28. B. Goulard et al., "Beta Decays and Related Processes in the A = 14 Nuclei," *Physical Review,* Series *C,* 16 (1977):1999–2009.

29. See M. D. Kamen and S. Ruben, "Studies in Photosynthesis with Radio-Carbon" and "Synthesis in Vivo of Organic Molecules Containing Radio-Carbon," *Journal of Applied Physics* 12 (1941):326,310, and S. Ruben and M. D. Kamen, "Nonphotochemical Reduction of CO_2 by Biological Systems," *Journal of Applied Physics* 12 (1941):321–22.

30. See Vannevar Bush, *Pieces of the Action* (New York: William Morrow, 1970), p.31 et seq.

31. The reaction was of the same kind as that in which slow neutrons are made ^{14}C from ^{14}N; the target nucleus was ^{35}Cl and the reaction $^{35}Cl + n \rightarrow {}^{35}S + {}^{1}H$.

32. M. D. Kamen, "Production and Isotopic Assignment of Long-lived Radioactive Sulfur," *Physical Review* 60 (1941):537–41.

33. M. D. Kamen, "Radioactive Sulfur from Neutron Activation of Chlorine Isotopes," *Physical Review* 62 (1942):303.

8. Pearl Harbor and the Manhattan Project (1941–1943)

1. See C. B. van Niel et al.,"Radioactive Carbon as an Indicator of Carbon Dioxide Utilization. VIII. The Role of Carbon Dioxide in Cellular Metabolism," *Proceedings of the National Academy of Sciences (USA)* 28 (1942):8–15. In the same issue of the *Proceedings,* Sam and I were co-authors with van Niel and his student J. O. Thomas in a study of CO_2 fixation in the protozoan *Tetrahymena gelii* (p.157).

2. See *Leo Szilard: His Version of the Facts. Selected Recollections and Correspondence,* ed. S. P. Weart and G. W. Szilard (Cambridge, Mass.: MIT Press, 1978), in which I. I. Rabi is quoted as saying to Fermi in early 1939, "What do you mean by remote possibility [of a bomb]?" and Fermi replied, "Well, ten percent!"

3. See G. E. Gordon, "Obituary: Charles Dubois Coryell," *Journal of Inorganic and Nuclear Chemistry* 34 (1971):1–11. Coryell's remarkable personality and his scientific contributions are well set forth in this obituary.

4. Deposition of M. D. Kamen, district court, District of Columbia, Civil Action No. 4461–51 (December 19, 1952), pp. 117–18.

5. For example, J. von Halban, Jr., et al., "Evidence of a Nuclear Chain

Reaction Inside a Uranium-Bearing Mass," *Journal de Physique et le Radium* 10 (1939):428–29, of which the abstract provides the following summary:

> A copper sphere 50 cm. in diameter had a source of photoneutrons (1 gram radium and 160 grams beryllium) at the center. The neutron density, inside and around the sphere, is measured with a dysprosium detector, when the sphere is filled with water, with dry uranium oxide, (or) with uranium oxide and various proportions of water. A consideration of the number of neutrons detected at various distances from the source, in the light of the average life of thermal neutrons, leads to the conclusion that in the presence of uranium some of the neutrons are secondary, tertiary, or higher origin. This indicates the presence of a converging chain reaction in such nuclear processes. At least 8 non primary neutrons are produced by each primary process.

9. Tragedy and Transition Again (1941–1945)

1. F. Lipmann, "Metabolic Generation and Utilization of Phosphate Bond Energy," *Advances in Enzymology* 1 (1941):99–162, and H. M. Kalckar, "The Nature of Energetic Coupling in Biological Systems," *Chemical Reviews* 28 (1941):71–178.

2. The central idea was that in living systems energy for biosyntheses could be supplied from a common reservoir in the form of adenosine triphosphate (ATP), a compound made by linking an organic base to triphosphate residues through a five-carbon sugar, d-ribose. If one symbolized the organic base (adenine) as A, the ribose as R, and the phosphates as P, the ATP could be written simply as ARPPP. There was excess energy in ATP as compared with products made by splitting it with water to ARP and PP or ARPP and P. This energy could be stored in other compounds by transferring one or two P's through the action of special enzymes, called kinases. In this way the energy was not lost as heat, as happened if water split the ATP to free phosphate. Moreover, the necessary specific placement of the P in ARP groups for carrying on the subsequent reactions of biosynthesis could be assured by using these enzymes. If a particular enzyme had to be energized, or activated, this could be done by transfer of P or ARP to make the active "phosphorylated" or "adenylated" form.

Applying this concept to the dark fixations of CO_2 in photosynthesis, we could imagine that light energy was used to make ATP, which then phosphorylated RH to supply an energized form of RH—phosphorylated RH—that contained sufficient energy to permit the carboxylation to proceed with good yield. In fact, the extra energy obtained for each P transferred from ATP was close to ten kilocalories, just the amount needed. The reaction could be written: Phosphorylated RH + $CO_2 \rightarrow$ RCOOH + P.

3. I mentioned in chapter 6 that in my encounter with Warburg, we dis-

cussed an example of the phosphate bond energy concept—phosphorylation of a three-carbon intermediate (pyruvate) found in glycolysis (the anaerobic breakdown of glucose) to phosphoenol pyruvate. Warburg and his associates had shown how the oxidation of the three-carbon precursor molecule, glyceraldehyde phosphate, was used to make ATP, which in turn phosphorylated the pyruvate. See O. Warburg and W. Christian, "Isolation and Crystallization of Proteins of the Oxidative Fermentation of Enzymes," *Biochemische Zeitschrift* 303 (1939):40–68.

4. F. Lipmann, "Analysis of the Pyruvic Acid Oxidation System," *Cold Spring Harbor Symposia in Quantitative Biology* 7 (1939):248–59.

5. S. Ruben, "Photosynthesis and Phosphorylation," *Journal of the American Chemical Society* 65 (1943):279. It is remarkable that while the actual process, worked out by Melvin Calvin, Andrew Benson, and their associates a decade later, showed Sam's proposals to be faulty in detail, his prediction of a charged, phosphorylated intermediate as a CO_2 acceptor was close to the mark. The phosphorylated RH turned out to be ribulose diphosphate.

6. I wrote an account of Sam's work on the presumptive aldehyde acceptor of cyanide and his deductions about the concentration of such an acceptor, which appeared in the publication of the Berkeley section of the American Chemical Society, *Vortex* 24 (1963):400–416. Some years later, Ed McMillan reported other unfinished work on which he had collaborated with Sam to observe formation of the long-lived isotope of beryllium (^{10}Be) in "Radioactivity of Be^{10}," *Physical Review* 70 (1946):123.

7. Deposition of P. S. Patterson in case of M. D. Kamen, district court, District of Columbia, Civil Action No. 4461–51 (1952), p. 290.

8. Some years later Barker sent me a page from his notes describing the samples we used. We had 1.829 grams of barium carbonate, with a total radioactivity of 1.4×10^6 counts per minute, corresponding to no more than a few microgram equivalents of radium.

9. To make the two molecules of CO_2 into acetic acid ($C_2H_4O_2$) some hydrogen is needed. In the fermentation of the glucose, this comes in the form of reductant(s) that supply the eight hydrogen atoms needed, i.e.,

$$2CO_2 + 8H \rightarrow C_2H_4O_2 + 2H_2O$$

Other possibilities existed. One might imagine that all the glucose was first fermented to CO_2—that is, to $6CO_2$—with the release of twenty-four hydrogen atoms, available from reductant(s) formed. Then, the $6CO_2$ could be recombined to make three acetic acids, i.e.,

$$6CO_2 + 24H \rightarrow 3C_2H_4O_2 + 6H_2O$$

Any variation between these extremes was possible. The reaction was ideal for investigation with tracers, because by conducting the fermentation in the presence of labeled CO_2, we could follow the flow of CO_2 in and out, i.e., observe how many molecules of CO_2 were made per molecule of glucose. This could be done by observing the rate at which the constant pool of labeled CO_2 lost label as unlabeled CO_2 entered from the glucose and labeled CO_2 left as acetic acid.

The situation was much like the classic problem posed to students in elementary calculus wherein a bath tub is filled with a quantity of water colored by a dye. This water in the tub is kept at a constant level by allowing it to run out at the bottom and replenishing it from above with fresh water containing no dye. The problem is to determine how fast water has to be supplied, given the rate at which color disappears. We could use a mathematical expression wholly analogous to the solution of this problem to determine which of the possibilities we envisioned was correct in the clostridial fermentation. Moreover, we could predict that both carbons of the acetic acid produced would contain closely identical amounts of labeled carbon because both would originate in calculable amounts from the labeled carbonate in the pool supplied. Our experiment was reported in 1945. See H. A. Barker and M. D. Kamen, "Carbon Dioxide Utilization in the Synthesis of Acetic Acid by *Clostridium Thermoaceticum*," *Proceedings of the National Academy of Sciences (USA)* 31 (1945):219–25.

10. See L. G. Ljungdahl and H. G. Wood, "Total Synthesis of Acetate from CO_2 by Heterotrophic Bacteria," *Annual Reviews of Microbiology* 23 (1969):515–38.

11. See H. A. Barker, M. D. Kamen, and V. Haas, "Carbon Dioxide Utilization in the Synthesis of Acetic and Butyric Acids by *Butyribacterium Rettgeri*," *Proceedings of the National Academy of Sciences (USA)* 31 (1945):355–60, and H. A. Barker, M. D. Kamen, and B. T. Bornstein, "The Synthesis of Butyric and Caproic Acids from Ethanol and Acetic Acid by *Clostridium kluyveri*," in *Proceedings of the National Academy of Sciences (USA)* 31 (1945):373–81. In these papers, we reported that another anaerobic fermenter, the butyric acid-producing microorganism *B. rettgeri*, made acetate from CO_2, followed by conversion of acetate to butyrate. In *C. kluyveri* we demonstrated that when anaerobic growth took place in ethanol- and carboxyl-labeled acetate, labeled butyric and caproic acids were formed. The additions of the carbon fragments from acetic acid were shown to occur by "beta" addition—that is, the label appeared only in alternative carbons, starting at the carboxyl end. Our observations were in accord with the suggestion that formation of butyric acid came about by simple addition of acetic acid and an active two-carbon product of ethanol. Further addition of this type resulted in production of the six-carbon caproic acid. We found no evidence for CO_2 utilization to make acetic acid in this fermentation. For Barker's reminiscences, "Explorations of Bacterial Metabolism," see *Annual Review of Biochemistry* 47 (1978):1–33. I wrote a popular account under the title *A Tracer Experiment* (New York: Holt, Rinehart and Winston, 1964).

12. M. D. Kamen and H. A. Barker, "Inadequacies in Present Knowledge

of the Relation between Photosynthesis and O^{18} Content of Atmospheric Oxygen," *Proceedings of the National Academy of Sciences (USA)* 31 (1945): 8–15.

13. M. Dole and G. Jenks, "Isotopic Composition of Photosynthetic Oxygen," *Science* 100 (1944):409–10.

14. Letter of H. C. Urey to H. A. Barker, dated September 29, 1944.

15. A. P. Vinogradov and R. V. Teis, "Isotopic Composition of Oxygen of Different Origins," *Comptes rendus de l'académie des sciences URSS* 33 (1941):490–93.

16. See A. S. Holt and C. S. French, "Isotopic Analysis of the Oxygen Evolved by Illuminated Chloroplasts in Normal Water and in Water Enriched with O^{18}," *Archives of Biochemistry and Biophysics* 19 (1948):429–35.

17. In January 1955, Chancellor Compton was examined by legal counsel in connection with a libel suit I had brought against the *Chicago Tribune* (see chapter 13). The deposition included his statement that "E. O. L. felt no doubt with regard to Kamen's essential loyalty to the United States and that the difficulties involving what he agreed were indiscretions in the action of Kamen should not be permitted to prevent Kamen from obtaining effective employment where his knowledge could be put to good use" (p. 31).

10. Postscripts on Lawrence and Oppenheimer

1. N. P. Davis, *Lawrence and Oppenheimer* (New York: Simon & Schuster, 1969).

2. R. Birge once observed that "Lawrence didn't like Bohemians" (ibid., p. 84 n. 1). I may add that Birge didn't either.

3. In *All in Our Time: The Reminiscences of Twelve Nuclear Pioneers,* ed. J. Wilson (Chicago: Bulletin of the Atomic Scientists, 1975), pp. 35–49, I have described the rise of Big Science as it was born in the Rad Lab.

4. In *In the Matter of J. Robert Oppenheimer: transcript of hearing before Personnel Security Board* (Washington, D.C.: Government Printing Office, 1954), p. 268, Colonel R. Lansdale is quoted as saying, "We had more trouble with Ernest Lawrence than any other four people put together."

5. See *Robert Oppenheimer: Letters and Recollections,* ed. A. K. Smith and C. Weiner (Cambridge, Mass.: Harvard University Press, 1980), p. 29.

6. Ibid., p. 220.

7. See *In the Matter of J. Robert Oppenheimer,* pp. 656–72.

11. A Career Builds Under a Cloud (1945–1948)

1. A letter from Hughes dated December 21, 1944, informed me that I was to do research in radiochemistry bearing on fundamental problems in

medicine and biology "working with scientists who have important problems to solve but have no experience whatever with the powerful new techniques made possible by the cyclotron."

2. See "Remarks on the Assay of Radioactive Phosphorus," *Journal of Laboratory and Clinical Medicine* 31 (1946):216.

3. See, e.g., C. J. Costello, C. Carruthers, M. D. Kamen and R. L. Simoes, "The Uptake of Radiophosphorus in the Phospholipide Fraction of Mouse Epidermis in Methylcholanthrene Carcinogenesis," *Cancer Research* 7 (1947):642.

4. See A. I. Lansing, T. B. Rosenthal and M. D. Kamen, "Calcium Ion Exchange in Some Normal Tissues and in Epidermal Carcinogenesis," *Archives of Biochemistry and Biophysics* 19 (1948):177–83.

5. W. I. Vernadsky published many works on isotopes in Russian in the late 1930s and an account later in English, see *Transactions of the Connecticut Academy of Sciences* 35 (1944):483.

6. M. D. Kamen, "Isotopic Phenomena in Biogeochemistry," in "Survey of Contemporary Knowledge of Biogeochemistry," ed. G. E. Hutchinson, *Bulletin of the American Museum of Natural History* 87 (1946):105–38.

7. Glycine, the simplest amino acid, had a single ("alpha") carbon attached on one side to an amino group (NH_2) and on the other to a carboxyl group (COOH). The formula was thus

$$\begin{array}{c} H \\ | \\ H_2N-C-COOH \\ | \\ H \end{array}$$

It was shown to be an efficient precursor of heme in the blood cell. See D. Shemin and D. Rittenberg, "The Biological Utilization of Glycine for the Synthesis of the Protoporphyrin of Hemoglobin," *Journal of Biological Chemistry* 166 (1946):621–25.

8. See M. Grinstein, C. V. Moore, and M. D. Kamen, *Journal of Biological Chemistry* 179 (1949):359–69. Later, Grinstein wrote an account in Spanish (my only publication in that language) in *Cienciae Investigación* 5 (1949):83–84.

9. In a letter to Arthur Hughes in May 1946, I described research using radio-iodine by I. L. Chaikoff at Berkeley; by I. Dziewiatkowsky using radiosulfur at Vanderbilt University; by C. W. Sheppard and P. F. Hahn using radiomanganese and radioiron at Rochester University; by a group studying peripheral circulation in humans using radio-sodium at Tulane University; by S. M. Seidlin using radio-iodine at Montefiore Hospital in New York; and by E. S. Gordon using radio-iodine at Wisconsin. We had spent $1,500 supplying materials for these projects, as well as for others on campus devoted to the study of phosphorus metabolism in yeast, with which I was personally involved.

10. See *Edward William Fager: In Memoriam* (University of California, 1978), p. 63 et seq. The competition between the Berkeley and Chicago groups generated one confrontation I found myself forced to attend. A symposium, organized by W. E. Loomis of Iowa State College and James Franck at the University of Chicago was held some time later in 1947 or early 1948, in which the relative merits of the two research programs and their results were debated. I contributed a paper summarizing and evaluating the work Ruben and I had done in Berkeley in the late thirties. The symposium monograph was published under the title *Photosynthesis In Plants* (Ames: Iowa State University Press, 1949).

11. We began by assaying the specific ^{32}P content of phosphate fractions that could be distinguished by such chemical criteria as differences in water and alcohol solubility, resistance to mild acid hydrolysis, precipitability by metallic and alkaline earth ions, and extractability in various mixtures of alcohol and ether. These fractions were compared for ^{32}P content with the original labeled content of inorganic phosphate added as nutrient. We compared cells merely "resting"—that is, not supplied with nutrient nitrogen, or inhibited in growth by addition of compounds such as sodium azide, which permitted yeast fermentation but no growth, or dinitrophenol, which uncoupled phosphate ester hydrolyses from oxidation—and cells induced to make protein and enzymes by provision of growth nutrients. Further procedures probed the composition of such mixtures as were obtained by the fractionation methods described, using known compounds that were intermediates in fermentation.

12. S. Spiegelman and M. D. Kamen, "Genes and Nucleoproteins in the Synthesis of Enzymes," *Science* 104 (1946):581–84.

13. See Sewall Wright, *Physiological Reviews* 21 (1941):487–527 and *American Naturalist* 79 (1945):289; C. D. Darlington, *Nature* (London) 154 (1944):164–69.

14. S. Spiegelman and M. D. Kamen, "Some Basic Problems in the Relation of Nucleic Acid Turnover to Protein Synthesis," *Cold Spring Harbor Symposia in Quantitative Biology* 12 (1947):211–23.

15. See *Time*, January 20, 1947, p. 49.

16. Letter of M. Delbruck to M. Kamen, November 23, 1945.

17. See "Radioelements as a Tool in Scientific Research," *Nature* 161 (1948):456.

18. The original edition was entitled *Radioactive Tracers in Biology* (New York: Academic Press, 1947). In 1951 I rewrote and expanded the text, under the title *Isotopic Tracers in Biology*, to include a treatment of rare stable isotopes as well as radioactive isotopes. Obsolete material was removed in large chunks, but still the newer material caused the text to expand by over a hundred pages.

19. The proceedings of the symposium appeared the following spring; see "Preparation and Measurement of Isotopes," Suppl. *United States Naval Medical Bulletin*, March–April 1948, pp. 115–121.

20. *Isotopic Tracers in Biology,* 3d ed. (New York: Academic Press, 1957), ch. 7, pt. 2.

12. Times of Discovery and Dismay (1948–1951)

1. The expression "splitting of water" as used in those days was not meant to be taken literally. It was a convenient peg, or locution, on which to hang mechanistic schemes using H to symbolize the electron-donating (reducing) system, and OH for the corresponding electron-accepting (oxidizing) system in photosynthesis. The energy needed for green plant photosynthesis is equivalent to that required to move electrons back from the oxygen atom to the hydrogen atoms in water. The elements of water are thereby reformed as hydrogen gas and oxygen gas, schematically $2HOH \rightarrow 2H_2 + O_2$. The energy required for this process is the amount spontaneously released in the reaction between hydrogen and oxygen when hydrogen gives up its electrons to oxygen, making the unreactive stable molecule water.

Little was known about the chemical nature of the molecules actually involved in the photosynthetic process when water was "split." One might use x to indicate the electron acceptor that is part of the reducing complex associated with H, and y to indicate the electron donor associated with OH. Schematically: $HOH + x + y + \text{light} \rightarrow Hx + yOH$. At least four water molecules must participate in this process, because four electrons are needed to form one molecule of O_2 from yOH, viz., $4yOH \rightarrow 4y + 2H_2O + O_2$. Simultaneously, the 4H$x$ made in this process can be used to give electrons to the carbon of CO_2, completing the cycle, viz., $4Hx + CO_2 \rightarrow CH_2O + H_2O + 4x$. Four electrons are required in this process to reduce the carbon of CO_2 back to its condition in carbohydrate, CH_2O.

In summary, electrons originally on the oxygen of water are transferred to the oxidized carbon of CO_2, viz., $CO_2 + H_2O \rightarrow CH_2O + O_2$. In terms of this model of the photosynthetic oxidations of water in the presence of CO_2, all cellular carbon, symbolized as CH_2O, is considered to be derived from the carbon of CO_2.

A general definition of photosynthesis does not require that CO_2 be the ultimate acceptor of electrons. The basic characteristic of green plant photosynthesis is that the energy of radiant light be converted to chemical energy. This chemical energy drives the reaction that "splits" water; further transfer of electrons to CO_2 occurs in the process of green plant photosynthesis.

2. A good description of chromatophores as isolated particles will be found in the first report of such preparations by H. K. Schachman, A. H. Pardee, and R. Y. Stanier, "The Macromolecular Organization of Microbial Cells," *Archives of Biochemistry and Biophysics* 8 (1952):245–60.

3. If the general hydrogen donor in bacterial photosynthesis were writ-

ten HA, the corresponding reaction could be symbolized à la van Niel as: $4yOH + 4HA \rightarrow 4H_2O + 4y + 4A$. Contrast this with the plant systems in which $4yOH$ simply broke down to $4y$, O_2, and $2H_2O$. If A happened to be sulfur, as in the case of certain "green" and "purple" species of photosynthetic bacteria, the reaction became specifically: $4yOH + 2H_2S \rightarrow 2S + 4H_2O + 4y$.

4. In terms of a chemical reaction:

$$\underset{\text{isopropanol}}{2\,\overset{\displaystyle CH_3}{\underset{\displaystyle CH_3}{HC\!-\!OH}}} + CO_2 + \text{light} \rightarrow \underset{\text{acetone}}{2\,\overset{\displaystyle CH_3}{\underset{\displaystyle CH_3}{C\!=\!O}}} + \underset{\text{"cell material"}}{CH_2O} + H_2O$$

see J. W. Foster, "Oxidation of Alcohols by Non-Sulfur Photosynthetic Bacteria," *Journal of Bacteriology* 47 (1944):355–72.

5. See L. S. Tsai and M. D. Kamen, "On The Exchange Of Cyanide With Nitriles," *Journal of Chemical Physics* 17 (1949):585–86.

6. Our paper was eventually published after being presented at the meeting in Paris by Haissinsky. It dealt with some general considerations of isotopic exchange reaction, and included a detached discussion of some experiments by Kennedy and Myers on the mechanism of reaction between arsenious acid and iodine. They measured incorporation of labeled arsenic into the oxidation product, arsenic acid, when the reactant, arsenious acid, was labeled initially with 26-hour ^{76}As isotope. The results, together with discussions they provoked, can be found in our paper. See M. D. Kamen, J. W. Kennedy, and O. E. Myers, "Kinetic Studies on Isotope Exchange Reactions with Remarks on Their Relevance in Classification of Types of Chemical Bonding," *Journal of Chemical Physics* 45 (1948):199–204.

7. M. D. Kamen, "Tracers," *Scientific American* 180 (1949):30–41.

8. For a representative roundup of the smear campaign stories, see the *St. Louis Post-Dispatch,* September 3, 1947, and the *St. Louis Globe-Democrat,* the *San Francisco Chronicle,* and the *Washington Star,* September 4, 1947. The stories were also carried in most of the newspapers belonging to the big syndicated press organizations, such as the Associated Press.

9. Dr. Condon, who made genuinely original and important contributions to nuclear physics in its early years, was a target of the committee and of lobbyists seeking to discredit scientists nearly all his life after he left the Manhattan Project in 1944, having incurred the dislike of General Groves. He was later questioned by HUAC on September 15, 1952, with allegations that he was "the weakest link in atomic security," based mostly on associations with various leftist groups proscribed by the infamous attorney general's list. This was concocted by President Truman and the Democratic administration in the early

years of the Cold War in the hope that it would help counter the fire of Republicans accusing Franklin D. Roosevelt and the Democratic party of harboring communist sympathizers, and even spies.

10. Deutsch's stories appeared in the *New York Star,* September 20, 21, and 25, 1948, headlined "The Case of Professor Kamen," "Professor Kamen: Tartar to Thomas and Co.," and "Scientists Confined by Atomic Curtain." Stone's articles were published in successive issues of the same paper, beginning September 29, 1948, immediately after the distribution of the HUAC report on the alleged espionage activities of atomic scientists. There were five in all, ending October 4. Stone concluded that the report was a "fizzle," promising much and producing little, if anything, to back up its promise of "shocking revelations."

11. Hydrogenase catalyzes the reversible reaction between molecular hydrogen and its dissociation products, protons and electrons, i.e.,

$$H_2 \rightleftharpoons 2H^+ + 2e^-$$

12. For the first report of nitrogen fixation in *R. rubrum*, see M. D. Kamen and H. Gest, "Evidence for a Nitrogenase System in the Photosynthetic Bacterium, *Rhodospirillum Rubrum,*" *Science* 109 (1949):560. Later a complete description appeared; see H. Gest, M. D. Kamen, and H. M. Bregoff, "Photoproduction of Hydrogen and Nitrogen Fixation by *Rhodospirillum rubrum,*" *Journal of Biological Chemistry* 182 (1950):153–70. Confirmation of our findings appeared in an article by E. S. Lindstrom, R. H. Burris, and P. W. Wilson, "Nitrogen Fixation by Photosynthetic Bacteria," *Journal of Bacteriology* 58 (1949):313–16. They also commented on the general occurrence of the phenomenon. The correspondence with this Wisconsin group makes interesting reading and is available for those interested in documentation.

13. See M. D. Kamen and H. Gest, "Serendipic Aspects of Recent Nutritional Research in Bacterial Photosynthesis," *Phosphorus Metabolism,* ed. W. D. McElroy and B. Glass, (Baltimore: Johns Hopkins University Press, 1952), pp. 507–19.

14. My article "Discoveries in Nitrogen Fixation," *Scientific American,* 188 (1953):38–42, may be of interest to readers wishing to probe further into the history of our research on nitrogen fixation.

15. *Chicago Sun-Times,* November 7, 1948.

16. See, for example, a second-page story in the *St. Louis Star-Times,* September 3, 1948, the issue featuring Compton's defense of me on the front page.

17. See report in *Newsweek,* science section, September 27, 1948.

18. A. D. Hershey, M. D. Kamen, J. W. Kennedy, and H. Gest, "The Mortality of Bacteriophage Containing Assimilated Radioactive Phosphorus," *Journal of General Physiology* 34 (1951):305–19.

19. They were edited by J. L. Magee, R. L. Platzman, and myself; see *Physical and Chemical Aspects of Basic Mechanisms in Radiobiology,* Nuclear Science Series, no. 305 (Washington, D.C.: National Academy of Sciences, 1953).

20. M. D. Kamen, "Hydrogenase Activity and Photoassimilation," *Federation Proceedings* 9 (1950):543–49.

21. See W. A. Swanberg, *Luce and His Empire* (New York: Charles Scribner's Sons, 1972), p. 256 et seq. for a discussion of the editorial methods and debasement of researchers at *Time.*

22. Letter to the author dated October 6, 1950.

23. See writ of certiorari prepared by Nathan David, acting as attorney for the Federation of American Scientists, and S. H. Bolz, as attorney for the American Jewish Congress, filed in the Supreme Court as no. 621 during the October term of 1957 in the case of Dr. W. B. Dayton vs. J. F. Dulles and the Department of State. This writ, based on Nate's experience in my case as well as others, cogently states the basis for labeling the passport policy of the government as invalid. Numerous articles, mostly critical of the passport policy, appeared in newspapers and magazines all through the Cold War years. See, in particular, W. H. Saltzman, "Passport Refusals for Political Reasons: Constitutional Issues and Judicial Review," *Yale Law Journal* 61 (February 1952):171–203. Occasionally, apologists for the State Department entered the fray; see, for example, H. A. Philbrick and N. H. Fulbright, "The Red Underground—Bosses Order Speedup in Attack on State Dept. Passport Policy," *New York Herald Tribune,* June 1, 1952, sec. 2, p. 3.

24. 22 *Code of Federal Regulations,* Executive Order No. 7856, sec. 5175 (1949).

25. The tendency of bureaucrats to grab power wherever possible has been a matter of concern to government ever since the Founding Fathers first tried to contain it with the Bill of Rights. A recent instance is the "misinterpreting" of a memorandum from the president on the part of the Internal Revenue Service in order to arrogate to itself the right to deny tax exemptions to certain non-profit educational institutions. In the resultant flap, President Reagan stated: "I am opposed to administrative agencies exercising powers that the Constitution assigns to Congress. Such agencies, no matter how well intentioned, cannot be allowed to govern by administrative fiat" (*Los Angeles Herald Examiner,* January 13, 1982, pt. 1, p. 16).

13. War with the Establishment As Science Moves Forward (1951–1953)

1. A typical protocol is given in the paper by R. Glover, M. D. Kamen, and H. Van Genderen entitled "Comparative Light and Dark Metabolism of Acetate and Carbonate by *R. rubrum,*" *Archives of Biochemistry and Biophysics* 35 (1952):384–408.

2. See M. D. Kamen et al., "Nonequivalence of Methyl and Carboxyl Groups in the Photometabolism of Acetate by *Rhodospirillum rubrum,*" *Science* 113 (1951):302.

3. See S. Ajl et al., "Studies on the Mechanism of Acetate Oxidation by *Micrococcus lysodeikticus*," *Journal of Biological Chemistry* 189 (1951): 859–67.

4. The first in a series of articles was by J. M. Siegel, "Metabolism of Acetone by the Photosynthetic Bacterium, *Rhodopseudomonas gelatinosa*," *Journal of Bacteriology* 60 (1950):595–606. Others with M. D. Kamen appeared in the same journal, vols. 59 (1950):693–97, and 61 (1951):215–28.

5. See H. M. Bregoff and M. D. Kamen, "Photohydrogen Production in *Chromatium*," *Journal of Bacteriology* 63 (1952):147–49.

6. See H. M. Bregoff and M. D. Kamen, "Quantitative Relations between Malate Dissimilation, Photoproduction of Hydrogen and Fixation of Nitrogen in *Rhodospirillum rubrum*," *Archives of Biochemistry and Biophysics* 36 (1952):202–20.

7. The story appeared on the front page of the *Chicago Tribune* headlined "Atom Expert Who Gave Reds Data Named," and on the front page of the *Washington Times-Herald* with the heading "Atomic Expert in Photo with 2 Reds Named."

8. Laurence Burd, "Red Spy Helped Create A-Bomb, Senators Told," *Chicago Tribune*, July 1, 1951, pt. 1, p. 14, and "Red A-Bomb Spy Shielded in U.S.—Hickenlooper," *Washington Times-Herald*, July 1, 1951, pt. 1, p. 2.

9. See *Congressional Record*, 82d Cong., 1st sess., 1951, 97, no. 118, p. 7693.

10. See *The Shameful Years*, a report released by HUAC on December 30, 1951, and *Soviet Atomic Espionage* (Washington, D.C.: Government Printing Office, 1951).

11. Some time later, during the preparation of its defense in our libel action, *Tribune* lawyers got a court order to force a deposition from Teeple, hoping thereby to unearth some evidence upholding their case. This action excited the interest of Warren Unna, a reporter for the *Washington Post*, which had acquired ownership of the *Times-Herald*. Unna did a remarkable bit of investigation. He wrote that Teeple had figured only a few weeks previously in a story reporting his "resignation" from the AEC, where he had been one of three "special assistants" to Chairman Lewis L. Strauss. Teeple's tour of duty with Strauss appears to have coincided with the preparation of the case Strauss brought against J. Robert Oppenheimer. He was hired in October 1953 and left a year later. The AEC refused to say what Teeple was doing in Strauss's office. Unna pointed out that it was shortly after Teeple was hired that the famous letter removing clearance from Oppenheimer was sent, barring him from further participation in nuclear work. After the war ended, Teeple went to work for Senator Hickenlooper. As coincidence would have it, this was the period when the senator charged David E. Lilienthal with "incredible mismanagement" of the AEC. He accused Lilienthal of blocking FBI checks of scientists with communist "leanings" who were being given AEC fellowships. Teeple was still working for the senator when the

speech about me was delivered in the Senate, complete with the picture (which Hickenlooper waved about as he spoke). Later, Unna reported, he had spoken with Willard Edwards, who said that the picture was back with Teeple and that Teeple had told him he (Teeple) had "discussed methods" of ending my activities by physical violence. The story ended with the statement that Teeple was back working for Hickenlooper. (See "Senator Hickenlooper Aide Is Ordered by Court To Give Testimony in Libel Suit," *Washington Post–Times Herald,* November 19, 1954, p. 26.)

12. In green plant chloroplasts the separation of the oxidizing and reducing systems was still maintained despite the removal of the chloroplasts from their original location in green leaves or algal cells. This was manifest because they still produced molecular oxygen while reducing the Hill oxidant. In the bacterial chromatophore, there was no mechanism for separating the oxidizing system by making oxygen from it. Either it did not exist to begin with, or it had been lost in the isolation of the chromatophores from the bacteria. A reagent would be needed to replace the chloroplast system for making oxygen. Otherwise, the oxidizing system in the bacteria would back react with the reducing system and only heat would result, so that our attempt to demonstrate that Hill-type reactions could take place in chromatophores would fail.

13. To assay for the cytochrome enzyme systems, Leo used ordinary cytochrome *c* prepared from beef heart muscle mitochondria, and available commercially. He also used a chemical combination such as ascorbate (vitamin C) with a small quantity of a dye intermediate, the molecule dichlorophenolindophenol (DCIP), known to substitute for the physiological reductant in these enzyme systems.

14. Van Niel had confirmed this phenomenon in *R. rubrum* (see C. B. van Niel, "The Bacterial Photosyntheses and Their Importance in the General Problem of Photosynthesis," *Advances in Enzymology* 1 (1941):263–328). Later, R. K. Clayton expanded these observations with an exhaustive quantitative study of the competition between photosynthesis and respiration in *R. rubrum* (see R. K. Clayton, *Archiv für Mikrobiologie* 22 (1955):180–94).

15. See L. P. Vernon, "Cytochrome *c* Content of *Rhodospirillum Rubrum,*" *Archives of Biochemistry and Biophysics* 43 (1953):492–93.

16. See R. Hill and R. Scarisbrick, "The Haematin Compounds of Leaves," *New Phytologist* 50 (1951):98–111.

17. Biochemists were just beginning to use the first primitive versions of these instruments—in particular, the famous old Beckman "DU" model. Even with this instrument, it required laborious point-by-point checking from 400 to 700 millimicra to do a spectrum analysis. Running two such spectra, one under oxidizing conditions (addition of ferricyanide) and another under reducing conditions (addition of sodium hydrosulfite), as needed for cytochrome *c,* could take several hours.

Cytochrome *c* is a protein in which the central iron atom of its active heme group can gain or lose one electron reversibly. When it is in its reduced form, it loses an electron. Oxygen can accept four electrons in being reduced to water, which also requires four hydrogen positive ions (protons) to complete the process, viz.:

$$4 \text{ reduced cytochrome } c + 4\,H^+ + 0_2 \rightarrow 4 \text{ oxidized cytochrome } c + 2H_2O,$$

that is, four cytochrome *c* molecules would be needed to react with one oxygen molecule. This is what we observed. The same ratio of cytochromes *c* to oxygen ("stoichiometry") is seen when cytochrome *c* is oxidized by oxygen in mitochondria from aerobic tissues.

18. L. P. Vernon and M. D. Kamen, "Photoauto-oxidation of Ferrocytochrome *c* by Sonic Extracts of *Rhodospirillum Rubrum,*" *Archives of Biochemistry and Biophysics* 44 (1953):298–311.

19. See E. I. Rabinowitch, *Photosynthesis and Related Processes* (New York: Interscience Publishers, 1945) pp. 168–69. In later research on the primary processes of photosynthesis, the first reaction to be established was an anaerobic oxidation in the light of chlorophyll, either as chlorophyll *a* in green plants or of bacteriochlorophyll *a* in bacteria. The electron produced was transferred to acceptors, which differed in the two types of photosynthesis. The following reactions in bacteria did not involve "water splitting" as in plants, but instead a removal of hydrogen from the carbon compounds provided for growth.

20. See *Psychiatric Quarterly* 27, no. 1(1953):170. The book, entitled *The Menopause,* was published by Random House in 1952, and later reissued as *Women Needn't Worry.* It went through several translations (Dutch and Japanese among them). Beka subsequently wrote another book with Dr. J. Hirsh for the Mt. Sinai centenary celebration, entitled *Saturday, Sunday and Everyday* (New York: United Hospital Fund, 1954).

21. Our approach followed that originated by Alan Mehler, as described in his two articles on "Studies on Reactions of Illuminated Chloroplasts," *Archives of Biochemistry and Biophysics,* 33 (1951):65–77, and 34 (1951):339–51. Mehler introduced the use of reductants, rather than oxidants as in the classic Hill system. Such additions resulted in an uptake, rather than evolution, of molecular oxygen when chloroplasts were illuminated. The results were consistent with the notion that the oxygen normally produced could itself act as a Hill oxidant. We confirmed these observations in chloroplasts.

We used reagents in chromatophore suspensions similar to those in Mehler's experiments, in particular a combination of vitamin C (ascorbate) with small amounts of the dye dichlorophenol-indophenol (DCIP) as in assaying for cytochrome enzyme systems. This reducing system reacted with the oxidizing system of the illuminated chromatophores in the presence of air to produce an uptake of oxygen, just as we had observed previously using reduced mitochondrial cytochrome *c* as the reducing reagent. If we added alcohol (ethanol) in the presence of large quantities of the enzyme catalase, we found that there was an additional oxidation (with uptake of oxygen) of the ethanol to its oxidation product, acetaldehyde. This combination had also been used by Mehler in his experiments with chloroplasts, and indicated that some substance such as a peroxide might be made by chloroplasts and chromatophores in the light-induced reactions.

We could conclude that with air present, chromatophores acted just like chloroplasts. See L. P. Vernon and M. D. Kamen, "Comparative Studies on Simultaneous Photooxidations in Bacterial and Plant Extracts," *Archives of Biochemistry and Biophysics* 51 (1954):122–38.

22. An example of this general attitude was the reaction of my old friend Roger Stanier, an expert on bacterial physiology at Berkeley, to whom I had written when we first became aware of the *R. rubrum* cytochrome *c*. I had inquired whether he, or anyone in van Niel's laboratory, had ever noticed anything peculiar about the color of *R. rubrum* extracts. He replied that often they were pinkish. It was assumed that the color came from acidic degradation of bacteriochlorophyll to bile pigments. No one had ever examined the extracts with a simple visual spectroscope, or they would have seen there was a cytochrome *c* present.

23. See S. R. Elsden, M. D. Kamen, and L. P. Vernon, "A New Soluble Cytochrome," *Journal of the American Chemical Society* 75 (1953):6347.

24. See M. D. Kamen and L. P. Vernon, "Existence of Haem Compounds in a Photosynthetic Obligate Anaerobe," *Journal of Bacteriology* 67 (1954): 617–18.

25. See J. R. Postgate, "Presence of Cytochrome in an Obligate Anaerobe," *Biochemical Journal* 56 (1954):xi, and M. Ishimoto, J. Koyama, and Y. Nagai, "The Cytochrome System of Sulfate-reducing Bacteria," *Journal of Biochemistry* (Japan) 41 (1954):763–70; also see their paper "A Cytochrome and a Green Pigment of Sulfate-reducing Bacteria," *Bulletin of the Chemical Society of Japan* 27 (1954):564–65.

26. Her letter dated December 21, 1953, read in part, "The Department has given very careful consideration to your letter and affidavit and to the other information in regard to your case. However, the Department must disapprove your application for a passport on the ground that the granting of passport facilities to you is precluded under the provisions of sub-sections (b) and (c) of Section 51.135 of the Passport Regulations. In this connection it has been alleged that you participated in meetings of the Joint Anti-Fascist Refugee Committee, including meetings held for the benefit of "The Peoples' World," a Communist publication, and that you associated over a period of several years with Communists including Louise R. Bransten, an active Communist in the San Francisco area, and Ralph Shaw, State Chairman, District No. 21, Communist Party, U.S.A. Further, it is alleged that you were a member of the American League Against War and Fascism which has been designated by the

Attorney General as an organization falling within the purview of Executive Order No. 10450. It is also alleged that while you were employed on classified work on the Manhattan Project you furnished certain classified information to Soviet officials: that as a result thereof, your services were terminated; and that you discussed with Soviet officials and other persons the question of your travel to the Soviet Union after the termination of World War II for the purpose of engaging in similar work in the Soviet Union."

These allegations were a tired reassertion in much distorted form of material already discredited in the HUAC affair of 1948.

27. Much of the burden of preparing literature and writing articles supporting the federation's fight was effectively borne by a noted biologist, Dr. Clifford Grobstein.

14. The War Is Won (1953–1955)

1. By contrast, the process of burning carbon compounds in non-biological systems is uncontrolled. The energy that would be stored in ATP in cells is lost wholly as heat when sugar, for instance, is burned to form CO_2 and water. The temperature becomes too high to permit cells to survive and the rate of CO_2 and water production is large and erratic. In cells, the rate of combustion is such as to be consistent with the need to keep the temperature within bounds. The energy of combustion is conserved in ATP rather than being lost as heat.

2. L. P. Vernon and M. D. Kamen, "Hematin Compounds in Photosynthetic Bacteria," *Journal of Biological Chemistry* 211 (1954):643–62. In these papers we also included studies extending observations to another of the non-sulfur photosynthetic purple bacteria, *Rhodopseudomonas spheroides,* in which we also found heme proteins like those in *R. rubrum.*

3. See Peter Goodchild's *J. Robert Oppenheimer: Shatterer of Worlds* (London: British Broadcasting Corporation, 1980), p. 195.

4. I learned later that Nate had been correct in his surmise. According to a letter I received from the chief of the Freedom of Information–Privacy Acts Section, U.S. Department of Justice, dated September 15, 1982, the Appeals Board advised the Passport Office on April 16, 1954, that "after consideration of all evidence including Kamen's hearing on March 10, 1954, it was not believed that a preponderance of evidence existed to indicate that Kamen was an individual within the purview of Section 51.135, subsections *b* or *c,* Passport Regulations." Just five days later, Walter Bedell Smith, the acting secretary of state, reversed this finding.

5. Excerpts from letter of Nathan H. David to Chancellor E. A. Shepley, May 12, 1954:

The question is what to do about the two alternatives which are Congressional pressure and litigation. Frankly, my view is that in these trying times it would be most unlikely that the kind of pressure which would do Kamen any good could be generated on Capitol Hill. From where I sit there is little hope for aid in that quarter.

With reference to a possible lawsuit, however, I have a feeling of optimism. It will probably strike you as unbelievable, but the fact is that there is only one case in the books in which the discretionary power of the Secretary of State to issue passports has been squarely raised. That was the *Bauer* case in which Mrs. Shipley sought to revoke a passport without a hearing. A three judge court here in the District of Columbia held that this could not be done. The Department of Justice did not petition for *certiorari.*

In analyzing the legal problems presented, it is necessary, therefore, to turn to related fields. As I see it, the joint effect of the basic statute, the executive order pursuant thereto and the regulations of the Secretary of State, is to give the Secretary of State power to limit the movement of American citizens to within the borders of the United States and to do this without affording due process of law. If any one of the states assumed to limit the movement of its domiciliaries in a similar way, the shocking nature of such an action would be apparent to everyone.

It is true that after many years of naked, unexplained refusals to issue passports, a kind of hearing is now offered, but it is not a proceeding in which any evidence is produced against the applicant. You are aware, I assume, the hearing on Dr. Kamen's application was the first formal hearing on a passport application which has ever been held in the history of the country. In view of the way the Department of State handled the matter, it seems to me there are several avenues open for approaching the problem in the courts.

I believe it is possible to construe the Secretary's regulations as requiring that the evidence against the applicant be produced at the hearing. You will note, for example, that in Section 51.151(a) there is a provision that the Board "shall be convinced by a preponderance of the evidence, as would a trial court in a civil case." This language, it seems to me, is consistent only with the theory that evidence on both sides is expected to be introduced at the hearing.

On the other hand, since the regulations also provide in Section 51.137 that the reasons on which a refusal is based shall be stated only "as specifically as in the judgment of the Department of State security considerations permit," it may well be argued that a limited, as opposed to a full and open hearing, is contemplated by the regulations at least in some cases. Such a limited hearing would be practically identical with that which is afforded government employees in loyalty and security cases. In the *Bailey* case involving the loyalty procedures, the Supreme Court several years ago split 4 to 4 with Mr. Justice Clark not participating, on the question of the validity of a hearing procedure in which no evidence was submitted against the employee. It seems to me that a passport case would be a much easier case to win because, while one can accept the view that government employment is a privilege, free mobility is a right, a facet of "liberty" under the due process clause. If this is so, the law must be that under certain circumstances the government has the power to forbid exit to its nationals, but only after a proceeding in which enough evidence is brought forward *on the record* to permit a court to test whether the government's conclusion is arbitrary or based on reasonable evidence.

With reference to whether the *Kamen* case is one in which "security considerations" preclude full disclosure, it should be noted that thus far no such claim had been made. I think that if we got this case into court and that claim were then made, we could urge that the judge at least should be entitled to look at what the Department of State has to determine whether it is substantial. All of us have reason to believe that there is nothing substantial in any file, closed or open.

Another point which might prevail in court is that if we in fact won before the Board, the Secretary of State would have the duty of examining all the evidence before he rejected the Board's recommendation. I am certain that was not done in this case. In this connection, it would be relevant to argue that even though the statute providing for the issuance of passports vests discretion in the Secretary, it is not intended to be an arbitrary discretion but a reasonable discretion to be exercised on the basis of substantial evidence.

Under these circumstances I am definitely of the view that a suit should be brought on behalf of Kamen. If a complaint is filed, I think it must be anticipated that the case will wind up in the Supreme Court. It is clear to me that the government will resist the case strongly and will take it all the way up if we win below. This, of course, presents a serious problem of financing the litigation. I have reached a stage at which I feel that some of the burden of this problem might well be shared by others. As I see it, the protection and advancement of some basic liberties are involved, and this should concern all of us.

6. See M. D. Kamen and Leo P. Vernon, "Comparative Studies on Bacterial Cytochromes," *Biochemica Biophysica Acta* 17 (1955):10–22.

7. See H. E. Davenport and R. Hill, "Preparation and Some Properties of Cytochrome f," *Proceedings of the Royal Society* (London) 139 B (1952): 137–45.

8. See *F.A.S. Newsletter,* Member's Bulletin No. 17, December 7, 1954, reporting actions of the council.

9. Miller later achieved considerable prestige as counsel in the capital and was Nixon's lawyer in the Watergate case.

10. See *The Cosmos of Arthur Holly Compton,* ed. M. Johnson (New York: Alfred A. Knopf, 1967), p. 441 et seq.

11. Listed originally in Trial Records, U.S. District Court, D.C. CA 4461–51.

12. Front-page story in the *Washington Post,* April 14, 1955.

13. *Washington Evening Star,* April 14, p. A-20. The *Philadelphia Inquirer* also noted the episode in its issue of April 16, as did a New York paper, which also carried a disclaimer by Mrs. Shipley that there had been any "big fellow" in her office.

14. *St. Louis Post-Dispatch,* April 23, 1955.

15. *New York Times,* June 24, 1955, p. C-12.

16. FBI memorandum dated April 18, 1956, made available with the letter mentioned in note 4 above, after consultation with the Department of State, Passport Office.

17. *Washington Post–Times Herald,* July 15, 1955.

18. James Thurber, *My World—And Welcome To It* (New York: Harcourt Brace Co., 1942), p. 133.

Epilogue

1. The proceedings were published under the title *From Cyclotrons to Cytochromes,* edited by N. O. Kaplan and A. B. Robinson (New York: Academic Press, 1982).

2. A description of the apparatus is given in F. S. Brown et al., "The Biology Instrument for the Viking Mars Mission," *Reviews of Scientific Instruments* 49 (1978): 144–48.

Index

Compositor: Innovative Media
Printer: Vail-Ballou Press
Binder: Vail-Ballou Press
Text: 10/12 VIP Times Roman
Display: Goudy Bold